Comperia comperiana

By the same author:

Plant Breeding and Genetics in Horticulture
The Macmillan Press Ltd. 1979

A BOTANICAL TOUR ROUND THE MEDITERRANEAN

BY

CHRISTOPHER NORTH

NEW MILLENNIUM
292 Kennington Road, London SE11 4LD.

Copyright © 1997 Christopher North

All rights reserved. No part of this publication
may be reproduced in any form, except for
the purposes of review, without prior
written permission from the
copyright owner.

Printed and bound by Arm Crown Ltd., Uxbridge Road, Middx.
Issued by New Millennium*
ISBN 1 85845 075 6
*An imprint of The Professional Authors' & Publishers' Association

Acknowledgements

The witty remarks by A.E.Bowles on pages 118 and 273 come from A Handbook of Crocus and Colchicum (1952) and I am grateful to the 'Bodley Head' for allowing me to quote them.

Many of the friends and acquaintances who helped in the preparation of this book are mentioned in the Introduction on page 2.

To the memory of my late father-in-law,
CHARLES HENRI LAUENER,
from La Chaux-de-Fonds, Switzerland.
He was a very enthusiastic amateur plantsman.

CONTENTS

	Page
Foreword by Christopher Brickell C.B.E., V.M.H.	xiii
Chapter 1: Introduction	1
Chapter 2: The Mediterranean Environment	3
Chapter 3: Plant Identification	7
Chapter 4: Southern Portugal	15
Chapter 5: Spain West of Gibraltar	33
Chapter 6: Western Andalusia	49
Chapter 7: Eastern Andalusia	69
Chapter 8: Majorca	97
Chapter 9: Minorca	113
Chapter 10: South of France	123
Chapter 11: Corsica	159
Chapter 12: The Gargano	181
Chapter 13: Sicily	195
Chapter 14: The Maltese Islands	217
Chapter 15: Albania	231
Chapter 16: Northern Greece	247
Chapter 17: Southern Greece	275
Chapter 18: Corfu and Cephalonia	303
Chapter 19: Lesvos	323
Chapter 20: Crete	333
Chapter 21: Rhodes	363
Chapter 22: Cyprus	373
Chapter 23: Southern Turkey	403
Chapter 24: Israel	431
Chapter 25: Tunisia	451
Chapter 26: Conservation	469
Bibliography:	479
Alphabetical list of plant names:	483
Place Index:	491

LIST OF COLOUR PLATES

FRONT COVER *Narcissus gaditanus* growing among rocks on Sierra Cabrera near Mojacar, Andalusia, Spain (p.76)

FRONTISPIECE *Comperia comperiana* in pine woods near Mugla. A rare prize to be found in Turkey (p.429), Rhodes (p.369) and Lesbos (p.328)

111. *Senecio rodriguezii* Fornells, Menorca. (p.118)

Asclepias curassavica Barbate de Franco, Spain. (p.44)

112. *Iberis candolleana* Mont Ventoux, S. France. (p.141)

Iris pseudopumila Lentini, Sicily. (p.201)

193. *Ophrys garganica* Mattinata, Gargano, Italy. (p.187)

Pancratium illyricum Cap Corse, Corsica. (p.163)

194. *Cyclamen repandum* var. *rhodense* Mt. Profitis Ilias, Rhodes. (p.369)

Sternbergia sicula Sellia Gorge, Crete. (p.351)

287. *Campanula topaliana* Delphi ruins, Greece. (p.259)

Verbascum macrurum Mt. Ossa, Greece. (p.258)

288. *Viola graeca* Mt. Parnassos, Greece. (p.260)

Quercus alnifolia Platres, Cyprus. (p.390)

391. *Dorystoechas hastata* Termessos, Turkey. (p.407)

Tulipa armena var. *lycia* Baba Dag, Turkey. (p.423)

392. *Iris haynei* Mt. Gilboa, Israel. (p.440)

Centaurea pullata Nefza, Tunisia. (p.463)

More than 500 monochrome line drawings are listed in index pages.

Foreword

Since the pioneering efforts of Anthony Huxley and Oleg Polunin, who obtained a publisher after some considerable difficulty for their standard work *Flowers of the Mediterranean (1965)*, several very useful guides to the plants of this much visited area have been produced. These have proved invaluable to those travellers to Mediterranean countries who, for the most part, shun the beaches and the bustle and exuberance of the more popular tourist facilities and instead spend their time seeking and photographing plants in their wild habitats.

None of the authors of any guides to this area has, to my knowledge, approached an exploration of the flora in quite the same way that Chris North has done in this very personal, but extremely informative book, which is based on his intimate knowledge of all the areas described and is presented in a narrative fashion in a very methodical and helpful way for travellers to Mediterranean countries.

As the author indicates it would be impractical to produce a single portable, or even semi-portable, volume that could cover all the plants of this exceedingly rich and variable flora and wisely he has restricted himself to describing only the plants in the areas which he has explored thoroughly himself. The information provided, therefore, is a result of his personal observations over twenty-five years and is set out in an easily read and logical format which contains "tasters" of the plants found in the sites visited plus further details of selected plants giving salient diagnostic characteristics which are supplemented by line drawings, all done by Chris North from living specimens.

The aim is to provide an overall view of what is likely to be seen in particular seasons both for plant enthusiasts and for those visiting the areas who simply want to enjoy the countryside and flowers without the bother of too serious botanical identification. It covers, in broad terms and non-technical language the geography and terrain of each area specifying commonly found plants as well as describing specific sites within an area where plants of particular interest may be located. Very helpful information on travel and accommodation in the area and botanical references are all gently blended into a text packed with plant information that should prove a stimulus for casual visitors as well as naturalists, gardeners and botanists to explore and enjoy further the richness, variation and beauty to be found in the Mediterranean flora.

This book will undoubtedly prove to be an invaluable companion for visits to the Mediterranean as well as much winter armchair anticipation of what is likely to be seen when planning which area to visit next without the need for detailed study of the appropriate flora.

CHRISTOPHER BRICKELL C.B.E., V.M.H.
Pulborough 1996

1. Introduction

This book is intended as a travellers' introductory guide to the flora of specific parts of the Mediterranean. It is compiled especially for those visitors who feel a compelling urge to look on the ground and round about to see what plants grow wild. The reader is assumed to be a plant enthusiast, but not necessarily a trained botanist so scientific terminology has been kept to a minimum. Nevertheless, Latinized plant names are used throughout as English equivalents are frequently imprecise, confusing, and often non-existent.

The text indicates which species may be seen in selected areas; it does not aim to provide exact information for their identification. It is complementary to standard floras and popular books on plant recognition in this respect, and gives details of these works most likely to be of use to the reader. Nevertheless, the text will aid identification in some instances, for a certain amount of plant description is given and it is possible to identify some species by inference - as when a limited number of a genus occur in a given region. The line drawings will also be of help; all have been made from living specimens and, with very few exceptions, on the site where they were found growing in the wild. They have been done in such a way as to show not only their salient diagnostic features but also some attributes of their natural elegance. Most plant enthusiasts will appreciate the aesthetic characteristics of the material they search for as much as the urge to name it correctly.

It is self-evident that no single volume can cover all the interesting areas and all the different plants likely to be seen within the Mediterranean region. This book is no exception and, of course, its contents are influenced by personal choice. All the regions described here have been visited personally over a period of 25 years. Regrettably, there are other places which ought to have been explored. It is especially disappointing not to have visited the former Yugoslavia and other regions such as Lebanon and more parts of North Africa and Italy, but life is too short. However, I believe that this work covers a wider area in a single volume than anything of the kind that has been published to date.

The plants dealt with in the text inevitably represent a personal selection. Readers will soon observe a preference for orchids and the more showy bulbous and tuberous-rooted species. Some other plant families such as the grasses are given scant mention though they are also of great interest. I make no excuse - most 'plant enthusiasts' who have come to the subject through horticultural leanings like myself will feel the same.

Many experts and enthusiasts have helped with information and encouragement in the compilation of this work. I am indebted especially to the late Professor Peter Davis and his team of Dr. Kit Tan and Dr. John Mill of the Science Research Council's Flora of Turkey Unit and to Mr. Ian Hedge and Mrs. Jennifer Woods of the herbarium at the Royal Botanic Garden Edinburgh for help in identification of plant material. Other friends and advisers who have provided information and encouragement include; Dr John Akeroyd, Dr Mike Almond, Mrs Llyn Almond, Mrs Mona Fortesque, Mr Andrew Mayo and Mr Nick Turland. My wife Marie Louise has accompanied me on all but one of the journeys envolved in writing this book and become an expert at spotting interesting plant species from afar. Without her encouragement and tolerance of my periods of frustration in trying to find 'correct' names for the plant material I should not have been able to complete this work.

Mr Leslie Bisset (Dundee University Botanic Garden), Professor Pierre Quézel (Marseille University) and Mr Edwin Lanfranco (University of Malta) have kindly read through parts of the draft and made corrections and suggested amendments and Mr Calder Jamieson has checked the presentation. In spite of the care that has been taken, there will no doubt be some textual errors for which I accept full personal responsibility. Nevertheless, it is appropriate to quote, as Mr Stuart Thompson did when he wrote one of the first popular books on the Mediterranean flora in 1914, a comment from the great French natural historian M. Jean-Henri Fabre; "Il n'y a que ceux qui ne font rien qui ne se trompent pas" - It is only those who do nothing who make no mistakes.

2. THE MEDITERRANEAN ENVIRONMENT

2.1 Climate: The Mediterranean climate is typified by nearly frost-free winters, a warm and sunny spring and hot dry summers with rainstorms in the autumn and winter. Attempts have been made to define this in a mathematical formula combining seasonal rainfall and temperature and when this is done the adjective 'mediterranean' is applicable to the climate of other parts of the world such as that of California, South Africa, South West Australia and parts of Chile. The climate has also been defined in terms of the vegetation it supports and the usual indicator is the olive - the Mediterranean being typified by the region where olive trees grow to give a crop of ripe fruit. This, of course, precludes the area around the Chelsea Physic Garden in London where there is an olive tree that occasionally produces green fruits - ripe olives are black, green ones are unripe. However, the olive originates from Asia Minor and Syria and is not native to all parts of the Mediterranean, though it has been cultivated there for thousands of years and the domesticated form exists alongside the small prickly, wild species and numerous hybrid intermediates.

Plant enthusiasts will usually wish to go to the Mediterranean in spring when most of the interesting plants are in flower but one should not be deluded into thinking that the weather is invariably fine there during this season. It varies from year to year and from place to place. As a broad generalisation the southern islands with few high hills, such as Malta and Rhodes, have the earliest and most reliable spring weather. Some areas with high mountains, as in Turkey, Crete and Sicily, may have more variable weather, sometimes with torrential downpours at this time of the year so it is important to go prepared with waterproof clothing, and an umbrella may well be appreciated. In any case the onset of finer spring weather can vary from year to year by at least two weeks and planning the best time to go is not always straightforward.

The average rainfall varies considerably throughout the region, being generally lower in the south than in the north. Precipitation is especially high (1000-1500mm per annum) in the Alpes Maritimes of France and accompanying region of northern Italy and along the Balkans, stretching into northern Greece. Local topography may also have a profound influence on rainfall as in Cyprus where the northern slopes of the Kyrenia range are relatively moist and those facing south are dry and desert-like. A similar effect is noticible, though to a lesser extent, in the hills of Provence.

In nearly all places within the Mediterranean region one can stand looking out to sea and turn round to see hills or mountains clearly visible behind one. Many of these 'hills' are substantial heights. Compared with Snowdon 1085m and Ben Nevis 1342m, even some of the smaller mountains rise higher, such as Puig Mayor 1445m in Majorca, Monte Soro 1847m in Sicily; and others such as Monte Cinto 2710m in Corsica and Mulhacén 3482m in Southern Spain are twice as high. Here the climate is certainly not typically mediterranean. Snow falls on the peaks in autumn and often lasts up to the end of May with some patches surviving throughout the year. Much of northern Greece and the Balkans is entirely mountainous, except for a narrow region around the coasts. The mountains within the Mediterranean area often have isolated endemic plant species of very great interest to the plant enthusiast but these usually do not flower before late May or June.

2.2 Geology: During the Tertiary period, some 50 million years ago, much of the land discussed here was covered by an ancient sea called the Tethys which was much vaster than the present day Mediterranean. The animalcules that lived in this sea left a calcareous deposit giving rise to the predominately limestone areas which typify the region today. Only the highest peaks projected above the Tethys sea and retained their summits of primary rocks. Then the continent of 'Africa' began to move northwards exerting, enormous pressures on the land of 'Eurasia' and causing great folds which created the relatively young mountains of the Atlas, Pyrenees, Alps, Appenines, Balkans and further afield in the Carpathians and Himalayas. The folding process led to complicated intermingling of limestones with older primary rocks and was accompanied by volcanic activity in places. As a result much of the coastal areas of the Mediterranean today are of limestone ringed by mountains of complicated geological structure. This situation provides a great variety of soils and habitats which in part explains the richness of the flora of the region. A much fuller account of these geological processes and their resultant effect on the present day land structure is given in Polunin & Smythies (1973) and Polunin (1980) - see Chapter 3.

2.3 **Plant Communities:** The area around the Mediterranean is the birthplace of modern civilisation and it is not surprising that the wild flora has been considerably modified by human activities. However, a high proportion of the land is rocky and cannot be cultivated, so the complete clearing of regions for monoculture agriculture are uncommon and much of the changes brought about by human activity are indirectly through the grazing of domestic animals.

It is generally thought that woodland covered most of the area before humans cut down trees for fuel and building purposes. The Mediterranean Sea has been a very important means of communication and transport for the communities living around it and much of the best timber was felled for ship building a thousand or more years ago. On lower-lying areas the evergreen holm oak was probably dominant with some pines and higher up, especially in the wetter areas, deciduous trees predominated. Most of the pine forests now seen in the Mediterranean are planted for timber. On the very highest tree regions several species of abies existed but are now greatly reduced in numbers such as *Abies pinsapo* in Spain, *A. nebrodensis* in Italy though *Abies cephalonica* still holds its own in Greece and the Balkans. These may have been accompanied by cedars which still occur in North Africa, Cyprus and Turkey and have been reintroduced to the south of France.

Once the trees are destroyed for building purposes or charcoal the land is rapidly taken over by scrub comprised of species which can withstand the dry hot summers. The tallest shrub communities, about the height of a man, are usually referred to as maquis and it is generally assumed that further grazing reduces this to garigue. The terminology for these plant associations can be confusing. In Spain the word matorral covers both maquis and garigue (sometimes spelt as in French 'garrigue') and the words 'tomillar' and 'jaral' are used to designate special kinds of matorral dominated by thyme and cistus respectively. In Greece phrygana corresponds to garigue, though the word in Greek simply means common grazing land. True maquis usually occurs on acid soils and in areas where the rainfall is higher than usual, and many limestone areas that carry garigue have never been covered by maquis.

Areas of garigue are undoubtedly the best places to look for plants typical of the Mediterranean. Light grazing by sheep or goats helps to provide a habitat for many of the most interesting bulbous and tuberous rooted species and the maintenance of this type of plant association has greatly helped the spread of members of the orchidaceae. However, excessive grazing, as occurs in North Africa and Turkey, can destroy garigue and lead to pseudo-steppe conditions, especially in the drier areas. Large numbers of asphodels and spiny plants such as *Sarcopoterium spinosum*, which are unpalatable to animals, are indicators of this condition.

The higher mountains have mostly retained their vegetation on the summits more or less unchanged though grazing usually occurs during the summer but is not so severe as at lower levels. Nevertheless, grazing can be damaging even at these high altitudes and may destroy vegetation which has only a short summer in which to put on growth, as on Mount Ida in Crete. The summits of the Mediterranean mountains carry numbers of interesting endemic species which are hardly known to gardeners who seem at present more enthusiastic about the flora of China and the Himalayas than they are about these remote parts of the Mediterranean. There is still a chance for the plant enthusiast to find a new species in the more remote mountains of Greece and Turkey.

Mediterranean plant associations make an interesting study in their own right but are dealt with in a very cursory manner in this travellers' guide. Further information is given in some of the books reccommended in 3.2 and the one by Schönfelder & Schönfelder (1990) covers the subject rather well.

3. Plant Identification

3.1 Finding the correct name for a plant one has discovered in the wild may well be difficult or even seem impossible. It is a question of numbers. The bird watcher can use a single pocket book which accurately describes all the species to be met with in Europe and the Mediterranean but no equivalent volume that will account for the 25,000 or so plant species which grow around the Mediterranean basin can be produced. The task then of putting the correct name to a plant from amongst these thousands is often a daunting one and may well deter the faint-hearted. But it may also be a welcome challenge to those who enjoy an intellectual stimulus similar to that required for solving a whodunnit or a crossword puzzle.

3.2 **Handbooks:** Books weigh heavy in personal luggage and it would be convenient, when looking for plants, if one could take just a single volume on identification. For most readers this undoubtedly would be the new 'Mediterranean Wild Flowers' (Blamey & Grey-Wilson 1993). It describes 2,539 species and illustrates most of them by small drawings of the salient morphological differences. It also has 1,500 colour illustrations based on paintings.

A somewhat earlier book is 'Collins Photoguide to the Wild Flowers of the Mediterranean' (Schönfelder & Schönfelder 1990) (originally published in German in 1984) which describes some 1,000 species. It is well worth having if only for its 500 excellent colour photographs and fairly full introduction. An even earlier book, indeed the first of its kind, is 'Flowers of the Mediterranean' (Polunin & Huxley 1965) which is still available and also well worth having though it only deals with some 700 species. It includes over 300 useful colour photographs and additional excellent line drawings.

At best, all three of these works describe only some 10% of the total number of different vascular plants found around the Mediterranean. They are wisely concentrated on those species most commonly encountered growing in the true mediterranean climate below about 1,000m. However, many of the most interesting and exciting species with a limited distribution are to be found in mountains above this altitude and for their identification one must search through other literature. An indication of the best places to look for this information is given in section 3 under the different chapter headings.

Two other books which the serious enthusiast will need are 'Flowers of South-west Europe' (Polunin & Smythies 1973) and 'Flowers of Greece and the Balkans' (Polunin 1980). They describe many mountain species, as well as those from lower altitudes, and have illustrations and keys for identification purposes. They also give some details of where to look for plants, as in this present volume, but do not cover all countries around the Mediterranean. Neither deal with such important regions as Corsica, Sicily, Italy mainland, Turkey, Cyprus, Israel or North Africa.

Other books which are a valuable help to identification are those devoted to special groups of plants. A notable example is 'Orchids of Britain and Europe' (Pierre Delforge 1995), a transalation from the French written by a Belgian and originally published in Switzerland in 1994. This is the latest of a number of books on the subject, every one of which includes 'new' species by species-splitting and amateurs may wonder whether this latest has gone a little 'over-the-top'. Slightly older, but with especially good photographs is 'Field Guide to Orchids of Britain and Europe' (Karl Peter Buttler 1991). Other useful works include 'Wild Orchids of Britain and Europe' (Davis & Davis 1983); a delightful and helpful work which is still available. Another useful book is 'The Bulb Book' (Rix & Phillips 1981) which has excellent photographs of nearly all bulbous species found in the Mediterranean, including rare ones. There are also monographs on individual genera produced by or in collaboration with The Alpine Garden Society such as 'Narcissus, A Guide to Wild Daffodils' (Blanchard 1990) and 'Cyclamen, the Genus in the Wild and in Cultivation (Doris E. Saunders 1975).

3.3 **Floras:** These are definitive scientific works but they do become out of date with time. The most important is the 5 vol. 'Flora Europaea' by Tutin (1964-1980) though it does not cover the whole of the Mediterranean and tends to be a 'lumper' - ie. it does not recognise some of the generally acclaimed sub-species. It is also somewhat out of date and at present under revision. However it is the most useful single work of its kind for most readers and is available in some public libraries.

The excellent, and relatively new 10 vol.'Flora of Turkey and the East Aegean Islands' (Davis et al. 1965-1986) takes over to the east of the Aegean where Flora Europaea leaves off but it does not include Cyprus which is well covered with the 2 vol. 'Flora of Cyprus' (Meikle 1977-1985). The extreme east of the Mediterranean region is best covered by 'Flora Palaestina',in English (Zohary 1966-1978) and North Africa by the 12 Vol 'Flore de l'Afrique du Nord' (Maire 1952-1980). In addition to these regional floras there are national ones for most countries and some of these are quoted, when appropriate, under the various chapter headings.

A start has been made on a check list of Mediterranean plant species that will eventually be the standard for naming. It includes information on which species have been recorded in the countries of the region.

3.4 Plant Names: Scientific latinized names of plants have been used throughout this book. Not only are they more precise than anglicised versions, they are universal. Botanists from all over the world, even though they speak a language that does not use the Roman alphabet such as Russian, Japanese or Arabic will understand what is meant by *Taraxacum officinale* though they may never have heard of a dandelion. The correct naming of a plant will include the generic and specific names, usually written in italics followed by the abbreviated name of the author who originally described the plant ie.*Taraxacum officinale* Weber. Sometimes studies show that it is more logical to change the original generic name but retain the specific name and then one presents it thus:*Misopates orontium* (L.) Raf. which was originally called *Antirrhinum orontium* by Linneus (abbreviated as L.). To be valid, a botanical name has to be published with an accurate description in Latin. As a general rule the authors of the names are not included in this book, solely for the sake of brevity.

Species are not infrequently divided into subspecies which adds a third latinised word to the name as *Ranunculus ficaria* ssp. *ficariiformis* - the word for subspecies being abbreviated as ssp. or subsp. In this book the abbreviation has been left out for the sake of brevity, but this is not a recognised convention. Another smaller taxonomic category than a species is a variety which often implies a segregating form, as when seedlings produce flowers of different colours. An example is *Capparis spinosa* var. *inermis* which has no thorns or small ones that drop at an early stage of development whereas the typical species is distinctly thorny. In such cases the abbreviation var. has been retained in the text to distinguish between subspecies and varieties.

As a result of continuing research, plant names are revised and changed from time to time. It may be that an earlier described name has come to light and takes precedence over the current one by the rules of nomenclature. It may also be that further studies show that the genus should be split into several new genera to provide a more logical classification. This gives rise to synonyms which remain in use for a time. Some of these synonyms are included in the text here, especially as readers may at times be using some of the older books. They are mostly indicated in parenthesis with an equals sign (=). In places where old and new names are still current usage the synonyms are indicated with O = or N = (meaning old and new respectively) as in *Buglossoides purpurocaerulea* (O = *Lithospermum purpureocaeruleum*) or *Chinodoxa forbesii* (N = *Scilla forbesii*).

Recently some of the names of the plant 'families' have been changed to bring them in line with the naming of typical genera. This means that up-to-date books should change to the following new family names:

Compositae	is now	*Asteraceae*
Cruciferae		*Brassicaceae*
Graminae		*Poaceae*
Labiatae		*Lamiaceae*

Leguminosae   *Fabaceae*
Umbelliferae   *Daucaceae*

Presumably we will continue to talk about composites, crucifers, labiates, legumes and umbellifers, at any rate for a few years to come.

3.5 Herbarium specimens: Taxonomists often can not name plants with certainty from photographs and drawings, they need fresh or pressed specimens. Pressing is not difficult and requires little equipment - all that is needed for the process is a simple lattice-wood press and absorbent paper. Failing this, two pieces of hardboard and sheets of newspaper will suffice to produce good quality specimens. It is important to change the paper frequently so that the material dries quickly. However, pressing should not be undertaken without prior arrangement with a botanic garden, specialist university department, or museum. Many plants are protected by law and customs officials may confiscate the material if an attempt is made to import it without the required authority.

3.6 Personal Records: Most readers will wish to keep a record of species they have seen and undoubtedly the best way to do this is to take colour photographs. For this, one will need a reliable reflex camera with a macro lens that will give a 1:1 image if necessary and it is strongly recommended that it should be used whenever possible on a short tripod to reduce the risk of camera shake. Good cameras are relatively cheap nowadays but they do go wrong from time to time. The macro lens (preferably F50 or F100mm, some zoom lenses are suitable) may well cost more than the camera body and if one takes the matter seriously it is advisable to carry a spare camera body. There is nothing more irritating than to have found some very interesting plant in a remote region only to discover that the camera has developed a fault. One should also always carry spare batteries. The usual 35mm camera system is the most convenient and there is a wide choice of films including some which are very fast and suitable for poor light conditions. Flash photography is sometimes used and is good for poor light conditions or when a very fast exposure is needed to counteract high wind conditions. It tends to give a sharp, but somewhat unnatural picture.

As an alternative to photography, or perhaps as an addition, one may wish to make drawings of the plants where they are found. This can conveniently be done in a pocket-sized drawing book. The results may, if required, be traced and then put through a photocopier, as in this present work.

3.7 OPTIMA: This is the Organization for the Phyto-Taxonomic Investigation of the Mediterranean Area. It is primarily for professional taxonomists but welcomes non-professional members and can be of considerable help to serious enthusiasts. It publishes a newsletter and makes available literature on current taxonomical works as well as organising meetings and collecting trips. Those interested should contact:

OPTIMA Secretariat
Botanischer Garten und Botanisches Museum
Königin-Luise. Str. 6-8
D-1000 Berlin 33

It has members in most countries bordering the Mediterranean and thus provides a ready way of contacting other enthusiasts.

3.8 MGS - The Mediterranean Garden Society: This very new organisation exists, as its name suggests, essentially for gardeners in the Mediterranean, many of whom have settled there from outside the region. It may be a useful means of communication with visitors from abroad who come to see the wild flora and are themselves gardeners. It produces a journal called appropriately 'The Mediterranean Garden'. The address of the secretariat is:

Sparoza, Box 14,
Peania 19002, Greece.

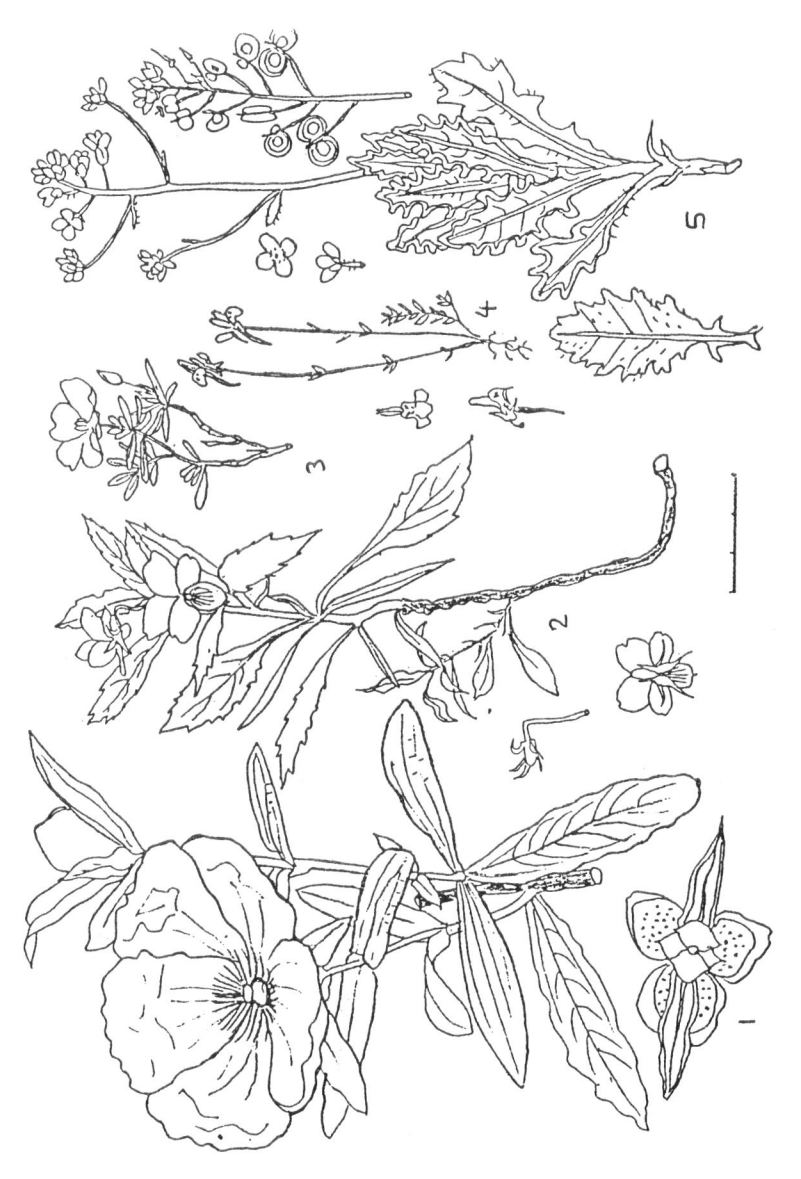

Algarve 1. Cistus palhinhae 2. Viola arborescens 3. Halimium commutatum
4. Linaria algarviana 5. Biscutella vincentina

Algarve 1. Cotula coronopifolia 2. Bellevalia hackelii 3. Fedia cornucopiae 4. Bellis annua 5. Teucrium pseudochamaepitys 6. Scilla monophyllos

4. Southern Portugal

4.1 No part of Portugal borders the Mediterranean Sea but the flora of its southern region is typically Mediterranean in character. The weather is also similar though the wind from the Atlantic can be especially strong and persistant. The area bordering the south coast, usually referred to as the Algarve, is very popular with sun seekers so plenty of holiday accommodation is available and as a bonus the Atlantic fish makes for especially good eating.

In this chapter, the area around Cape St. Vincent in the extreme south west of Portugal is treated in some detail and little direct information is given about the flora further eastwards towards the Spanish border which is essentially the same. The region described here is a very good place to visit for somebody embarking on an interest in Mediterranean plants. There are substantial stretches of typical garigue which is not overgrazed and most of the more exciting species can be seen within walking distance of Sagres or some of the other small towns nearby. It is not essential to have the use of a car and there seem to be no fences, barriers, or restrictions to make access difficult. However, as in most other parts of the Mediterranean coastal regions, 'development' for tourism is slowly eroding the native flora.

Maps which include the south of Portugal are easily obtained and the 1:200,000 scale Mapa Turistico do Algarve published by the Automovel Club de Portugal is especially useful and readily available at most resorts there.

4.2 **The Terrain:** There is a broken line of low limestone hills roughly parallel to the south coast called in Portugese the 'barrocal'. The rock forms impressive high cliffs along the coast with some fine beaches at their base. Strong winds have blown sand over the west facing cliffs north of Cape St. Vincent to form dunes, some of which are now well colonised with vegetation. Immediately to the north of the barrocal there are cultivated fields and orchards and still further north the land rises to the Serra de Espinhaço de Cao in the west and the higher Serra de Monchique north of the coastal town of Portamao. These northern ranges mostly have acid volcanic soils with a different and less varied, flora to those of the limestone areas nearer the coast.

4.3 **Literature:** Polunin and Smythies (1973) book has a very helpful description of the flora of the Algarve but it by no means covers all the species to be found in the Cape St. Vincent area. Furthermore, many of the plants mentioned in their section on the Serra da Arrabida, which lies just south of Lisbon, can also be seen around Cape St. Vincent.

4.4 **Around Sagres:** Just to the north of Sagres there is an area of garigue which, from the plant enthusiasts point of view, must at present be one of the richest in the Mediterranean and deserves some degree of protection. Here, at the western end of the barrocal, amongst outcrops of limestone and with excellent views of the sea, grow many small shrub species of which the following are the most common :

Cistus albidus   *Juniperus phoenicea*
C. ladanifer   *Lavandula stoechas*
C. monspeliensis   *Phlomis purpurea*
C. palhinhae   *Rosmarinus officinalis*
C. salvifolius   *Ulex minor*

Cistus palhinhae is the glory of the area and found only in this small corner of the world. It forms low, neat rounded bushes that have dark green leaves covered with a sticky exudate so that they glisten in the sunshine as though they had been newly varnished. The flowers are large - up to 9cm diameter - and usually pure white though a few specimens may have dark blotches at the base of the petals. It closely resembles the widespread *C. ladanifer* but is much shorter growing and more branched. *Cistus ladanifer*, often called the gum cistus, also occurs near here and can reach 3m tall, bearing large white flowers but these usually have a dark marking at the base of the petals though pure white-flowered forms do

Algarve 1. Armeria pungens 2. Calendula suffruticosa 3. Astragalus massiliensis 4. Lobularia maritima 5. Romulea clusiana 6. Ruta chalepensis

Algarve 1. Phlomis lychnitis 2. Lavandula stoechas 3. Erica lusitanica 4. Erica arborea 5. Lathyrus ochrus

occur. It seems that these two species may hybridise for intermediates can sometimes be found. Other less widespread shrubby species which occur scattered amongst the above include:

Chamaerops humilis   *Ononis sp.*
Cistus crispus   *Phillyrea angustifolia*
Daphne gnidium   *Pistacia lentiscus*
Genista hirsuta   *Quercus coccifera*
Jasminum fruticans   *Teucrium fruticans*
Lithodora diffusa lusitanica

The jasminum resembles the winter jasmine of our gardens but is shorter growing in the wild and the flowers have practically no scent. It is also sometimes cultivated in home gardens and, as a bonus to the flowers, it produces decorative, shiny, black, berries in autumn. The lithodora (O = *Lithospermum diffusum*) sub-species is a plant special to the area and in some places here it is abundant both on basic and acid soils. It is easy to recognise at a distance in the garigue because of its brilliant blue flowers. The ononis needs checking. It is not in flower with most of the other species in March and it is not mentioned by Polunin and Smithies (1973). Amongst the shrubs there are a number of herbaceous and sub-shrubby species:

Anagallis arvensis   *Cerinthe major*
Misopates orontium   *Linaria algarviana*
Astragalus lusitanicus   *Lobularia maritima*
Bellis annua   *Salvia verbenaca*
Calendula suffruticosa   *Tuberaria guttata*
Campanula lusitanica   *Viola arborescens*

The anagallis is our scarlet pimpernel but it nearly always has blue flowers in the Mediterranean. The astragalus is rather tall growing with large white flowers followed by inflated pods - more resembling a broad bean than an astragalus. As its name suggests it is rather special to the region, for Lusitania is the old Latin name for

Portugal. However, it has a curious distribution and can be found far away in the Peloponnese (17.7) and in Cyprus (22.10) though the plants there are classed as a sub-species of those in Portugal. Near Sagres it grows in a number of different sites and is not infrequently seen by the roadside. *Campanula lusitanica* is another special endemic plant of the area. It is a fragile annual with bright mauve-blue flowers which are not produced in any quantity until the end of March. The viola is a sub-shrub which at times forms clumps that look at a distance rather like a pale flowered aubretia. It grows also in southern Spain near Grenada and as far east as Sardinia. Amongst all these shrubs and herbaceous species there are many bulbous and tuberous-rooted plants of special interest to the enthusiast:

Arisarum vulgare	*Orchis italica*
Asphodelus ramosus	*Ophrys bombyliflora*
Anemone palmata	*O. vernixia*
Bellevalia hackelii	*O. tenthredinifera*
Fritillaria lusitanica	*Romulea clusiana*
Gynandriris sisyrinchium	*Serapias lingua*
Leopoldia comosa	*Tulipa sylvestris australis*
Narcissus obesus	*Valeriana tuberosa*
Orchis morio	

The anemone, which forms small pea-like tubers, typically has beautiful yellow flowers about the size of our wood anemone. Here the population is unusual in that about a third of the plants produce pure white flowers. The species is reputed to be perfectly hardy and adaptable in British gardens but is rarely seen at home. The bellevalia is a grape-hyacinth like plant resembling a small form of the tassel hyacinth *Leopoldia comosa* (=*Muscari comosum*) and is endemic to southern Portugal. The fine fritillaria is fairly abundant in places here and sometimes produces two or three flowers per stem. It is a variable species and many of the specimens here have narrow leaves suggesting that they could be ascribed to the variety *stenophylla*. In spite of its name it is frequently seen also in parts of southern Spain. The romulea may be the widespread and variable species *R. bulbocodium*. Its flowers are mauve-pink or white and sometimes of two distinct sizes. The smaller flowered forms are often pollen sterile, ensuring that their cross fertilisation takes place. Of the orchids, *Ophrys tenthredinifera* is the most apparent in early March followed then by *Orchis morio* and the mirror orchis *Ophrys ciliata* (O = *O.speculum*). A sub-species *lusitanica* of the last of these is special to the extreme south-west of the Iberian peninsula. It is a more vigorous plant than the type and has a more elongated and narrower lip with longer and more pronounced side lobes. It is now usually referred to as *Ophrys vernixia*. *Orchis italica* is less common here and there are only a few scattered specimens of *Ophrys bombyliflora* and the serapias.

Walking along the top of the south-west facing cliffs near the garigue, several other interesting plants may be found. There are patches of *Omphalodes linifolia*, an annual member of the Boraginaceae with pale blue forget-me-not like flowers. Another annual here is the tiny *Ionopsidium acaule* with masses of mauve flowers - it is a crucifer (or member of the Brassicaceae) which is native to Portugal but grown for ornament in other parts of the Mediterranean and sometimes becomes naturalised. In the shelter of bushes, protected from cattle, one may also find the Spanish bluebell *Hyacinthoides hispanica* (O = *Scilla hispanica* = *Endimion patulus*) and the brown bluebell *Dipcadi serotinum*. The first of these is a beautiful plant with flowers that are more cup-shaped than those of our own bluebell *Hyacinthoides non-scriptus* and hybrids between these two species are frequently offered in Dutch bulb catalogues and grown in our gardens. The so called Spanish bluebell and the brown bluebell are primarily plants of North Africa.

The species described above can be found throughout the barrocal as far as the Spanish border. Other plants which might be encountered include the winter-flowering *Iris planifolia* and the two rare narcissus species *N. gaditanus, N. willkommii* - both small members of the jonquil group.

4.5 Sagres to Cape St. Vincent: This walk along the road takes about an hour and in March and April is botanically very rewarding. There are few barriers and in most places it is easy to step off the road into the barrocal. As one leaves the roundabout joined by the main road to Faro, and before reaching the telegraphy station, there is an interesting flora on the east side. Here one can see *Narcissus obesus, Ophrys tenthredinifera, Scilla monophyllos.* The narcissus is the charming hoop-petticoat daffodil in which the corolla tube is more cup-shaped rather than bowl-shaped as in the typical *N. bulbocodium* of which it is sometimes classed as a subspecies. The scilla is a small squill which has, as its name suggests, usually only one rather broad leaf. It is special to the region but, like many of the other plants found here, also occurs in North Africa. By the roadside the endemic crucifer *Biscutella vincentina* is quite common - a yellow alyssum-like plant with typical fruits of the genus. The generic name 'two shields' refers to the shape of the fruits which resemble bucklers or targes.

Past the radio station grows another scilla, *S. verna*, easily distinguished from the one previously mentioned by having several leaves from each bulb and two, instead of one, small bracts at the base of each flower stem. It is a form found in this south-west corner of the Iberian peninsula and sometimes referred to as *S. ramburei* or *S. vincentina.* Here also is *Halimium commutatum* a yellow-flowered, rock-rose-like plant which is common on the barrocal near the coast. About half-way from Sagres to Cape St. Vincent the road takes a sharp bend to the west and the garigue is again full of interesting plants. There may be large groups of *Anemone palmata* and *Fritillaria lusitanica.* Other species include *Allium roseum, Convolvulus althaeoides, Iberis crenata* and *Teucrium pseudochamaepitys.* The iberis is an annual candytuft with brilliant white flowers and leaves with comb-like margins (crenate means with rounded or convex flattish teeth). It is found in other parts of southern Portugal and Spain but is rare outside this region. About a kilometer further along the road, on the south side, there are ruins of the small Beliche fort perched on the sea cliffs. The most noticeable species here is *Antirrhinum majus ssp.*

Algarve 1. Hyacinthoides hispanica 2. Fritillaria lusitanica 3. Tulipa sylvestris australis 4. Scilla verna

linkianum - a showy, red-flowered form of the garden antirrhinum which has the habit of climbing through the bushes by twisting its petioles around the branches. *Allium subhirsutum* is also common here; it is a small garlic with white flowers and leaves that are sparsely hairy. Another plant of some interest found growing around the fort is the low, spiny shrub *Lycium intricatum* with fleshy leaves and small, dusky-mauve, trumpet-shaped flowers. *Ophrys scolopax* has also been recorded from here, though it is not common in this area.

On the north side of the road, opposite the fort of Beliche, it is somewhat surprising to see colonised sand dunes. The flora is different here. *Cistus palhinhae* is absent, the main shrubs being *Cistus salvifolius, Halimium commutatum, Helichrysum stoechas, Lithodora diffusa*, and the wild olive. There is also the curious shrub *Corema album* which resembles a tall form of our cowberry *Empetrum nigrum*. Like the latter, it is also a member of the Ericaceae but prefers the hot dry conditions here rather than the cold wet climate of the Scottish and Welsh Highlands.

From the Beliche fort it is only a short distance to the lighthouse on Cape St. Vincent where the shrubs include:

Cistus palhinhae	*Pistacia lentiscus*
C. salvifolius	*Teucrium polium*
Helichrysum stoechas	*Rosmarinus officinalis*
Juniperus phoenicea	*Thymus camphoratus*

Cistus palhinhae, juniper and teucrium are dominant. The thymus is an uncommon species of southern Portugal with whitish flowers and, as its name suggests, a strong camphor-like smell. Amongst these shrubs grow:

Allium ampeloprasum	*Cerinthe major*
Armeria pungens	*Calendula suffruticosa*
Asteriscus maritimus	*Lobularia maritima*
Astragalus lusitanicus	*Silene littorea*
Bellis annua	*Narcissus obesus*

Biscutella vincentina   *Urginea maritima*

The allium is the wild form of our cultivated leek and the armeria a tall, robust thrift with spiny-tipped leaves. The calendula is a relatively large flowered perennial marigold that tends to be shrubby at the base. It is a variable species which is widely distributed in the southern part of the Mediterranean. In addition to these, one may see scattered plants of the pink-flowered *Centaurium erythraea ssp. grandiflorum* and the white-flowered perennial candytuft *Iberis procumbens*. The lighthouse area is remote and evocative and one frequently hears the calls of choughs which nest on the cliffs.

4.6 Sagres to Vila do Bispo: A twenty-minute bus ride from Sagres takes one to this small village just to the west of which an unsurfaced path (1256 on the map) leaves to join up with the Cape St. Vincent road near Beliche. It passes through well-watered farmland where the paper-white narcissus *Narcissus papyraceus* is fairly abundant in places though it often finishes flowering by the middle of March. The moist habitat here favours *Vinca difformis*, the small *Muscari neglectum* (O = *M. atlanticum*), and an occasional plant of the spectacular *Scilla peruviana*.

Another road from Vila do Bispo leads to the coast at Castelejo Aguia - an easy walk there and back. Here are fairly extensive plantations of the umbrella pine *Pinus pinea* which on this site grow not much larger than the sunshades of seaside cafés and are frequently damaged by a large wood-boring insect. They are accompanied in places by wind-stunted eucalyptus trees. Under the trees here it is possible to find plants of the unassuming small green-flowered orchid *Gennaria diphylla* growing with *Orchis morio*. The gennaria is rare in Europe and only found in the south west of the Iberian peninsula and first discovered in this area in 1985. It is particularly common in the Canary Islands. In a field near here one may sometimes see thousands of plants of the tiny endemic *Linaria algarviana* which colour the ground purple in places.

4.7 Around Carrapateira: The N125 road northwards out of Sagres joins the N268 for Lisbon after passing through Vila do Bispo. Some 25km from Sagres one comes to the village of Carrapateira which boasts substantial sand dunes at the top of the impressive sea cliffs. On the stony ground there is typical *Cistus palhinhae* garigue. The well-colonised sand dunes here are home to several species:

Asparagus acutifolius	*Corema album*
Allium subvillosum	*Helichrysum stoechas*
Antirrhinum majus linkianum	*Linaria sp.*
Armeria pungens	*Malcomia littorea*
Carpobrotus edulis	*Scilla monophyllos*
Centaurea sphaerocephala	*Silene colorata*

The *centaurea* is the sub-species *polyacantha* and one of the most beautiful of the genus with large reddish-mauve cornflowers that have spiny involucral bracts. The linaria had small yellow flowers and is probably *L. spartea*. In the slacks between dunes there are colonies of *Cryptostemma calendulacea* (= *C. calendula*, = *Arctotheca calendula*), a showy annual composite introduced from South Africa. It has rosettes of deeply-cut leaves and flower heads some 4cm across with yellow ray florets that are white towards the tips and dark on the underside, as well as blackish disc florets. One may also see it by the roadside and near buildings in some local villages where it was probably introduced as a garden plant.

4.8 Sagres to Portimao: Driving along the main N125 road from Sagres through Lagos to Portimao one will note that the road verges are planted in places with mimosa *Acacia longifolia* and *Myoporum tenuifolium* both Australian species - and occasional bushes of the beautiful native *Lygos monosperma* which has scented white broom-like flowers on hanging branches. Large groups of orchids can be seen on the banks in places, especially *Orchis morio*, *Ophrys fusca* and *Ophrys lutea*. Clumps of the common fennel

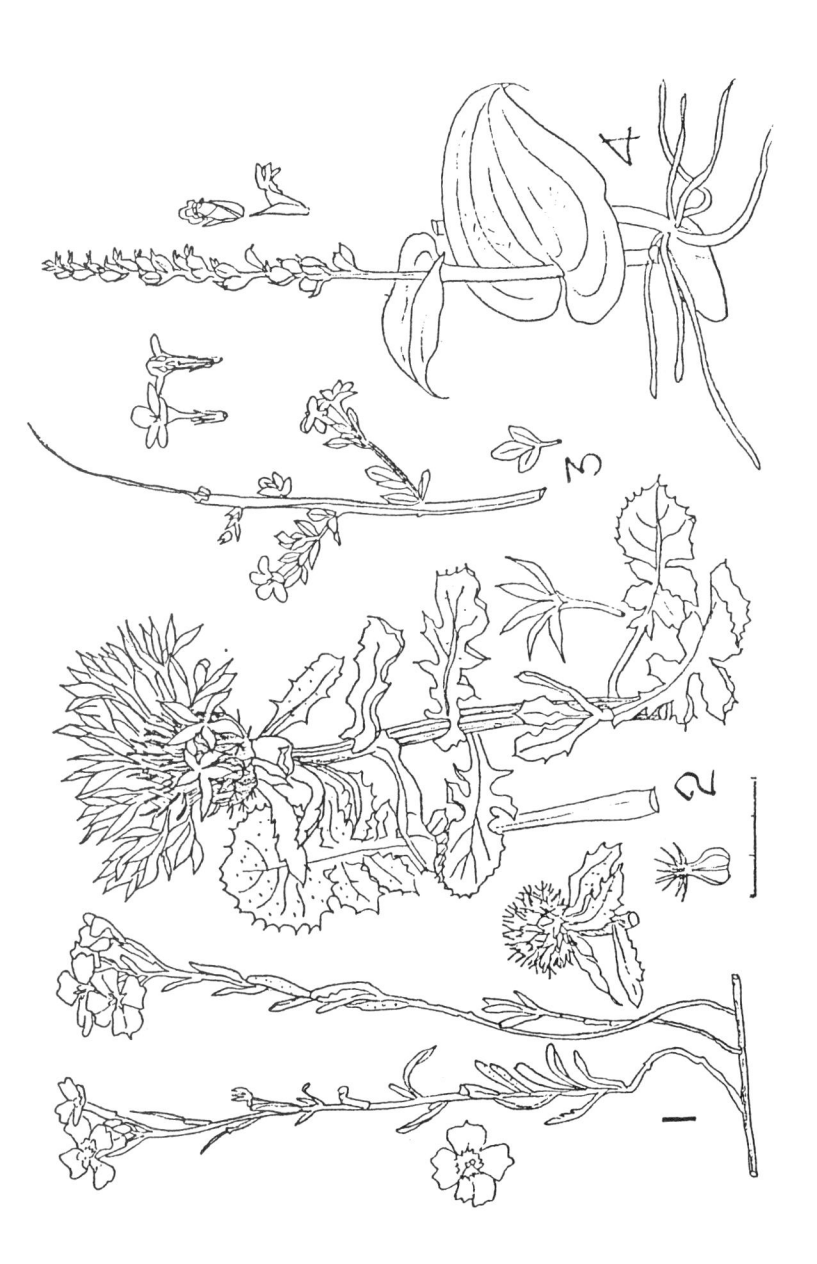

Algarve 1. Malcomia littorea 2. Centaurea polyacantha 3. Jasminum fruticans 4. Gennaria diphylla

Foeniculum vulgare are just coming through the ground in March but not usually flowering until July or later. Seeds of this wild form are sometimes used for flavouring but the leaves tend to be bitter; the cultivated sweet fennel is simply a form of this species which has leaves that are more suitable for culinary purposes. Another plant to be seen here is the rather strange vetch *Lathyrus ochrus* that has winged, leaf-like stems, tendrils and pale yellow flowers. It has been cultivated as a fodder crop and as a vegetable for its seeds since ancient times, but is rarely grown now.

Amongst the mixed, and often geriatric, orchards of figs, almonds, peaches and carobs that line the road in places, there are drifts of the ubiquitous yellow *Oxalis pes-caprae* which contrast with sheets of the purple *Fedia cornucopiae*. The last of these is like a small annual valerian and the form here may be a sub-species with two distinct horn-like appendages which give it the name *F. caput-bovis* (ox head). Later in April and May these plants are joined by the yellow *Chrysanthemum coronarium* and *Coleostephus myconis* (= *Chrysanthemum myconis*).

Keeping on the main N125 road one passes through the pleasant small town of Lagos and then to Portimao from where there is a secondary road that goes south towards the lighthouse at Ferragudo. The top of the cliffs here is a charming environment on a sunny day. A short distance out to sea there is a large nesting colony of cattle egrets on a rock stack they share with a few cormorants. These birds are fairly common inland accompanying cattle but have the habit of disappearing behind the cows when one approaches to photograph them. The terrain here is typical barrocal with cistus, gynandriris, lithodora and rue *Ruta chalepensis*. Many butterflies can also be seen, including the cleopatra, Chapman's green hairstreak, wall brown, painted lady, red admiral, clouded yellow, swallowtail, small copper and Spanish festoon.

4.9 Serra de Espinhaço do Cao: This is a low range of hills that stretch some 35km parallel to the west coast but never exceed 250m in height. They are mainly composed of acid rocks so the flora differs somewhat from that of the limestone barrocal. There is an artificial lake dammed by the Barragem da Bravura (Barragem de Odiáxere on some maps) about 15km north of Lagos. One can approach it either from Odiáxere on the N125 road and then northwestwards along the 125-9 to the barragem or avoid Lagos and leave the main road at Vale de Boi to Bensafrim to join the 125-9 at Cotifo.

The route via Vale de Boi takes one through dense stands of *Cistus ladanifer* with very few plants growing below except *Scilla verna* and *Arisarum vulgare*. Approaching the dam one also sees a beautiful form of *Erica australis* the Spanish heath which grows 2m tall and has pink flowers in early summer. It is accompanied by other shrubs including *Lavandula stoechas, Myrtus communis, Ulex parviflorus*, our own native heather *Calluna vulgaris* and *Gynandriris sisyrinchium* grows in profusion in places. Around the lake there is a monoculture of eucalyptus and one feels somewhat apprehensive at the way these trees are taking over from the native flora in parts of the Mediterranean. However, eucalyptus is favoured for planting in wet areas since it is reputed to deter mosquitoes. Under the trees there is little growing except for groups of *Scilla monophyllos* in a range of colours from white through pale to dark blue. But it is calm and peaceful here in spring with honey-scented air and the sound of millions of bees working the eucalyptus flowers.

Another way over the Serra de Espinahço do Cao is to take the 120 road out of Lagos or to join it at Bensafrim. This crosses the range and leads to Alfambras and on to Lisbon. At the higest point of this road (248m) there are cork oak plantations with the heathers *Erica arborea, E. lusitanica, E. australis* accompanied by *Cistus ladanifer*. In places there are specimens of the strawberry tree *Arbutus unedo* called the 'Medronheiro' in Portugal where the fruit is used to produce a fiery spirit of that name. Another plant to be seen round here is *Bellis sylvestris*, like an extra large perennial daisy that flowers from September to May and contrasts with the tiny annual *Bellis annua* found in thousands in garigue and by the roadside near Sagres.

The road leads on from Alfambras to near the coast at

Arrifana and around here one may see the attractive Mediterranean buckthorn *Rhamnus alaternus* that has bright red berries in late summer and somewhat resembles a holly, though the leaves are not prickly. Other plants that may be found here include the buttercup *Ranunculus paludosus*, which just gets into the British flora as it is found in Jersey and the beautiful large, blue pimpernel *Anagallis monelli* but much of the land here is now being taken over for housing.

4.10 **Monchique:** This is the highest point in southern Portugal and rises to 902m. It is most easily approached by driving north out of Portimao along N124 to Porto de Lagos. A right fork in the road here takes one to the interesting old Moorish capital of Silves and in mud flats near here the ground is covered with the curious little camomile-like *Cotula coronopifolia*. The capitulae have no ray florets and the disc florets are yellow, so the plant aptly deserves its name 'brass buttons'. It is an introduced South African plant that is rapidly spreading throughout the Mediterranean, especially in wet areas but it will also grow on dry ground.

If, instead of taking the turning to Silves, one continues northwards along N266 passing through well-watered country with orange orchards one eventually arrives at Caldas de Monchique. This is a quaint small spa which was originally used by the Romans and now has an old pump house that reminds one of a small version of the Brighton Pavilion. There is a kind of 'fairy glen' through which the spa waters flow and where *Viburnum tinus* bushes thrive, sometimes festooned with the curious climbing *Aristolochia baetica*. The viburnum is the 'laurustinus' of our gardens and a native of the Mediterranean where it flowers profusely during the winter and early summer.

After Caldas de Monchique the road passes between the two peaks of Foia to the west and Picota to the east before arriving at the town of Monchique. The whole of this area was of great botanical interest at one time with forests of *Quercus canariensis* and *Q. faginea* but these have nearly all been cleared and replaced by plantations of pines and eucalyptus. There is a convenient road

(263-3) to the summit of Foia which is rather spoiled by telecommunication masts and is now practically treeless. *Cistus salvifolius* is common here and in the spring the ground is decorated with the tiny mauve flowers of *Romulea bulbocodium*. There are also groups of *Paeonia broteroi* which do not come into flower much before May. The Spanish bluebell *Hyacinthoides hispanica* flowers here in March and April but does not grow in large numbers.

The summit of Picota, to the east of the main road, is less accessible and one must climb it on foot, but it is more interesting than Foia. On the way up, *Saxifraga granulata* and the local asphodel *Asphodelus morisianus* (which resembles a rather robust form of *A. aestivus* and is not recognised by Flora Europaea), grow in chestnut plantations. Near the rocks on the summit there are *Erica australis*, *Cistus crispus* and *Halimium commutatum*. A speciality of the region is *Myrica faya* a member of the family Myricaceae and closely allied to our bog myrtle (*Myrica gale*). It is a large bush resembling the strawberry tree but has narrower leaves and produces catkins. Like other members of the family it prefers moist or wet acid soils. Polunin and Smythies (1973) also quote that *Campanula primuliflora*, *Centaurea longifolia*, *Leucoium longifolium* and *Senecio grandiflorus* may be found here. The leucoium is an uncommon species which may be extinct now on Picota. It resembles *L. trichophyllum* but has smaller flowers and more rounded tepals. It is also found in Corsica and Pignatti (1982) claims it to be endemic there! In woods lower down one may see *Rhododendron ponticum* which has become a troublesome 'weed' in some parts of Britain but is probably genuinely native to this part of the world though its special home area is the Pontus region of Turkey to the south east of the Black Sea around Trabzon.

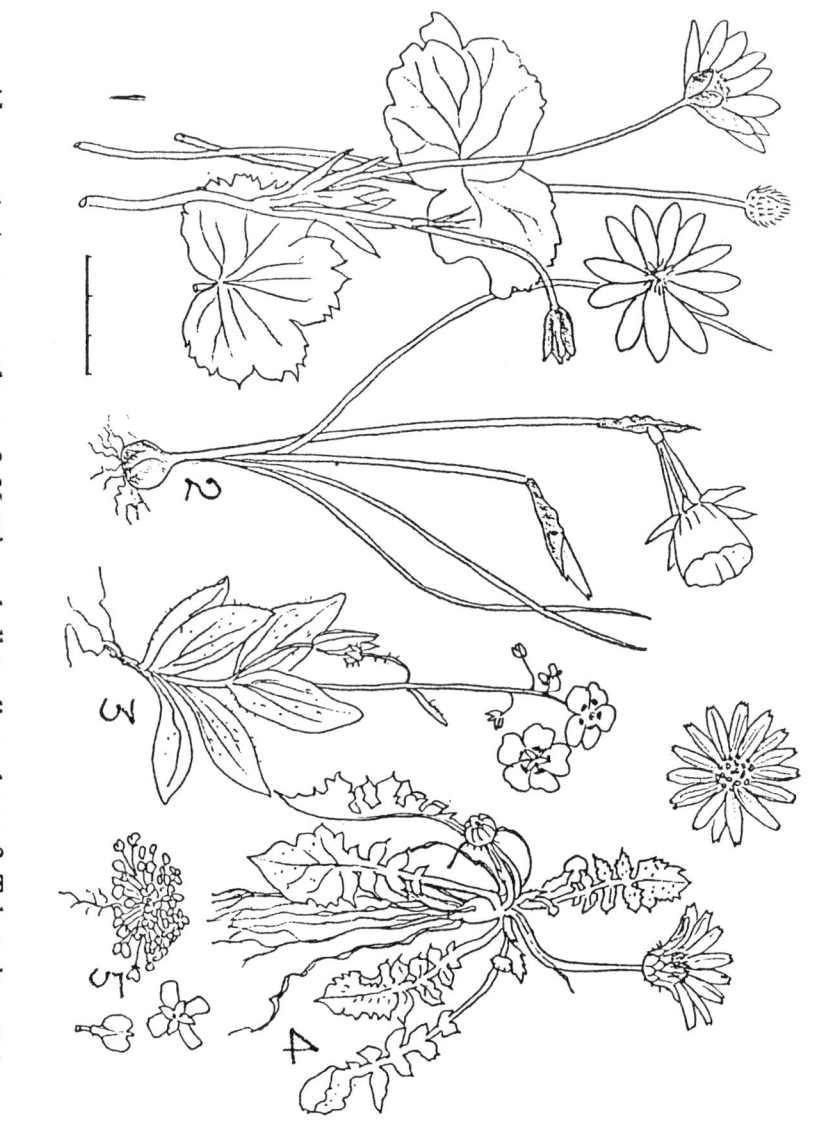

Algarve 1. Anemone palmata 2. Narcissus bulbocodium obesus 3. Tuberaria guttata 4. Cryptostemma calendula 5. Inopsidium acaule

5. Spain West of Gibraltar

5.1 As one travels westwards along the Spanish coast past Gibraltar it is sometimes possible to detect a change in the weather. The effect of the Atlantic may become evident. The rainfall is slightly higher here than further east and a strong wind frequently blows from the south west. Although the coastline is strictly Atlantic, the climate and vegetation are still typically Mediterranean.

The sandy coastal strip known as the Costa de la Luz is a holiday region used mainly by the Spaniards themselves and by those tourists who prefer camping; package holiday firms do not very often include it in their brochures. However this part of the world is a good place to watch migrating birds in spring and autumn and the Coto de Doñana to the west of the region is a Nature Reserve world renowned for its animal life. To take advantage of this facility, several specialist holidays are arranged for ornithologists to visit the region at appropriate times. The flora differs somewhat from that to the east of Gibraltar and is well worth examining..

Algeciras, which is near to the airport at Gibraltar, is a convenient place to stay for studying the flora of the eastern end of this region. However, it is not a town planned with tourists in mind though the interesting Hotel Reina Cristina there may appeal to those who like a little extra comfort. For the Coto de Doñana region, further west, one will need to find accommodation at the resort of Matalascañas, Huelva, the Parador Cristobel Colon, Mazagón or close to the Spanish/Portuguese border. There is much to interest the plant enthusiast in this part of Spain, especially the flora of the sandy coasts between Tarifa and Cadiz which is probably the finest of its kind in the whole Mediterranean region but, alas, rapidly being taken up for building. The best time to come is in April or early May but there are still a few interesting plants to see in flower during the autumn.

5.2 **The Terrain:** At the eastern end of the region, the Sierras del Niño, Blanquilla and Algibe rise to over 600m (with the highest point at 1092m) and drop to the sea as cliffs between Algeciras and Tarifa. They are composed of sandstone overlying limestone and carry an interesting flora of heathers and other acid-loving species.

The coastline between Tarifa and Cape Trafalgar and further on to Cadiz is relatively flat and has long sandy beaches which constitute the Costa de la Luz. North-west of Jerez lies the delta of the river Guadalquivir and the sands of the Coto de Doñana with wetlands and moving sand dunes.

5.3 **Literature:** Polunin and Smythies (1973) and Stocken (1969) (6.3) give some information on the plant hunting areas of this region. The most detailed descriptions of species from here are to be found in the new, three volume Spanish flora of Andalusia (Valdés et al 1987) which is well illustrated with line illustrations. For those who like drawing-room table books, there are lavishly-illustrated works on the Coto de Doñana which deal principly with the fauna there.

5.4 **Coastal region between Algeciras and Tarifa:** It is a fair day's walk out of Algeciras southwards to the lighthouse on the Punta del Carnero. The land to the west and south of the Hotel Reina Cristina, which was originally isolated 'in the country', has now been used for housing, nearly as far as the beach at Getares. However there are a number of interesting roadside 'weeds' and other plants en route including the cocklebur *Xanthium strumarium*, the mandrake *Mandragora autumnalis* and the stemless thistle *Atractylis gummifer* which produces its impressive purple flowers in late summer and autumn. Continuing past the Playa de Getares it is not difficult to get on to the rough ground beside the road that leads to the lighthouse. The flora here is fairly typical of the rocky parts of the coastal area as far as Tarifa and includes:

Calicotome spinosa	*Pallenis spinosa*
Carduncellus caeruleus	*Phlomis purpurea*
Chamaerops humilis	*Rubia peregrina*
Daphne gnideum	*Scolymus hispanica*

Delphinium gracile *Smilax aspera*
Dittrichia viscosa *Urginea maritima*
Foeniculum vulgare *Verbascum sinuatum*
Linum maritimum

The delphinium is a well-branched annual or perennial with few leaves and small, dark-blue, long-spurred flowers in summer; it is fairly common in low-lying parts of the region. *Dittrichia viscosa* (O = *Inula viscosa*) is a very common species throughout the Mediterranean and often the first plant to be seen when arriving at a holiday airport. It frequents the roadsides and covers waste places with its yellow flowers in autumn. In addition to these plants there are 2-3m tall specimens of wild olive. These differ mainly from the cultivated varieties by having some thorns, narrower leaves and smaller fruit but they hybridise readily with cultivated trees and intermediate forms occur. Occasionally one may also see the beautiful chaste plant *Vitex agnus-castus* here.

5.5 Sierra del Niño and Sierra de Algibe: The approaches to this region are mainly via the C-440 road, from Algeciras to Alcala and Jerez, and the small roads which turn off it in a westerly direction. Unfortunately for plant enthusiasts, this highway has been improved and the surface raised leaving deep ditches on either side to take away water from flash floods so it is now difficult to find safe stopping places. Furthermore it is designated as a 'Ruta del Toro' on account of the greatly increased cattle raising and the fields are very securely wired in; judging from the appearance of some of the bulls, this seems a wise precaution. The notices 'Coto de Caza Privado' also mean more than they did in the past when one just needed to keep clear of the area if hunting was in progress. Now they have been very effectively fenced, perhaps to keep the game from straying on the road as much as to keep poachers out, but it is often difficult for plant enthusiasts to get over them.

The whole region is covered with scattered groups of cork oaks which are indicative of acid soil conditions. Amongst them grow tall maquis shrubs including an extraordinary number of heather species:

Arbutus unedo
Cistus populifolius
Cytisus villosus
Calluna vulgaris
Erica arborea
E. australis
E. ciliaris

Erica erigena
E. scoparia
E. umbellata
Halimium lasianthum
Quercus fruticosa
Rhododendron ponticum
Viburnum tinus

The ericas are not easy to identify but are dealt with rather well on pages 286-287 in Polunin and Smythies (1973). The quercus is a suckering semi-deciduous shrub growing to 2m. The rhododendron is the subspecies *baeticum* which is special to the Iberian Peninsular, and found also further west in Portugal (4.10).

If one can get through the wire netting defences there are some special plants to be found in the undergrowth, notably the mouse plant *Arisarum proboscidium* which has its only station outside Italy here. Its mauve-grey spadix has a long fine projection on the tip and looks like a small mouse dissappearing under the soil. It is perfectly hardy in Britain and sometimes grown as a garden curiosity. *Fritillaria lusitanica* can be found here and the fine, blue-flowered milkwort *Polygala microphylla* is fairly common. In any damp places there is a chance to see *Sibthorpia europaea* the Cornish moneywort which is a rare British native. It belongs to the family Scrophulariaceae and is a creeping plant with rounded leaves and tiny, axillary, white or cream flowers. One may also be fortunate to see the curious sundew-like plant *Drosophyllum lusitanicum* (5.7).

The main coast road N-340 (also referred to as E-25) skirts the southern edge of this region and Stocken describes sites for two interesting plants along here. Some 10km west of Algeciras there is a viewpoint at Puerto del Bujeo and in cork woods to the north of the road he found *Narcissus viridiflorus* which is a small jonquil with scented green flowers in the autumn. It is primarily a Moroccan native which has a precarious foothold here in Europe and should

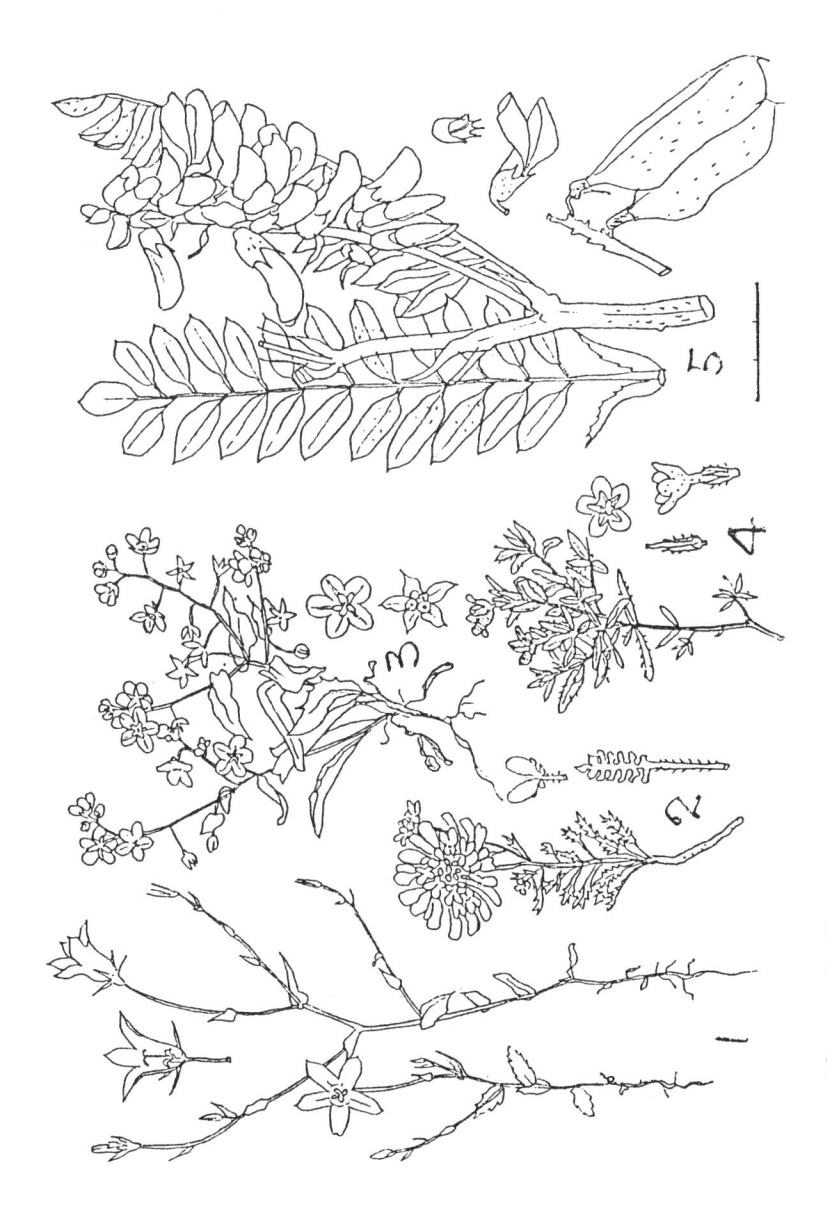

Algarve 1. Campanula lusitanica 2. Iberis crenata 3. Omphalodes linifolia
4. Lithodora diffusum lusitanicum 5. Astragalus lusitanicus

Costa De La Luz 1. Daphne gnidium 2. Gomphocarpus fruticosus 3. Scilla autumnalis 4. Linum maritimum 5. Urginea maritima

be looked for in damp areas with an acid soil. There seems to have been some building in the area and the site is difficult to locate with certainty. Another recording described for this narcissus is by a large rock a few kilometers west of Alcala de los Gazules on the C-440 road. Further on, towards Tarifa, one comes to a viewpoint at the Puerto del Cabrito close to a group of wind generators set up to take advantage of the strong winds that blow through the Straits of Gibraltar. Near here, in autumn another rare narcissus is said to occur. This is *Narcissus cavanillesii* which used to be known as *Tapeinanthus humilis* and sometimes *Narcissus humilis*. It also is a small plant that has its main site in North Africa and grows with rush like leaves and upright-facing yellow flowers in autumn.

Some 20km along the N-340 westwards out of Tarifa a road leads eastwards across the region to join up with the C-440 at Puente de las Cañas. It has a rough surface and one is forced to drive slowly but it is worth making the journey for the eastern half of this route passes through cork woods where it is fairly easy to get access to the countryside. Plants to be found here include *Gladiolus illyricus* and fine specimens of *Colchicum lusitanum* (= *C. autumnale ssp. algeriense*) during the autumn. The latter is rather variable in size, shape and degree of chequerboard markings (or tessellation) of the perianth. Some specimens probably have the largest flowers of the genus. By a stream here, with a degree of good fortune, one may see *Lobelia urens* which is not a common plant in this part of the world. From a farm called Cortijo de Ojén along this road a track leads up the Sierra de Niño through cork oak woods to areas with *Quercus fruticosa, Arbutus unedo* and *Chamaespartium tridentatum*. A number of ferns grow here including the royal fern *Osmunda regalis* and the hare's foot fern *Davallia canariensis*. *Fritillaria hispanica* and *Romulea bulbocodium* may also be found in the vicinity. The fritillaria is often considered to be simply a form of *F. lusitanica*.

5.6 Shore flora between Tarifa and Cabo de Trafalgar: Access to this very interesting shore line is limited in places due to military restrictions and camp sites. Tarifa is an interesting old town with a reputation for having a very windy climate. Along the beach here grow the usual shore plants such as *Crithmum maritimum, Eryngium maritimum,*and *Xanthium strumarium.*

Some 10km westwards, along the N-340 out of Tarifa, there is a turning to the left for Punta Paloma. One arrives near the shore at the Playa de Valdevaqueros which has a car park. By the river bank here are large groups of the showy, red-flowered *Centaurium erythraea* and near to the car park there is an enormous sand bank with large groves of the broom-like *Lygos monosperma* growing to 2m high. The hanging branches with scented white flowers make a beautiful sight in March and April when the ground below is carpeted with *Anagallis monelli, Echium gaditanum, Iberis linifolia* and *Malcomia littorea.* Growing in damp areas there are plants of the attractive silkweed *Gomphocarpus fruticosus* which was originally introduced from South Africa as a potential alternative to cotton and has established itself in parts of the Mediterranean. It has heads of hanging white flowers followed by curiously-shaped and inflated hairy fruits. In late summer it is possible to see the rare 'plain tiger' butterfly *Danaus chrysippus* (tigre mariposa in Spanish) which is bright orange brown with black and white markings on the leading edge of the forewings. This butterfly is a rare migrant from Africa and the larvae feed on members of the the Asclepiadaceae. I have watched it here laying its eggs on the gomphocarpus - perhaps, with the changing climate this elegant butterfly may be moving further north.

Near the giant dune a road leads off in a north westerly direction through woods of stone pine and there is a promise of interesting plants but it is a military restricted area and entrance is not permitted. Fortunately a similar area is to be found further along the coast. The next turning to the left from Tarifa leads to the Roman archeological site of Bolonia where excavation is still in progress. *Iris xiphium* is especially common here and along a side road that leads off in a westerly direction one can find *Leucoium trichophyllum* in February to March and *Triguera ambrosiaca* and *Viola arborescens.* The triguera is an annual member of the Solanaceae with large deep-mauve flowers. The viola is a low shrubby species, common near Cape St. Vincent in Portugal (4.4) and also found further east (7.10). Unfortunately there are more military restrictions further along the road so one has to return the way one came.

Back on the main N-340 and heading towards Vejer one passes through farmland which becomes flooded at times and where crops of cotton are grown. At certain times of the year, when the ground was partly under water, one could see many storks and an occasional marsh harrier and in places there were very large plants of *Scilla peruviana* with rosettes 30cm across and correspondingly large flower heads. Much of this land has been drained and these wet lands now hardly exist. There is a turning to Zahara de los Atunes to the south but, unfortunately for the plant enthusiast, this is yet another restricted area and one cannot get off the road before reaching the unattractive fishing port of Barbate de Franco. Approaching this town there is much *Eryngium maritimum* and *Pancratium maritimum* growing on the beach with some *Otanthus maritimus*. A new bridge crosses the Rio Barbate and a road leads towards the interesting old hill town of Vejer de la Frontera. There is also a road along the foreshore which, at the west end of Barbate, enters an area of pine woods and sand dunes which extend to Los Caños near to Cabo de Trafalgar. This is a Parc Nacional and of extreme interest to botanists. Plants growing in the sandy soil under the pines include:

Armeria macrophylla	*Iris xiphium*
Clematis flammula	*Leucoium trichophyllum*
Crocus clusii	*Lygos monosperma*
Cyperus kalli	*Malcomia littorea*
Delphinium gracile	*Ononis natrix*
Dianthus broteri	*Romulea clusiana*
Iris filifolia	*Ruscus aculeatus*

The crocus flowers in autumn with pointed perianth segments that are violet-coloured with a white throat; it is a form or subspecies of *C. serotinus*. There are short hairs at the base of the corolla but these can only be seen with the aid of a pocket lens. There are also a number of orchids in flower here between March and April.

The rocky edge of these woods leading down to the sea shore carries an extraordinarily rich flora. The upper slopes are mostly densely covered with shrubs which thin out toward the sandy beach. They include:

Cistus libanotis	*Lavandula dentata*
C. populifolius	*L. stoechas*
C. salvifolius	*Lygos monosperma*
Corema album	*Ononis variegata*
Halimium lasianthum	*Pistacio lentiscus*
Juniperus oxycedrus	*Rosmarinus officinalis*
J. phoenicea	*Thymelaea hirsuta*

The most uncommon of these is *Cistus libanotis* which resembles *C. clusii* but the leaves are larger, broader and hairless. The sticky calyx has 3 instead of 5 sepals as in the other species. The corema is an interesting plant that strangely resembles the cowberry *Empetrum nigrum* of our moorlands at home and belongs to the same family. It is not very common here but grows in quantity near Cape St. Vincent in Portugal (4.7).

In between these shrubs there is a colourful carpet of flowers in early summer, including the following species:

Allium stramineum	*Iberis linifolia*
Anagallis monelli	*Malcomia littorea*
Antirrhinum barrelieri	*Romulea gaditana*
Armeria macrophylla	*Ruta chalepensis*
Centaurea polyacantha	*Viola arborescens*
Dipcadi serotinum	

The allium is a rather uncommon species with yellow flowers. The anagallis is a very beautiful pimpernel with large blue flowers. It is occasionally grown as a half-hardy annual in gardens at

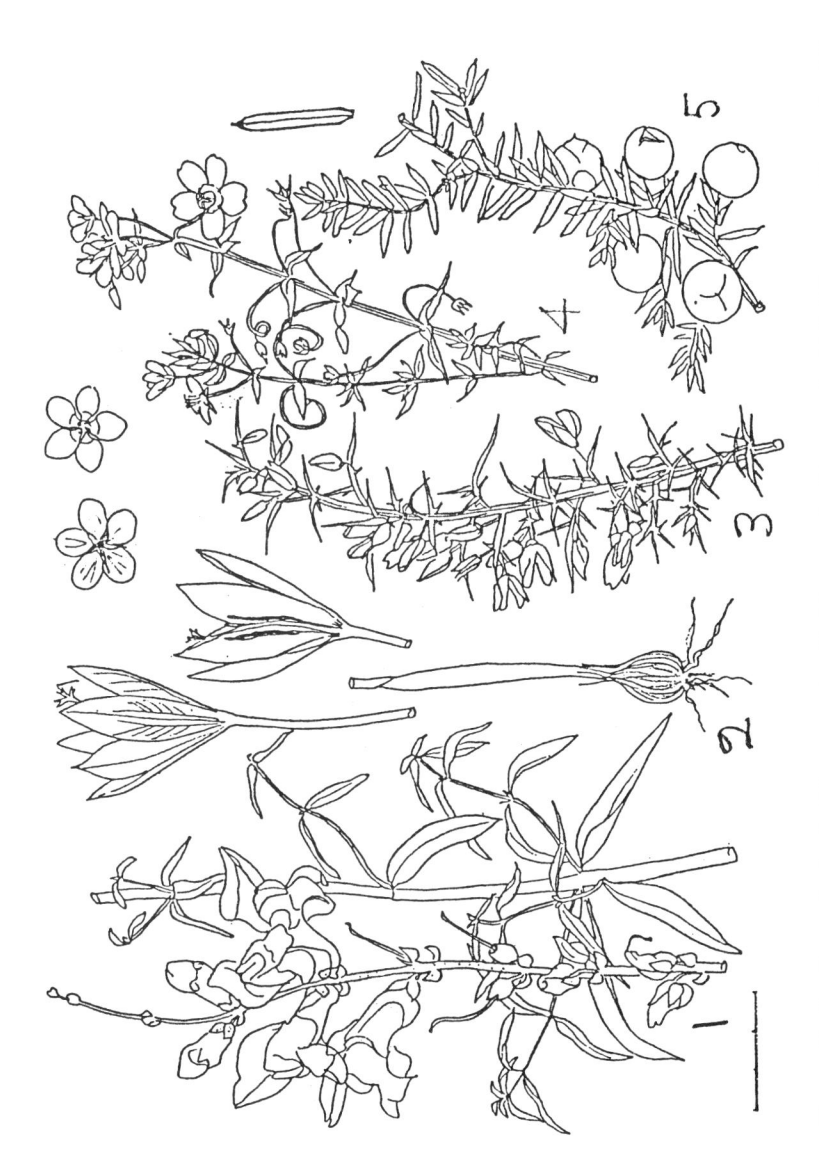

Costa De La Luz 1. Antirrhinum majus linkianum 2. Crocus clusii 3. Ulex parviflorus funkii 4. Anagallis monelli 5. Juniperus oxycedrus macrocarpa

home though in the wild it is a short-lived perennial. The antirrhinum is a rather tall growing species which is able to climb through bushes and has red flowers but it is possible that some of the plants here may be forms of *A. majus* or *A. granaticum* - the genus is complicated. The centaurea is one of the finest of the genus with large pink flowers. The romulea has relatively large purple flowers with white stigmas which are shorter than the stamens. It is more or less confined to the Cadiz province as its specific name suggests and is included in Flora Europea as a sub species of *R. ramiflora*.

At the Barbate end of the pine wood there is a small damp area leading to the shore and here *Asclepias curassavica* has become naturalised. This species, which Stocken (1969) also records as growing by Maro, near to Nerja east of Malaga, is a North American species belonging to the Asclepiadaceae. It has brilliant orange and red flowers in late summer and like the gomphocarpus attracts the rare plain tiger butterfly seen near Punta Paloma (5.6). Other rare butterflies worth looking out for in this region are the cardinal, two-tailed pasha and the Spanish festoon.

5.7 **The Coto de Doñana:** Based on the delta of the river Guadalquivir, this is the largest 'wetland' in Europe - about the same size as the Isle of Wight. It is an especially important area for different kinds of herons and other water birds and was made a state National Park in 1967. Sadly, it is under considerable pressure from a lowering of the water table for domestic purposes and crop irrigation, from agricultural chemicals and from sewage and industrial waste. Permission for access is difficult to obtain and almost entirely arranged in organised parties carried through the reserve in motor vehicles - a mode of transport that is not very conducive to botanizing. Indeed, it is hardly worth the effort to make arrangements to look for plants there for most of the species can be seen elsewhere in the vicinity as around the relatively new tourist centre of Playa de Matalascanas, the Parador Cristobal Colon near Mazagón and around the coast west of Huelva which borders Portugal.

The Monastery of La Rabida, just to the east of Huelva, is a good place to start. In the spring exciting plants such as *Narcissus bulbocodium, N. gaditanus* and *Leucoium trichophyllum* are fairly plentiful. *Narcissus gaditanus* is a rather rare small jonquil which is distinguished by having a slightly bent corolla tube (7.5). The shrubs that grow around here include:

Cistus salvifolius   *Halimium commutatum*
Erica erigena   *H. halimifolium*
E. scoparia   *Lavendula stoechas*

Erica erigena (= *E. mediterranea*) is closely allied to *E. carnea* but it grows somewhat taller and flowers later in March and April. It manages to get into the flora of the British Isles for it is found in boggy heaths of Mayo and Galloway in Ireland and forms of it are often grown in gardens in Britain. Non-shrubby species of interest to be found in the reserve and elsewhere in the vicinity include:

Armeria baetica   *Iris xiphium*
A. gaditana   *Limoniastrum monopetalum*
Astragalus lusitanicus   *Triguera ambrosiaca*
Drosophyllum lusitanicum

Both of the armerias are uncommon species limited to the southern part of the Iberian peninsula; the second is particularly handsome and tall with large pink flowers. The drosophyllum is also special to this area and well worth searching for - a sub-shrubby carnivorous species, allied to the sundews, with long leaves, coiled when they are young, and bearing red-tipped glandular hairs. The flowers are bright yellow, some 25mm diameter, and borne on stalks up to 30cm tall. Unlike our sundews, to which it is closely allied, it prefers sandy soils near the sea rather than wet heathland bogs. Limoniastrum is a shrub with silvery leaves that grows in salt marshes and sands near the coast. It has bright pink flowers some 15mm in diameter and is sufficiently attractive to be cultivated in some seaside resorts. The triguera is a handsome annual with bell-shaped deep violet flowers - a member of the Solanaceae (5.6).

The Playa de Mazagon is a long sandy beach with the Parador Cristobal Colon near the eastern end. From here the road continues eastwards through a wood of umbrella pines to Torre de Higuera and the relatively new resort of Matalascañas. The wood and beach are home for several of the interesting plants one sees further east around Barbate de Franco (5.6).

East of the road that leads from Matalascañas northwards to El Rocio is the Coto de Doñana reserve. Here are three types of habitat; seashore, sand dunes and salt marshes called 'marismas' - similar to what is found in the Camargue of southern France (10.8). The salt marshes which flood in winter and dry out during the summer have a number of small islands known as 'vetas'. Amongst the more interesting species which grow here are: *Cotula coronopifolia* an introduced plant from South Africa with composite heads of yellow disc, but no ray florets and aptly described as batchelor's buttons, *Lathyrus palustris* which is a fairly typical vetch with three pairs of leaflets and mauve flowers and the rather strange *Scorzonera fistulosa*. The last of these has typical yellow flower heads of the genus but hollow, cylindrical, ribbed leaves as its specific name suggests (fistulosa means hollow or pipe-like) and it thrives in wet areas of Spain and Portugal. However, the enthusiast may well find all three of these plants growing in the surrounding country outside the reserve. This is helpful for it has been said that it is easier to get out of Alcatraz than into the Coto de Doñana!

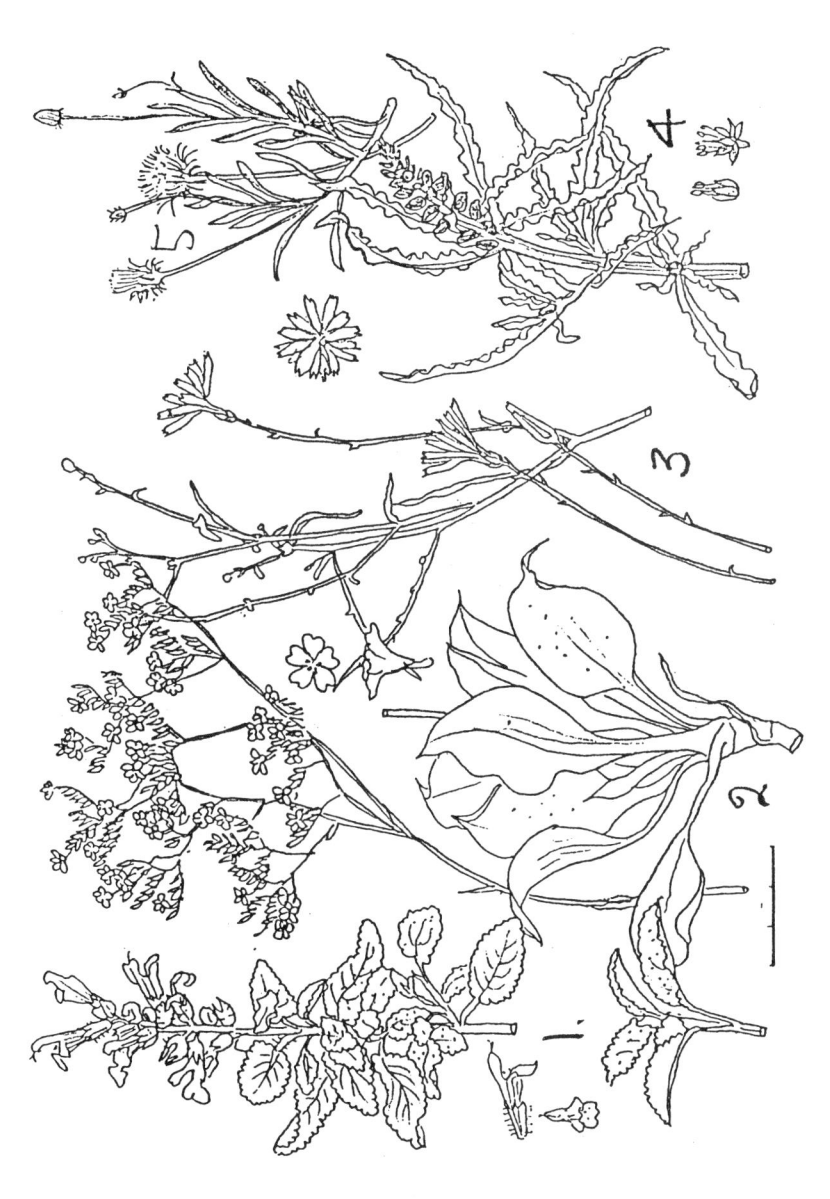

Andalusia 1. Salvia fruticosa 2. Limonium ovalifolium 3. Launea resedifolia 4. Lysimachia atropurpurea 5. Phagnalon rupestre

Andalusia 1. Kickxia commutata 2. Solanum bonariense 3. Dittrichia viscosa 4. Polygonum equisetiforme 5. Antirrhinum granaticum

6. Western Andalusia

6.1 The well-known tourist feature of this region, is the 'Costa del Sol' which stretches along the coast between Malaga and Gibraltar. It was given this name by General Franco to encourage tourism and, having started the 'costa' craze, it has probably attracted more sun-seekers than any other part of the Mediterranean. Behind its facade of hi-rise hotels, pubs, discos and restaurants there is a back-cloth of mountains aptly described in a Spanish poem as 'triste y callando' (sad and silent). It is easy to get to these wild places from the main holiday centres such as Torremolinos, Fuengirola, Marbella and Estepona and they have much to offer the plant enthusiast. Here, is the region par excellence to see cistus and allied species such as halimium, which are at their best in April and May. A number of interesting bulbous plants flower in the autumn and winter but the time to see them varies somewhat from year to year according to the onset of autumn rains. However a few, such as the colchicums, flower regularly in October irrespective of rainfall.

Package holidays to any of the major resorts of the Costa del Sol provide convenient stepping-off places to examine the flora but if one wishes to make one's own arrangements then there are regular flights to Malaga and Gibraltar. The Firestone C9 map covers this area and extends as far west as Cape St. Vincent in Portugal. The Hallwag map of the Costa del Sol is on a slightly larger scale (1:200,000) and carries some useful additional details.

6.2 **The Terrain:** Although the coastline is mainly 'developed' and of relatively little importance botanically, all the mountains behind are of great interest, especially those in the west of the area. Just north of Estepona and Marbella lie the Sierra Bermeja ('bermeja' means red or vermillion coloured) which have an overlay of acid rocks and a red soil due to the abundance of iron oxide. North of this range is the Serrania de Ronda, comprised of hard dolomitic limestone that has been compressed to marble in places and carries a special flora with several endemic species. The lower

Sierra de Mijas, of similar limestone composition, also has interesting plants, notably in the north-east. Further inland from Mijas are the fantastic limestone shapes of El Torcal which resemble a ruined city from a distance and provide a habitat for several special plants.

In the extreme west of this region Gibraltar rises at the limit of the Mediterranean proper. It is a limestone outcrop with a flora that shows the influence of its proximity to North Africa.

6.3 **Literature:** Perhaps the most interesting and useful small book about the region is that by Stocken (1969). The author was a naval officer stationed on Gibraltar and studied the plants of western Andalusia over a period of years. He was unfortunately killed in an accident in Greenland and the book was produced from his notes. It is a charming and very informative publication and the reader is made to feel that the author is standing at one's elbow as a guide. Unfortunately it is out of print and not easily available. However, Polunin and Smythies (1973) deals with the region fairly thoroughly. For those who wish to make a more detailed study there is the excellent new three-volume Spanish flora of western Andalusia (Valdés et al 1987) which is well illustrated with line drawings and distribution maps..

6.4 **San Pedro to Ronda:** All visitors to the Costa del Sol will wish to visit the attractive old town of Ronda and the direct, and botanically most interesting, way there is from San Pedro de Alcantara along the C-339. This has recently been widened and carries a fair amount of traffic but there are a reasonable number of stopping places en route and few fences.

The first half of the journey is through the Sierra Bermeja which is covered with forests of *Pinus pinaster* of varying density and some cork oak plantations shortly after leaving the main road. Shrubs under the pines and oaks include:

Arbutus unedo	*Erica erigena (5.7)*
Calicotome spinosa	*E. scoparia*
Cistus albidus	*Halimium atriplicifolium*

C. populifolius
C. salvifolius
C. ladanifer
Erica arborea

Lygos sphaerocarpa
Pistacia terebinthus
Viburnum tinus

Cistus populifolius is one of the less common species of the genus, limited in distribution to the south of Spain and a few parts of the south of France. The plants here grow to some 2m tall, with white flowers and heart-shaped leaves, and probably belong to the sub-species *major*. The halimium is perhaps the most charming of the genus and found only in southern Spain. It also is fairly tall growing and has panicles of yellow flowers some 4-5cm diameter; the leaves are silver-grey and three nerved. A number of herbaceous species and sub-shrubs which grow beneath the above include:

Aphyllanthes monspeliensis
Coronilla juncea
Digitalis obscura
Iberis pruitii

Linum narbonense
L. suffruticosum
Valeriana tuberosa
Verbascum giganteum

The aphyllanthes is an interesting and decorative member of the liliaceae without leaves and rush-like stems carrying bright blue flowers. It is not common outside its 'homeland' in the south of France where it grows in profusion (10.14)- hence its name *monspeliensis* (from Monpellier). The digitalis is generally referred to the sub-species laciniata. It is a partly shrubby perennial with rusty red flowers and is occasionally grown in British gardens. The iberis is an annual candytuft with white or mauve flowers and somewhat fleshy leaves. *Linum narbonense* has blue, and *L. suffruticosum* white flowers. The verbascum is also a sub-species special to the area. Careful searching may also reveal the linaria-like *Anarrhinum laxiflorum, Alyssum serpyllfolium, Saxifraga dichotoma* and *Tolpis barbata*. In February to April a rather fine form of *Narcissus cantabricus*, said to be stoloniferous, can be seen in rock crevices.

About half way between San Pedro and Ronda the road

passes from the peridotite rocks of the Sierra Bermeja to the hard limestones of the Serrania de Ronda and there is a fairly abrupt change in the vegetation. Relatively few trees grow on the limestone which carries a garigue containing the following shrubby species:

Cistus albidus   *Lavandula stoechas*
C. crispus   *Phlomis purpurea*
C. ladanifer   *Pistacia lentiscus*
C. monspeliensis   *Ulex parviflorus*
Lavandula lanata

Lavandula lanata is a typical lavender of the higher regions of Andalusia including the Sierra Nevada. It has rather broad, grey, felted leaves, heads of dark purple flowers and a distinct camphor-like smell. The ulex resembles a dwarf form of the common gorse and is the sub-species *funkii* which is special to the region.

Alyssum serpyllifolium, Saxifraga granulata and the white-flowered *Iberis saxatilis* grow amongst the shrubs. A common plant along the roadsides here is *Ptilostemon hispanicus*, (O = *Chamaepeuce hispanica*), a thistle with rosettes of fishbone-like leaves having pale midribs and beautiful long, amber-coloured spines. During the autumn and winter, a few interesting bulbous and tuberous rooted species can be found in flower. They include *Biarum carratracense, Crocus clusii, Narcissus papyraceus, Romulea bulbocodium*. The biarum closely resembles the more common *B. tenuifolium* but it is somewhat larger and can be distinguished by the additional rows of awl-shaped swellings between the male and female flowers. It is a fascinating plant and, like many other members of Araceae, it is perfectly hardy and multiplies well in gardens at home. The crocus, which is one of some five species special to southern Spain and North Africa, is sometimes classed as a sub-species of *C. serotinus*. It has pale mauve pointed flower segments which fade to near white towards the centre of the flower. Belonging to the meadow saffron group it has stigmas which are large, feathery and orange-yellow. A close examination of the flower with a pocket lens will reveal another diagnostic feature of the species - short hairs at the base of the corolla. The romulea is a rather fine, large-flowered form and can be seen here in spring with *Orchis morio* and *O. lactea*.

About two-thirds of the way from San Pedro to Ronda there is a signposted turn to the west leading to the village of Igualeja and nearly opposite, in a easterly direction, an unsurfaced road that takes one to the Coto Nacional de la Serrania de Ronda. It is well worth walking the 10 kilometers along here to see the stands of the native Spanish fir *Abies pinsapo*. It is an attractive tree with short blunt needles arranged all round the branchlet, instead of in rows as with most other species, and bearing purplish brown cones. It can be grown successfully in Britain and is especially useful on limey soils which do not suit most firs. Underneath the abies, and in clearings, *Paeonia coriacea* flowers in May and *Paeonia broteroi* is also found in the vicinity. From a nearby farm a track leads northwards to the boundaries of the Coto Nacional and along this route several interesting plants may be found:

Echium albicans	*Narcissus assoanus*
Fritillaria lusitanica	*Omphalodes brassicifolia*
Helianthemum apenninum	*Ranunculus gramineus*
Helleborus foetidus	*Senecio doronicum*
Jasminum fruticans	*Staehelina baetica*

Climbing still higher, one may encounter:

Aethionema saxatile	*Genista hispanica*
Arenaria montana	*G. triacanthos*
Armeria villosa	*Iberis pruitii*
Erinacea anthyllis	*Narcissus hispanicus*

The narcissus is a typical and rather large-flowered daffodil. In addition to the above species, in rock fissures one may encounter *Prunus prostrata* which has pink flowers in May and June, and the endemic *Saxifraga boisseri* - a cushion forming plant with leaves deeply cut into threes and clusters of white, yellow-centered flowers. They are often accompanied by the fern *Ceterach officinarum* and the pennywort *Umbilicus rupestris*.

Back on the C-339 and heading towards Ronda one enters a valley which is especially rich in orchids, particularly *Ophrys lutea* and *O. fusca* but also the less common *Orchis pallens*. Other plants worth looking for here include *Narcissus assoanus* and *Lapiedra martinezii* (7.6). Finally one crosses the spectacular ravine to enter the town of Ronda itself.

6.5 San Roque to Ronda: If one is staying at Algeciras or Gibraltar, this is the most direct route to take to Ronda and follows the C-333 as far as Jimena. Shortly after leaving the main N-340 there are groups of *Solanum bonariense*; an attractive, autumn-flowering weed from South America with mauve potato-like flowers often met with in southern Andalusia. The road at first follows the river Guadarranque and there are large groups of the giant reed *Arundo donax* which, in spite of being widespread in the Mediterranean is, like the solanum, an introduced plant - probably from India. The dried canes are used as wind-breaks and provide one of the sources of 'reeds' for Scottish bagpipes.

Some 10km from the N-340 one comes to the village of Almoriama where there is a corkworks and it is around here in marshland that Stocken found *Narcissus viridiflorus* - a small, green-flowered, scented jonquil native to northern Morocco which has its only foothold in Europe around San Roque and Algeciras. Some of the land has now been built on or used as a rubbish dump and I have not been able to locate the narcissus. It may be that timing is all important for although it usually flowers in October to November, this depends on the autumn rains. However, it seems that most of the sites for this interesting plant are now built on or 'developed' so it may not be long before it has to be excluded from the flora of Europe.

Just past the cork oak works a turning on the left leads to Castellar de la Frontera - an old Moorish fortified hill town, now

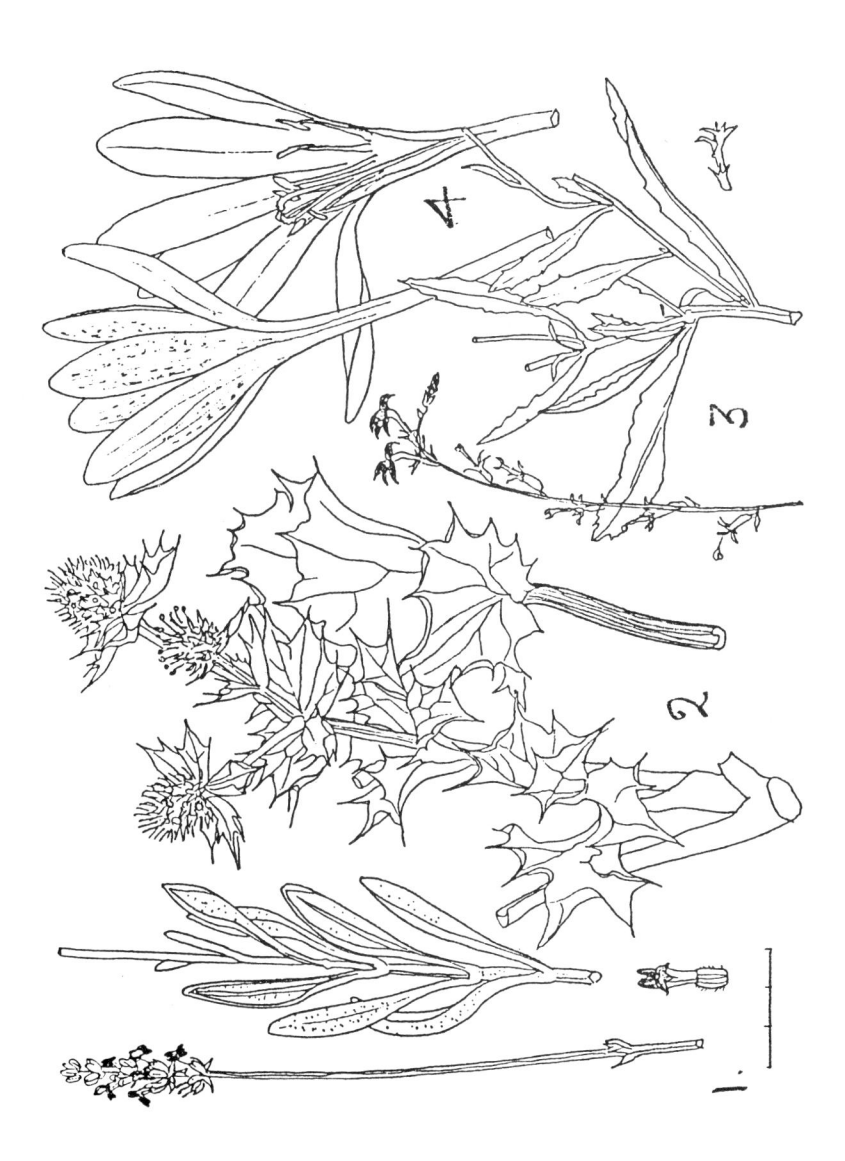

Andalusia 1. Lavandula lanata 2. Eryngium maritimum 3. Lobelia urens 4. Colchicum lusitanicum

more or less in ruins. The first part of the road passes through a delightful wooded area where nightingales, golden orioles and bee-eaters can be heard in spring and *Iris foetidissima* and *Colchicum lusitanicum* grow under the trees. At the beginning of the year *Narcissus cantabricus* flowers in cracks in the scattered outcrops of limestone rock.

Passing out of the wood the road climbs the hill and passes through areas with dense stands of heathers and cistus: *Calluna vulgaris, Erica arborea, E. australis, E. vagans* and *Cistus crispus, C.ladanifer, C. monspeliensis* with clumps of *Chamaerops humilis*. It is somewhat surprising to see the calluna which the same species as our heather but seems to grow much taller and to flower later, in October. Towards the top of the hill one has a good view of an artificial lake, the Embalse de Guadarranque. In grazed land by the roadside there are clumps of chamaerops and a few cistus amongst which one may find *Scilla autumnalis* and *Colchicum lusitanicum* in September and October. A treasure to be seen on rocks in the ruined town itself is the hare's foot fern *Davallia canariensis* which is a slender version of the type found in the Canary Islands.

Back on the C-333 one passes through country on the way to Jimena which is potentially very interesting botanically but it is difficult to get off the road. Roadside posters announce that it is a 'Ruta del Toro' devoted to cattle breeding. One sees plenty of the black bulls, reminiscent of the Vetterano advertisements. The fields are well-fenced but if one can get off the road to look around in damp places it may be possible to find the small bulbous *Narcissus cavanillesii* (O = *Tapeinanthus humilis*) which produces upright-facing, narcissus-like flowers in December. Like *Narcissus viridiflorus* it is another Moroccan species which has its only station in Europe near here. Its flowering often coincides with that of the much more widespread *Narcissus serotinus* that grows nearby and hybrids between the two species have been reported. During the spring the Spanish iris *Iris xiphium* may be seen and somewhat later *Nigella hispanica* which is a tall, attractive love-in-a-mist with especially large blue flowers, sometimes offered as an annual in seed catalogues at home and occasionally referred to as *Nigella papillosa*. Other plants which grow on the roadsides include *Convolvolus tricolor* and the figwort *Scrophularia sambucifolia* which is of limited distribution and has much larger orange-brown flowers than most other members of the genus.

Bypassing the interesting town of Jimena de la Frontera the road becomes the C-341 and begins to climb. On rocks by the roadside grow scattered plants of *Trachelium caeruleum*. There are fewer limiting fences here and the vegetation becomes more like that of the limestone of the Serrania de Ronda (6.4) As one approaches Algatocin, plants of *Rhus coriaria* can be seen by the roadside and past the village of Atajate there are large groups of the tall *Lygos sphaerocarpa* and what seems to be a dwarf, bushy form of the common broom *Cytisus scoparius*. The latter might be a useful plant for home gardens but when I have seen it in October all the seed was already shed. Around Atajate area grow large groups of *Iris planifolia* flowering in winter. As one descends from the Puerto de Encinas in further rocky areas one may see *Antirrhinum hispanicum*, *Erinacea anthyllis* and *Putoria calabrica* before coming in to Ronda.

6.6 Ronda and beyond: If one is fortunate enough to be at Ronda for Christmas or early in the New Year one may see the ground purple with the beautiful flowers of the juno iris *I. planifolia*. Later, during the summer, one can expect to find *Anemone palmata* (4.4), the purple cabbage *Moricandia moricandioides*, *Ranunculus rupestris*, *Iberis crenata*. The ranunculus is a rather fine buttercup with exceptionally large yellow flowers of 3-4cm diameter. It grows well in home gardens but is short-lived. A pine grove to the north of Ronda shelters *Prolongoa pectinata* and *Linaria amethystea*. The former is a tiny, anthemis-like plant with yellow flower-heads that hang down before opening and when they are past their best; the leaves have small tooth-comb like lobes. The linaria is a slender annual with white, cream or mauve flowers and and amethyst-coloured spurs; distinguished from most other species by the thickened wing on the seeds - if one is lucky enough to see both flowers and seeds! A number of orchids grow in the area including the less common *Orchis collina* (O = *O. saccata*) that flowers rather earlier than most other species and *Orchis olbiensis* (= *Orchis mascula ssp olbiensis)* which occurs in both mauve and white-flowered forms.

The country around Ronda is very rich botanically and it is well worth lodging for a night or two in the town itself. A celebrated hotel here is the Reina Victoria erected, like its sister hotel Reina Cristina in Algeciras, around 1890 by the British company Henderson Administration that built the first railway in Spain from Algeciras through Ronda to Bobadilla. The C-339 road leaves northwards out of the town and after crossing over the river Guadiaro and the railway there is a turning to the south to Montejaque and the Cueva de la Pileta. Along this road in summer grow the colourful annuals *Fedia cornucopiae* and *Centaurea pullata* with flowers varying through mauve to pink and white. Approaching Montejaque there are limestone outcrops and here one finds *Iris xiphium* and the rare *Ornithogalum reverchonii* carrying large white flowers on 30cm stems, somewhat resembling a white bluebell. Another interesting plant of the region is *Silene pseudovelutina* - a robust, woolly, species with heads of white flowers. Taking the rough road to Benaojan and then walking from the railway station to the Cueva del Gato one may have the good fortune to see the mistletoe *Viscum cruciatum* growing on olive trees. This resembles the common species but has bright red berries in December. It is a native of North Africa and parasitises a number of species though it seems to prefer olive trees. Understandably, it is unpopular with growers and is usually removed as soon as it is noticed.

Back on the main C-339, after about 3km, there is a turning to Grazalema which is well worth a visit. Rocks near the town are home for several interesting species:

Biscutella frutescens	*Helleborus foetidus*
Centaurea clementei	*Omphalodes brassicifolia*
Endymion hispanicus	*Papaver rupifragum*
Hesperis laciniata	*Pyracantha coccinea*

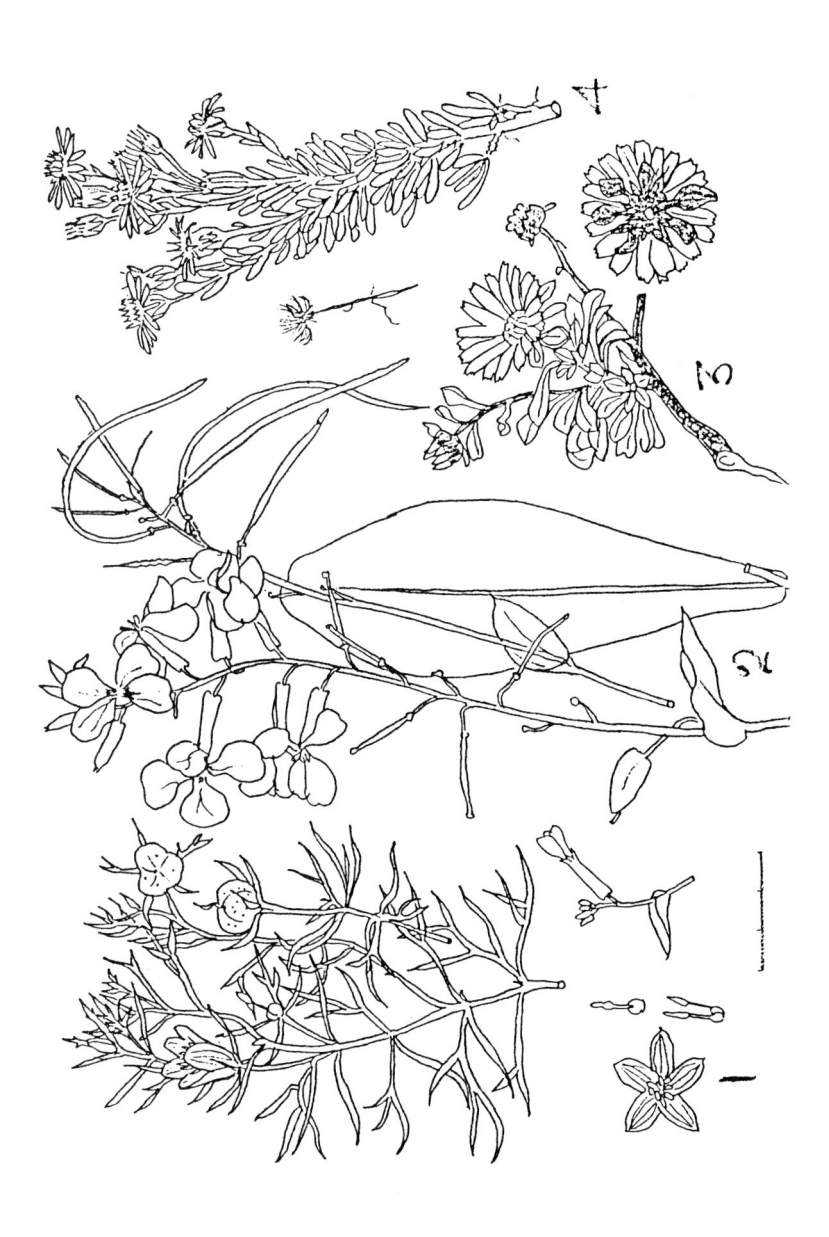

Andalusia 1. Peganum harmala 2. Moricandia moricandioides 3. Asteriscus maritimus
4. Inula crithmoides

The biscutella is covered with dense white hairs and has yellow flowers followed by typical fruits of the genus; it somewhat resembles *B. vincentina* of Cape St. Vincent (4.5) The centaurea is a handsome robust perennial with woolly, jagged-lobed leaves and large yellow flower heads. The hesperis is the cut-leaved dame's violet. *Papaver rupifragum* is only found in this part of Europe but also grows in Morocco; it is a perennial with rather small brown-red flowers and is sometimes grown in gardens where it has produced an attractive hybrid with *P. orientale*.

Three species of saxifrage limited to southern Spain grow on the rocks here - *S.boisseri, S.globulifera, S.haenseleri*. It is also a good area to look for narcissus species including *N. rupicola, N. jonquilla* and a form of *N. bulbocodium* described by Stocken as bearing "a resemblance to *N. hedraeanthus*". In addition to all these there are several orchids, especially *Orchis mascula var olbiensis*.

The Sierra del Pinar mountains, which rise to the north-west of Grazalema and reach a height of 1654m, are worth a visit but one will need to devote a whole day to the purpose. The lower parts are wooded but in clearings grow *Berberis hispanica* and *Ptilotrichum spinosum*; two plants also seen in the Sierra Nevada. Other plants to be found on the higher ridges include:

Arenaria aggregata	*Rupicapnos africana*
Draba hispanica	*Senecio minutus*
Erysimum grandiflorum	*Silene cretica*
Ionopsidium prolongoi	*Viola demetria*
Ononis saxicola	*V. parvula*

The arenaria is a cushion-forming plant that resembles *A. tetraquetra* (9.9) but the leaves are pointed and bent backwards. The draba has yellow flowers. The ionopsidium is a small, insignificant annual 'crucifer' with white or pale pink flowers and also grows in North Africa. *Rupicapnos africana*, as its name implies, is another African species; it resembles a prostrate fumitory with fleshy leaves and pink and white flowers. The senecio is a small annual with pale yellow ray-florets that are purple on the underside. The violas are both small annuals; *V. demetria* with attractive yellow, and *V. parvula* tiny white or yellow, flowers. The last of these is also found in the mountains of Greece (17.6).

6.7 North of Marbella, Fuengirola, Torremolinos: At a short distance from these resorts one can find many interesting plants. A profitable route is the C-337 from Marbella to Coin and from there eastwards along C-334 through Alhaurin to the coast at Benyamina. The rocks here are mainly alkaline like those of the Serrania de Ronda. The number of species to be seen is very numerous and includes the following:

Aphyllanthes monspeliensis
Asperula hirsuta
Centaurea prolongi
Cistus clusii
Coris monspeliensis
Crambe filifolia
Crocus salzmannii
Genista equisiteformis
Iberis linifolia
Iris filifolia
Leucoium trichophyllum
Linum narbonense
Matthiola fruticulosa
Narcissus gaditanus
Serratula flavescens
Valeriana tuberosa
Vinca difformis

The asperula has fine pink flowers. The crambe is quite unlike the ordinary seakale (*Crambe maritima*); it has a basal rosette of rough, hairy leaves and slender, wiry, branched flowering stems with small white flowers. It is not certain that the crocus is *C. salzmannii*, which is primarily a native of North Africa, but it resembles this species. Its pale mauve flowers with a yellowish centre appear in the autumn before the leaves appear. A similar species can be found in the woods near Castellar de la Frontera (6.5). The genista (= *G. umbellata*) is a distinctive species with almost leafless rush-like stems carrying tight, bun-like, terminal heads of yellow flowers. The iberis is a perennial with mauve, and sometimes white, flowers. *Iris filifolia* is a very beautiful bulbous species resembling a fine form of 'Spanish iris' (*I. xiphium*) with very narrow leaves and purple flowers in May. The leucoium (sometimes spelt leucojium) is somewhat of a speciality of the region - a very attractive small spring-flowering plant with nodding white flowers. The narcissus is another choice species resembling a small jonquil with a bent corolla tube. The serratula is like a centaurea with entire, toothed leaves and yellow florets. A number of orchids grow in this region including an attractive form of *Orchis papilionacea* with exceptionally large and brightly coloured flowers.

Some 30km north north-west of Fuengirola lie the Sierra de Alcaparain and Sierra Prieta, rising to 1505m, which can be reached either by the C-344 from Coin or the C-337 that leaves Malaga for Antequera. They are an extension of the limestone region of the Serrania de Ronda and home to a large number of orchid species. The less common ones to be found here include *Ophrys fusca ssp. durieui* (formerly *ssp. atlantica*). It closely resembles *ssp. iricolor* with large flowers that have a conspicuous bright blue speculum; it is primarily a Moroccan form of the species. The less common *Orchis champagneuxii* and *Orchis langei* may be found here: both are sometimes considered as sub-species of *O. mascula* by Valdes. *Gennaria diphylla* and *Spiranthes aestivalis* also grow in the vicinity. Other interesting plants here include :

> *Centaurea carratracensis*   *Salvia candelabrum*
> *Omphalodes linifolia*

The centaurea has mauve flowers and the salvia is an attractive species with large, violet-blue flowers that have a white lip. It somewhat resembles the garden sage *S. officinalis*.

To the north east of the Sierra de Alcaparin lies Antequera which can be approached along the C-337 and a little to the south of the town is El Torcal where the rocks resemble the ruins of an ancient city and remind one of Montpellier-le-Vieux in the south of France. There is a parador nearby if one should decide to stay in the vicinity. It is another area rich in orchids and also supports several other rare and interesting plants such as:

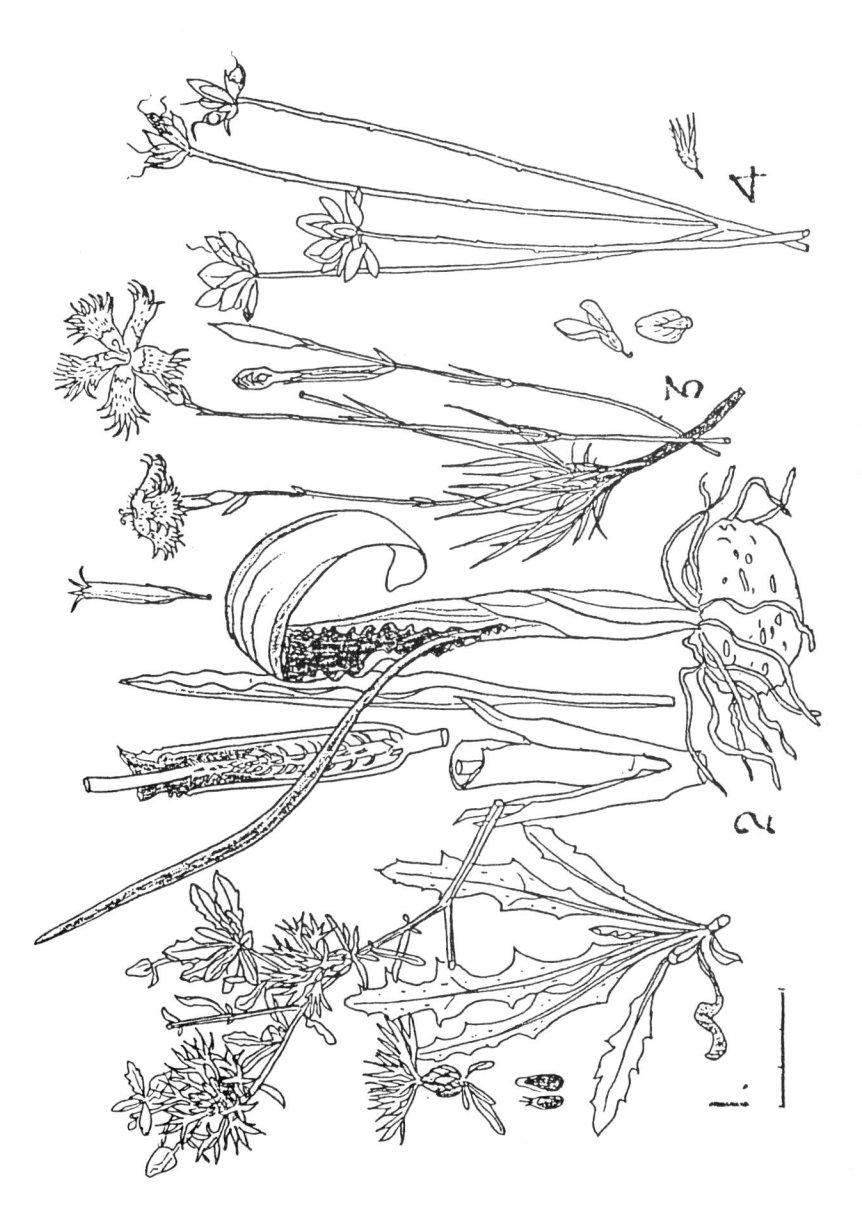

Andalusia 1. Mantisalca salmantica 2. Biarum carratracense
3. Dianthus lusitanus 4. Genista umbellata

Cynoglossum cheirifolium *Ranunculus rupestris*
Iris subbiflora *Saxifraga biternata*
Linaria anticaria *Viola demetria*
Paeonia broteroi

The cynoglossum has grey leaves and dark claret-red flowers. *Iris subbiflora* (= *I. biflora*) is classed in Flora Europaea as a sub-species or form of *Iris lutescens* (O = *I. chamaeiris*). It is somewhat taller than the last species and the stem carries one or two deep purple flowers with a violet coloured beard. The linaria is named after Antequera, where it was first discovered, but it has a limited distribution elsewhere in the Malaga region and hills west of Valencia. It is a glabrous perennial with white or pale lilac coloured flowers streaked with fine mauve lines and blooms from April to June. The saxifrage is only found here; it is low-growing with bulbil like growths in the axils of the lower leaves and produces relatively few, rather large bowl-shaped white flowers. The yellow viola has already been mentioned (6.6).

6.8 **Gibraltar:** All but two of the species that grow on Gibraltar can also be found on the adjacent Spanish mainland so it is hardly worth while spending a prolonged time to look for new plants there. However, many visitors on their way to Algeciras, Estepona and other places in western Andalucia land at Gibraltar airport and may decide to spend a few days on 'The Rock'. At present Gibraltar is not a good centre from which to study the Andalusian flora because it can, at times, be frustrating getting through the customs at the border although the situation may change before long. Most visitors spend much of their time in the town and make a short visit by motor vehicle or the cable car up the Rock to St. Michael's cave and the 'apes' den'. It is not easy to get off the road here and one may have to be content to accept that the flora can only be observed from a distance. During the summer the tall spikes of *Acanthus mollis* will be evident and in the autumn the flowers of the fine *Colchicum lusitanicum*. To make a more extensive study of the flora it may be necessary to obtain a permit to visit certain areas because of military restrictions. Some of the more interesting species to be seen by the persistent visitor include:

Aristolochia baetica	*Lavatera arborea*
Campanula velutina	*Lavandula dentata*
Calendula suffruticosa	*L. multifida*
Clematis cirrhosa	*Linaria tristis*
Crocus salzmannii	*Lobularia libyca*
Dianthus caryophyllus	*Narcissus papyraceus*
Dipcadi serotinum	*Ranunculus bullatus*
Echium boissieri	*Ruscus hypophyllum*
Gennaria diphylla	*Saxifraga globulifera*
Iberis gibraltarica	*Scilla peruviana*
Iris filifolia	*Spiranthes spiralis*

The dianthus is the wild form of the carnation and has strongly scented pink flowers that are, of course, single. It does not seem to occur wild on the mainland of Spain but is recorded from the Var department of South France. The iberis is a handsome shrubby perennial candytuft with white or reddish purple flowers - not endemic to Gibraltar since it also grows in North Africa. The lobularia closely resembles the common sweet alyssum *L. maritima* but is an annual and has 4-5 seeds in each cell of the fruit - not one only. It also is, primarily, a North African species.

A number of introduced plants have adapted to the conditions on 'The Rock' and now grow wild there, including the tree houseleek *Aeonium arboreum* from North Africa, *Aloe arborescens* from South Africa with brilliant red flowers in August and *Carpobrotus acinaciformis*.

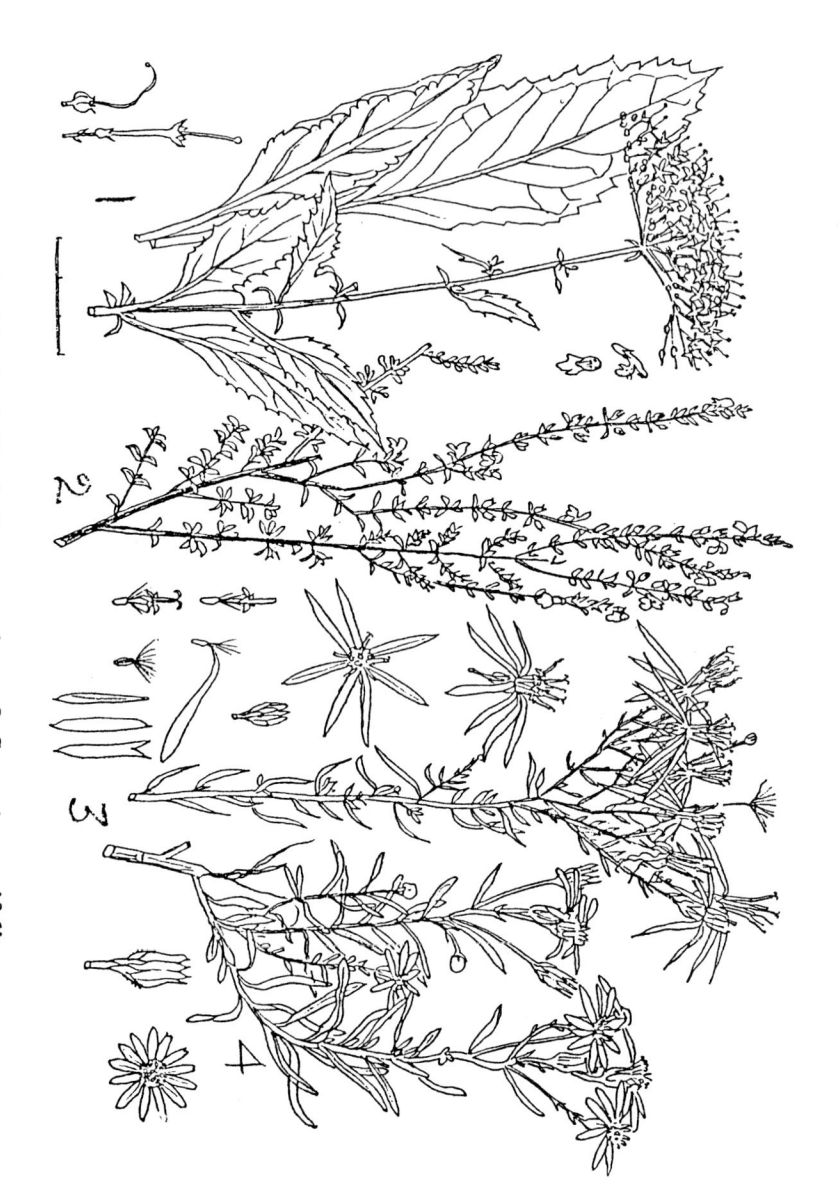

Andalusia 1. Trachelium caeruleum 2. Satureja cuneifolia
3. Aster sedifolius 4. Senecio linifolius

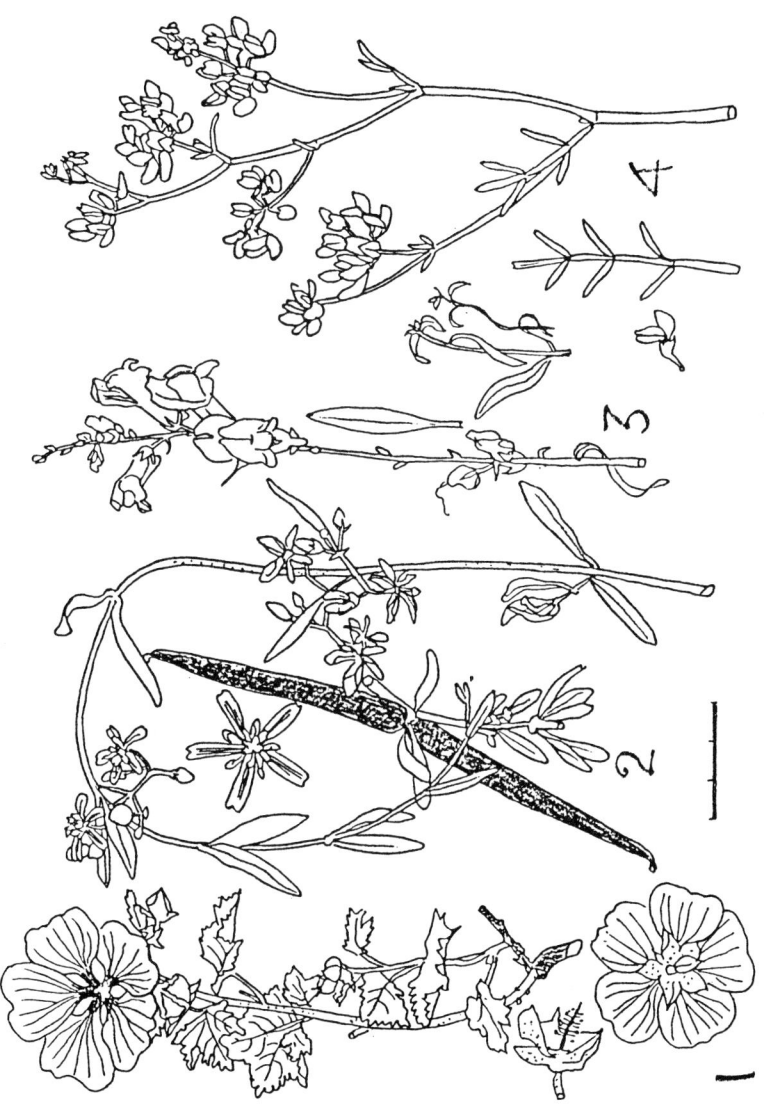

Andalusia 1. Lavatera maritima 2. Periploca laevigata
3. Antirrhinum barrelieri 4. Coronilla juncea

Andalusia 1. Lavandula multifida 2. Lycium intricatum 3. Reichardia tingitana 4. Narcissus gaditanus 5. Helianthemum cinereum 6. Fumana procumbens 7. Ononis speciosa

7. Eastern Andalusia

7.1 The region covered by this section roughly comprises the provinces of Granada and Almeria. It embraces the Sierra Nevada which rises to 3482m and is thus the highest mountain in Spain - indeed the highest in the Mediterranean Region. It also includes the Cabo de Gata which is the driest part of Europe and approaches desert conditions. Both of these areas have plants of special interest, and the Almeria province alone probably maintains the richest flora in Spain with over 2,500 species.

Tourism is not so developed here as in western Andalusia but year-round package holidays are available at Mojácar near the coast in the east of the region (not far from the fishing village of Garrucha) and, just west of Almeria town at resorts such as Aguadulce, Roquetas del Mar and Almerimar. These provide relatively inexpensive bases from which to explore the area. Those near Almeria are most suitable for visiting the Cabo de Gata. However, none of these resorts is sufficiently near to the Sierra Nevada and, if a prolonged visit to this area is the objective, it is probably best to stay at Grenada or at the attractive, modern parador at about 2,500m on the mountain itself. There is a good road to the parador and it is usually open in spring but the road that continues to traverse the mountain range may well be blocked by snow until July or August.

A suitable map for the area is the Firestone C9 and Hernando tourist maps of the individual provinces of Almeria and Granada on a larger scale are available at resorts.

7.2 **The Terrain:** There are extensive stretches of sand along the shore, especially to the south of the main coast road N340 between Balerma and Aguadulce. These support an interesting halophytic flora but much of this region is rapidly being exploited for early crops under plastic glasshouses and jokingly referred to as the 'Costa Plastica'. Backing the coast, east of Almeria, are the Sierra de Gata hills of volcanic origin which rise to only 388m and are included in the important Cabo de Gata Nature Reserve. Around and behind these hills in the Nijar region there is dry steppe land,

Andalusia 1. Phlomis purpurea 2. Narcissus cantabricus 3. Rhamnus lycioides 4. Limonium insigne 5. Matthiola lunata

noted as a locality for filming desert cowboy scenes. Further inland to the north the Sierra Alhamilla reach a height of 1,385m and they give way eastwards to the lower limestone hills of the Sierra Cabrera. Further inland still, is the more extensive range of the Sierra de los Filabres aligned approximately east to west.

To the west of Almeria town lie the Sierra de Gador, to the north of which rise the impressive Sierra Nevada with several peaks over 3,000m. All of these mountain ranges and other features have their special plants, to be discussed in the following sections.

The southern part of the region is important for agriculture. In addition to the crops of early tomatoes and carnations under plastic greenhouses, open-air crops include sugar cane (*Saccharum officinarum*) and custard apples (*Annona cherimola* sometimes spelt *Anona cherimolia* and called 'cherimiola' in Spanish), originally from Peru. In addition, the Mortil valley which leads from the coast to Grenada, has extensive vineyards and almond orchards with some crops of sisal (*Agave sisalana*) near Nijar.

7.3 Literature: Plants of the region are, of course, covered by floras of Spain. Polunin and Smythies (1973) treats the region fairly thoroughly but does not have much to say about the Mojácar region and Stocken (1969) (6.3) describes the plants of the Sierra Nevada. If one intends to look at the flora of the Sierra Nevada in detail then the 'Flora de la tundra de Sierra Nevada (Prieto 1975) is useful provided one understands enough Spanish. It is probably available in Grenada. There is an interesting small book in Spanish, available locally, called 'Cabo de Gata, Guía de la Naturaleza' (Rodriguez et al. 1982) which describes both the plants and animals of this fascinating region.

7.4 Mojácar: The small town of Mojácar is perched on a hill with whitewashed, flat-roofed buildings. Although the site is of ancient origin, many of the buildings are recent for much of the town has been rebuilt since it was almost entirely destroyed during the civil war. From the hill one has a magnificent view northwards over the wide valley of the river Aguas to the fishing village of

Garrucha. An earthquake, about 300 AD, lifted this valley so that there are extensive sandbanks now isolated from the sea; some have been flattened, cultivated and irrigated but a few remain untouched and covered with sparse vegetation. The earthquake was, presumably, accompanied by extensive volcanic activity and there are heaps of cinder-like volcanic slag forming extensive hills, especially to the north and east of Mojácar. They are reminiscent of coal-mining slag-heaps.

Walking down the road from the town towards the coast one passes a cultivated valley where oranges, broad beans and potatoes are grown. Here is a large patch of *Kundmannia sicula*, a showy umbellifer rather like our goutweed but with yellow flowers. Cutting across the hills by footpaths in a south westerly direction one passes through a desolate cinder-like area with ruined arab-style farmhouses surrounded by opuntia and agave plants. In spring the ground is covered in places by *Gynandriris sisyrinchium* (= *Iris sisyrinchium*) and *Fagonia cretica*. Other plants include:

Anthyllis cytisoides	*Iris lutescens*
Asphodelus tenuifolius	*Launaea spinosa*
Chaenorhinum villosum	*Lavandula multifida*
Helianthemum almeriense	*Ophrys tenthredinifera*

The asphodelus is an annual, like a scaled-down version of the common perennial *A. fistulosus* and can be clearly distinguished by its size when the two grow together. Nevertheless, it is not recognised by Flora Europaea as a distinct species. For those who understand the significance, it is interesting to note that *A. tenuifolius* is diploid ($2n = 28$) whereas *A. fistulosus* is a tetraploid ($2n = 56$). The endemic chaenorhinum is a tiny toadflax-like plant and the helianthemum is also endemic to this region but quite common in places. It has white flowers with a yellow blotch at the base of the petals. *Launaea spinosa*, a very spiny plant with small dandelion-like yellow flowers, is another species found throughout the region. Nearer the shore, growing in sand, one encounters:

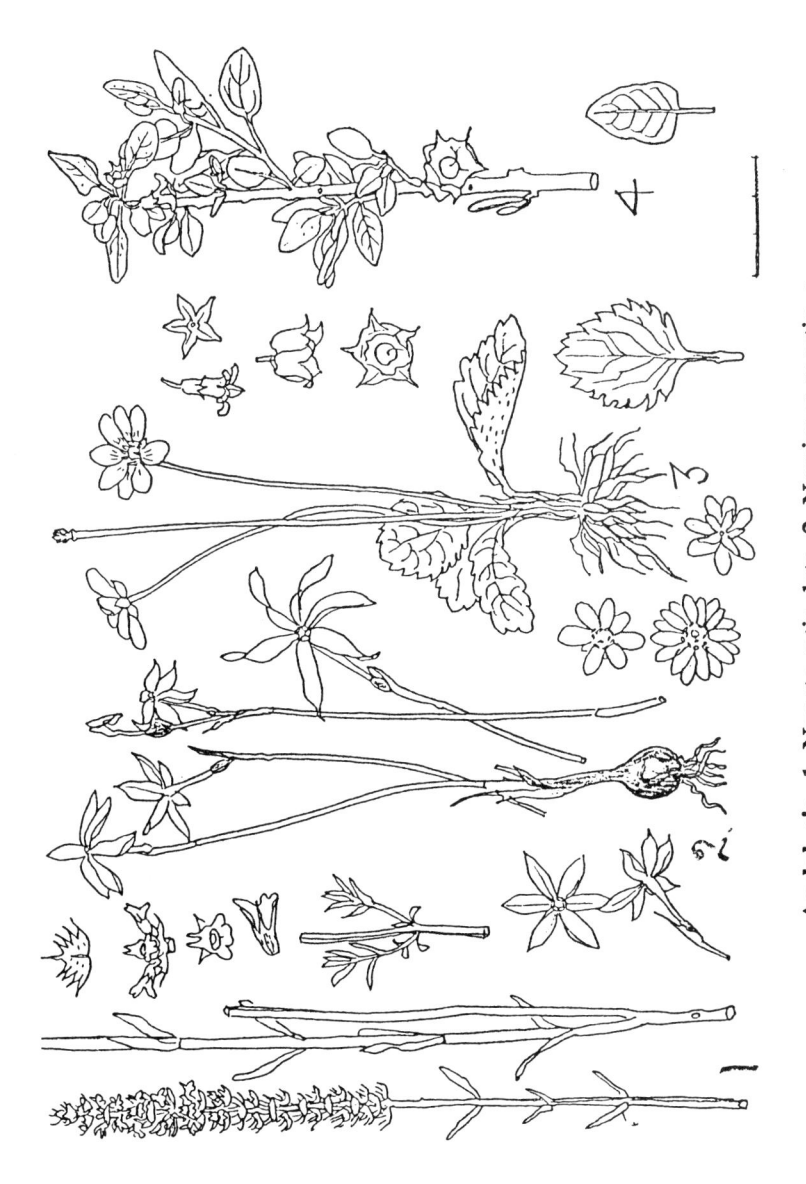

Andalusia 1. Nepeta reticulata 2. Narcissus serotinus 3. Ranunculus bullatus 4. Withania frutescens

Cakile maritima *Moricandia arvensis*
Centaurea sphaerocephala *Pancratium maritimum*
Convolvolus althaeoides *Reichardia tingitana*
Lobularia maritima *Silene littorea*

The moricandia is a common annual 'crucifer' in this region and known as the violet cabbage. It seems to flower at all seasons and during the autumn it sometimes colours the road verges mauve. The reichardia resembles a neat dandelion with a dark centre to the flower head.

Walking north-eastwards along the shore brings one to the mouth of the Rio Aguas and here, parasitising the atriplex plants, one may be lucky to see the giant broorape-like plant *Cistanche phelypaea* with robust spikes of attractive yellow flowers and no leaves. Following the coastal road through the resort in the opposite direction one comes to a corniche road leading to the small fishing port of Carboneras. There are some interesting plants to be seen along here if one can find a suitable place to park a car and look for them. They include:

Antirrhinum barrelieri *Lygeum spartum*
Cistus albidus *Limonium insigne*
Coronilla juncea *Nicotiana glauca*
Lathyrus clymenum *Ononis speciosa*
Lavandula stoechas *Periploca laevigata*

The antirrhinum has pink flowers and a curious habit of climbing by twining its petioles round obstacles. Both the ononis and coronilla are showy uncommon species with yellow flowers and the latter has a delicious scent. The lathyrus, with rather large mauve flowers, resembles the tangier pea *L. tingitanus*, and like it has been cultivated as a fodder crop.. The lygeum called in Spanish 'albardine' is an economically important plant of the region and will be mentioned later. The limonium is a very beautiful endemic species with horsetail-like foliage and 60cm branched inflorescences of small mauve flowers in summer. Periploca is a low-growing, or semi-climbing shrubby plant with thorns; it belongs to the Asclepiadaceae and the type here is a sub-species *angustifolia*. Its flowers, about 2cm in diameter, have five green petals striped with purple. These are followed by 7-8cm long pointed fruits held out like the horns of an old type draft oxen. Like other members of the family, the plant exudes a milky juice when the stems or leaves are broken.

7.5 Sierra Cabrera: From Mojácar it is easy to walk onto the Sierra Cabrera and an initial scramble to the broadcasting station, that can be seen from the town, may be very rewarding. In the stony limestone terrain there is a rich garigue embracing the following species:

Anthyllis cytisoides	*Lithodora fruticosa*
Fumana procumbens	*Lycium intricatum*
Globularia alypum	*Lavatera maritima*
Helianthemum almeriense	*Phlomis purpurea*
Launaea spinosa	*Rhamnus lycioides*
Lavandula dentata	*Ruta chalepensis*
Lavandula multifida	*Thymelaea tartonraira*
Lavandula stoechas	*Withania somnifera*

The lithodora (O = *Lithospermum fruticosum*) is a small shrub with beautiful gentian-blue flowers and, when grazed it has twisted bonsai-like trunks. The lavatera is truly beautiful with large flowers that are shell-pink in the form growing here. The lycium and rhamnus are superficially alike - small prickly shrubs with tiny succulent leaves. However the flowers are different; the lycium has dusky mauve tubular flowers and the rhamnus tiny, four-rayed greenish-yellow stars. The withania is a poisonous, evil-smelling, solanaceous shrub with yellow bell-shaped flowers followed by tomato-like berries. Growing amongst these shrubs one can find:

Arisarum vulgare	*Echium albicans*
Asteriscus maritimus	*Eryngium maritimum*

Calendula arvensis *Fagonia cretica*
Cynoglossum cheirifolium *Psoralea bituminosa*

In early spring there are large numbers of the rather uncommon *Orchis collina* (O = *O. saccata*) and the delightful small 10cm tall, jonquil *Narcissus gaditanus* recognisable by its curved corolla tube. Further searching may reveal such plants as *Gladiolus illyricus, Leopoldia comosa* and *Ophrys tenthredinifera*.

One can get into the Sierra Cabrera also via the road from Sorbas south to Saladar y Leche and this may well be interesting plant-hunting country. A rather uncommon species seen here in some quantity is *Centranthus angustifolius* (sometimes spelt *Kentranthus*), a spur-valerian with narrow, linear leaves and pink flowers in May and June.

7.6 Sierra de Filabres: An approach to this range can be made from Mojácar by driving west out of the town to the main N340 road. In moist cultivated land nearby the attractive *Solanum bonariense*, flowers in autumn - it is a native of Buenos Aires. Shortly after leaving one comes to Turre and then a bridge over the gorge through which the river Aguas flows (that is when there is some water). It is well worth looking in the moist soil here in late summer and autumn for *Narcissus serotinus, Ranunculus bullatus, Scilla obtusifolia* grow here in quantity. By careful searching one can find the curious endemic *Lapiedra martinezii*. It belongs to the amaryllidaceae, has a bulb like a daffodil, linear leaves with a distinct pale stripe running down the centre and an umbel of small white flowers in August. The flowers have a pleasant lilac-like scent.

Reaching the main road one turns westwards to the attractive small town of Sorbas and then in a north-west direction to Uleila del Campo to cross over the Sierra. At this point the soils are extraordinarily colourful with red, pink and mauve tints. There are crops of olives, peaches, apricots and vines. The highest point here is the Puerto de la Virgen where the ground is covered with a scrub of *Cistus albidus, Ulex parviflorus* and the taller *Lygos sphaerocephala*. Here in spring the ground between and amongst these shrubs is covered with thousands of the beautiful white petticoat daffodil *Narcissus cantabricus var monophyllus*. The flowers differ in size and degree of 'flatness' of the trumpet and some are really large. As a bonus, one may also find groups of *Barlia longibracteata*.

7.7 **Roquetas del Mar and the Sierra de Gador:** Roquetas del Mar is a rapidly expanding resort and, in its western part, has some exciting modern architecture. Unfortunately the approaches at the time of writing are rather squalid with a great deal of rubbish. One can walk westwards from the town to an extensive sandy area with an interesting halophitic flora and salt pans sometimes favoured by flamingos. Amongst the saltworts and seablites grow:

Asphodelus fistulosus	*Launaea spinosa*
Asteriscus maritimus	*Limonium ovalifolium*
Dittrichia viscosa	*Limonium thouinii*
Euphorbia parialis	*Lotus creticus*
Frankenia laevis	*Lycium intricatum*
Frankenia thymifolia	*Otanthus maritimus*
Glaucium flavum	*Ononis viscosa*
Helichrysum stoechas	*Teucrium spinosum*
Inula crithmoides	*Zygophyllum fabago*
Launaea resedifolia	

Launaea spinosa is very spiny whereas *L. resedifolia* has relatively few spines and larger flowers. *Limonium ovalifolium* is especially plentiful in places and colours the ground violet in summer.*Limonium thouinii* is, by contrast, an annual with yellow petals surrounded by a papery, mauve, calyx. The zygophyllum, known as the Syrian bean caper, is a curious-looking plant somewhat resembling mistletoe. It has leaves in pairs, small yellow flowers and flattened fruits. It is perennial and is said to be hardy in Britain.

The Sierra de Gador can be reached quite easily by car from Roquetas del Mar driving in a northerly direction out of the town, crossing over the main N340 and then climbing to Venta de la Mena. From here onwards, until one descends towards the valley of the river Aldarax, there are many interesting plants:

Anthyllis cytisoides *Lygos sphaerocarpa*
Arisarum vulgare *Nepeta reticulata*
Carlina racemosa *Ononis striata*
Cistus incanus *Phagnalon rupestre*
Coronilla juncea *Phlomis purpurea*
Daphne gnidium *Ptilostemon hispanicus*
Genista umbellata *Rosmarinus eriocalyx*
Lavandula lanata *Salsola sp.*
Lavandula multifida *Satureja cuneifolia*
Lavatera oblongifolia *Ulex parviflorus*

Lavandula lanata is one of the less common species usually found at higher altitudes. The lavatera is a 2m tall shrub with lanceolate leaves some 3cm long and covered with dense woolly hairs. The fine, lavender-coloured flowers are 4cm diameter. It is an endemic species with a very limited distribution. The nepeta is not a typical catmint though it smells like one. The leaves are narrow and linear and the inflorescence is cylindrical, up to 10cm long and 1cm wide; the flowers are small and pale yellow. The ptilostemon is a thistle with decorative, long, amber-coloured spines. The salsola is probably *S. vermiculata* or a related species. Plants of this genus usually grow near the shore but at the junction of the road to Enix they are plentiful. In the autumn the papery flowers are quite colourful in shades of cream and pink.

Much of the Aldarax valley is taken up with vineyards and other crops but there are some interesting roadside 'weeds'. The *Salsola vermiculata* or a similar species is again in evidence here. In September the plants look like 1-2m tall shrubs which range in colour from cream through pink to brilliant crimson - quite a sight. One can also see much *Zygophyllum fabago* and *Capparis spinosa* and in rocky areas the shrubby *Bupleurum fruticosum*. Travelling westwards one passes through the picturesque old town

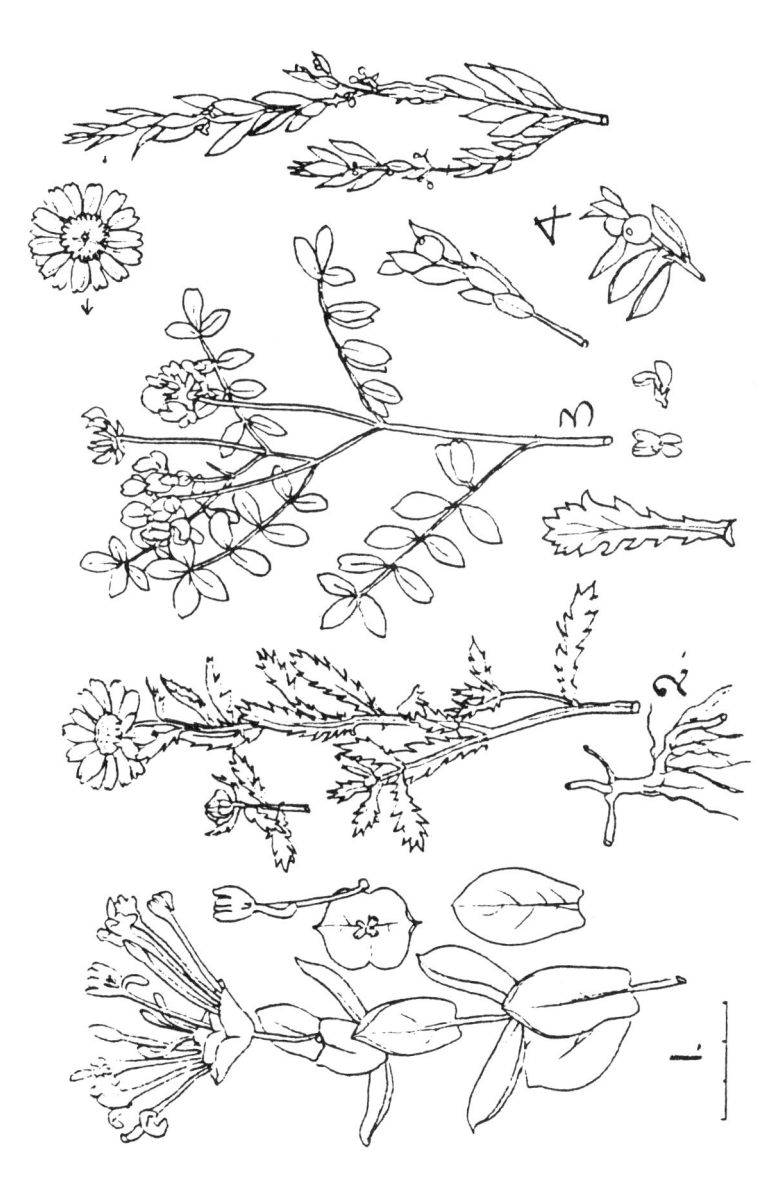

S.E. Spain 1. Lonicera implexa 2. Leucanthemum coronopifolium 3. Hippocrepis valentina 4. Osyris quadripartita

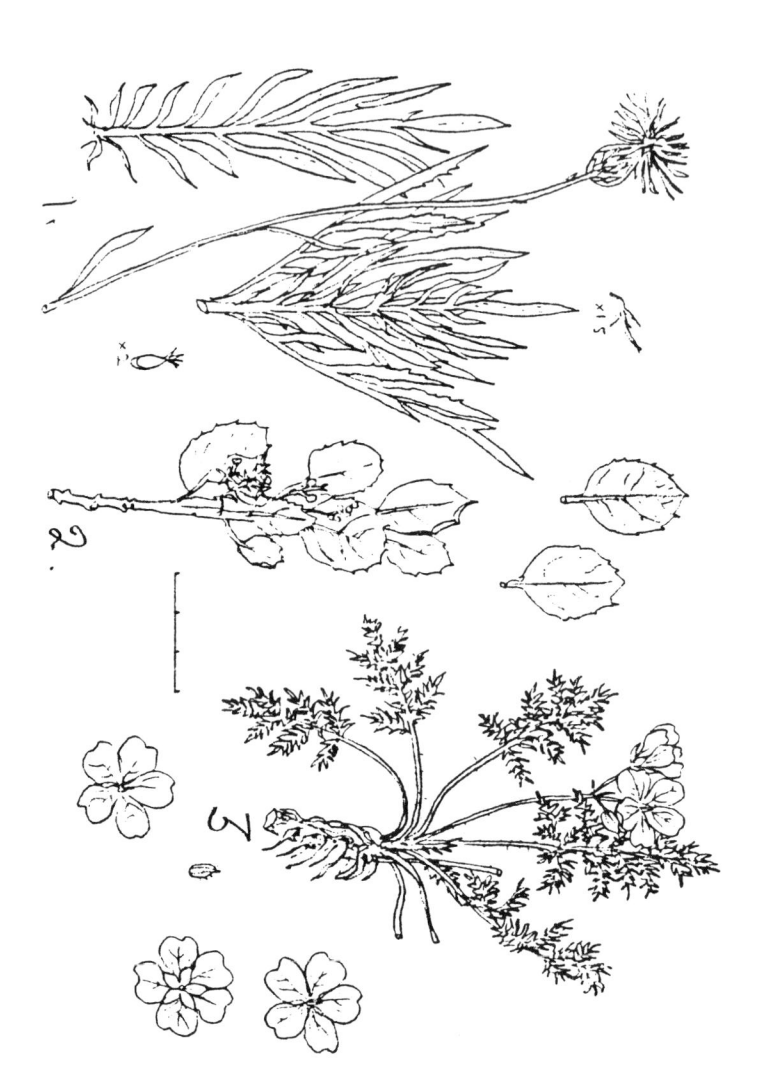

S.E. Spain 1. Centaurea boiseri spachi 2. Rhamnus alaternus 3. Erodium petraeum valentinum

of Fondón and then to Alcolea where the yellow-flowered thistle, or Spanish Oyster Plant, *Scolymus hispanicus* grows by the roadside. Its root may be eaten as a vegetable and tastes like salsify. From here one can return to Roquetas via Berja and Dalias to see more of the flora of the Sierra de Gador.

Somewhere in this region grows the interesting *Lafuentea rotundifolia* which is endemic and has a limited distribution. It is a member of Scrophulariaceae and an aromatic small shrub with 10cm high cylindrical spikes of white, purple-striped flowers in April and May. Polunin and Smythies (1973) describes it as growing by the corniche road approaching Almeria but nowadays it is practically impossible to park a car or walk from there.

7.8 **Cabo de Gata:** This very dry region includes an interesting nature reserve. One can reach there with ease from Almeria by following the signs to the airport and then turning south at a signpost announcing the reserve. Past El Aquián the roadside trees are *Schinus molle* the peppar tree or Peruvian mastic. It has neat pinnate foliage and small yellow flowers followed by rosy red fruits like peppercorns which are used in South America to make a peppary wine and sold in oriental markets as pink pepper. In this region there are old crops of sisal *Agave sisalana* which resembles the commonly grown *A. americana* but has no teeth on the leaves. One soon comes to sand dunes where the following plants may be seen:

Helianthemum almeriense	*Salsola vermiculata*
Launaea spinosa	*Thymelaea hirsuta*
Lavandula multifida	*Tribulus terrestris*
Lycium intricatum	*Ziziphus lotus*

The most obvious of these is the ziziphus which forms 2m high clumps of impenetrable spiny stems. It is in this sort of terrain that one may find the rare endemic *Androcymbium gramineum*, a member of the Liliaceae and resembling an ornithogalum with a rosette of leaves and white flowers striped with mauve in January and February. Bearing right at a road fork brings one to the fishing village of El Cabo de Gata. There are salt pans to the east of the road inhabited at times by hundreds of flamingos - the population is highest in the autumn. Continuing along here brings one to the lighthouse at Torre de Vela Blanca where the surfaced road ends. The rocks here have very sparse vegetation, which includes the native small palm *Chamaerops humilis* and the grasses *Stipa tenacissima, Lygeum spartum* (7.4). Other plants that manage to survive under these arid, windy conditions include *Asteriscus maritimus, Lavandula multifida, Limonium sinuatum, Lycium intricatum*.

Returning to the village of Cabo de Gata and, after passing through it, taking the turning to the right at the T junction puts one on the road for San José and the coast road to Playa de las Negras. This route leads through a rocky landscape with good sea views and an unique flora including the following species, many of which are rare or endemic:

Androcymbium gramineum　　*Helianthemum guiraoi*
Antirrhinum charidemi　　*Marrubium supinum*
Arisarum vulgare　　*Periploca laevigata*
Caralluma europaea　　*Phlomis purpurea*
Coris hispanica　　*Sideritis foetens*
Dianthus cintrans　　*Tamarix boveana*
Euphorbia dracunculoides　　*Teucrium charidemi*
Genista umbellata　　*Teucrium serranum*
Genista valentina　　*Thymus glandulosus*

The antirrhinum is a charming small plant with pink flowers and found only in the Sierras de Cabo de Gata. The caralluma is *var. confusa* and also a Gata endemic - a very strange plant belonging to the Asclepiadaceae. It has leafless, knobbly, succulent stems. The 12mm diameter flowers are dusky purple with a darker centre and these are followed by horn-like fruits resembling those of periploca but carried singly. The type species *Caralluma europaea* occurs on the small islands of Linosa and Lampedusa off the south coast of Sicily and in the drier parts of Israel around Jerusalem. The coris has flowers that are white to pale pink instead of the bright rosy-lilac colour of the more common *C. monspeliensis*. The helianthemum has yellow flowers and is thus easily distinguished from *H. almeriense* with white flowers that also grows there.

The above are just a selection of the interesting plants that one may see in this special area. Here are also roadside weeds such as *Heliotropum europaeum* and *Citrullus colocynthis*. The latter is the 'bitter apple' a small melon-like plant that produces spherical fruits about 8cm diameter, green-striped at first and ripening to yellow. It is an introduced plant, grown at one time for its purgative properties. *Narcissus serotinus* may be seen in flower in this area during the autumn.

7.9 **Sierra Nevada:** Anyone who is especially interested in the flora of the Sierra Nevada is strongly advised to purchase a copy of the book by Molero-Mesa and Pérez-Raya (1987) before making the ascent (7.3). It should be available in Granada as it is published by the university there.

The usual way to get up the Sierra Nevada is by road from Grenada. To see the spring-flowering bulbous rooted species it is best to make the journey in April or early May, but for the high alpines one must wait until August when much of the snow has melted. Shortly after leaving Granada, and around the village of Pinos Genil, one can expect to see flowering in spring a *Gagea species* quoted by Stocken (1969) as *G. hispanica* but plants of this genus are difficult to name correctly. With it grows *Orchis mascula olbiensis* in various colour forms and *Ranunculus rupestris* which is a 50cm tall buttercup with large (up to 4cm diameter) yellow flowers. The yellow-flowered form of *Narcissus triandus* has been recorded from the vicinity but it may take some time to find it. Later in the season the showy *Anthyllis vulneraria ssp. argyrophylla* with silvery foliage and red flowers may be seen.

As the road climbs to the viewpoint of Balcón de Canales other interesting plants include:

Daphne gnideum	*Lavandula lanata*
Daphne oleoides	*Paeonia broteroi*

Digitalis obscura *Phlomis crinita*
Geum heterocarpum *Sarcocapnos crassifolia*
Helleborus foetidus *Tulipa australis*

The digitalis is a sub-shrubby foxglove with rusty brown flowers. The geum has rather small bell-shaped flowers that tend to hang downwards and the phlomis is a decorative species with grey foliage and yellowish-brown flowers. Sarcocapnos grows in shady places amongst rocks and has white flowers - it is a member of the Papaveraceae and somewhat resembles a fumitory. The tulip is the only one of these that flowers in early spring but the paeony follows it in May.

The wide part of the road, which is usually kept free of snow, ends at Prado Lano 2,100m where there is an excellent modern parador. This is a good centre from which to study the high alpine flora when the snows clear in August. In any case it is probably worth spending a night there, if the weather is suitable, simply to witness the sunrise and sunset. Around the parador there are thick patches of dwarf shrubs, many of which are very prickly:

Adenocarpus decorticans *Lavandula lanata*
Astragalus granatensis *Ptilotrichium spinosum*
Astragalus sempervirens *Prunus prostrata*
Berberis vulgaris *Santolina virens*
Bupleurum spinosum *Ulex parviflorus*
Erinacea anthyllis *Vella spinosa*
Genista hispanica

The ptilotrichium and the vella are 'cruciferous' spiny shrubs, the first with pink or white flowers and the latter with larger white or yellowish purple-veined flowers, somewhat resembling a spiny wallflower. Erinacea is a rather beautiful small gorse-like plant with bluish violet flowers and sometimes called the hedgehog broom. The genista is the neat, dwarf, yellow-flowered 'Spanish gorse' of our gardens. All these shrubs flower in early summer but in patches of turf between them and amongst the shrubs themselves the ground

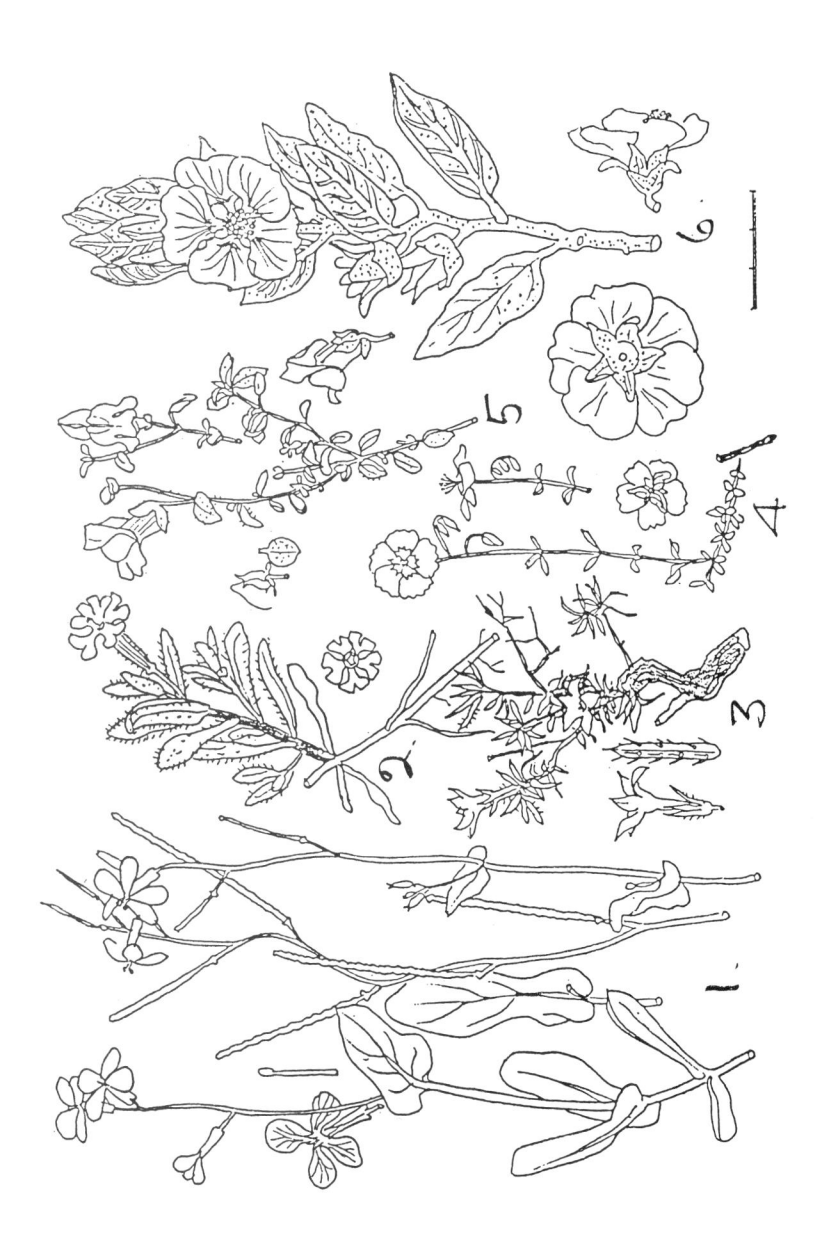

S.E Spain 1. Moricandia arvensis 2. Silene littorea 3. Lithodora fruticosa 4. Helianthemum almeriense 5. Antirrhinum hispanicum 6. Lavatera oblongifolia

Majorca 1. Arisarum vulgare 2. Fagonia cretica 3. Clematis cirrhosa 4. Cyclamen balearicum 5. Lavatera maritima

is coloured in spring with the flowers of *Colchicum triphyllum, Crocus nevadensis, Narcissus nevadensis*. The crocus has starry, pale lilac flowers with a greenish throat and is found in a few other parts of Spain and in Algeria. The colchicum is a small gem with narrow leaves that emerge at the same time as the crocus-sized, mauve pink flowers. The narcissus is a typical small daffodil but differs from most by having two flowers on a stem. It may need some searching for here as it is not so widespread as the crocus and colchicum.

A number of herbaceous plants also flower amongst the shrubs from June to August:

Arenaria tetraquetra	*Salvia lavandulifolia*
Chaenorhinum macropodium	*Saxifraga boisseri*
Draba hispanica	*Saxifraga dichotoma*
Jasione amethystina	*Sempervivum nevadense*
Myosotis alpestris	*Silene boryi*
Primula elatior	*Tanacetum radicans*
Ranunculus acetosellifolius	

The arenaria forms dense cushions of tiny quadrangular rosettes with small white flowers. It occurs in mountains in several parts of Spain including the Pyrenees and in the Maritime Alps. The chaenorhinum is a spreading, toadflax-like plant with lilac flowers that sometimes have a yellow or white throat-boss. The draba has yellow flowers and resembles *D.aizoides*. The jasione is a small scabious-like plant with mauve flowers. *Ranunculus acetosellifolius* is endemic to the Sierra Nevada where it grows in moist ground. It is an attractive species with unusual hastate leaves and fairly large white flowers. *Salvia lavandulifolia* is a kind of sage with rather narrow lance-shaped leaves. *Saxifraga boisseri* is endemic to southern Spain; it forms cushions of glandular, hairy leaves and reddish winter buds followed by white flowers with a yellow centre. The silene produces mats of greyish-green leaves and has pink petals that are red on the lower surface. The tanacetum is also mat-forming and has yellow inflorescences, usually borne singly.

About August, when the snow melts one can wander amongst the peaks of the Sierra Nevada National Park including; Mojón Alto 3,109m, Cuervo 3,191m, Vacares 3,149m, la Alcażaba 3,366m, Mulhacén 3,482m Cerro de Caballo 3,015m, Fraile del Valeta 3,210m and others. Here one can find many special alpines:

Androsace vandelli   *Gentiana brachyphylla*
Armeria splendens   *Linaria glacialis*
Artemesia granatensis   *Plantago nivalis*
Biscutella glacialis   *Saxifraga globulifera*
Eryngium glaciale   *Saxifraga nevadensis*
Fritillaria lusitanica   *Saxifraga oppositifolia*
Gentiana alpina   *Veronica repens*

The androsace, (= *A. imbricata*) has white flowers; it also grows in the Pennine Alps. *Artemesia granatensis* is a small cushion-forming species, endemic to the Sierra Nevada. *Saxifraga globulifera* and *S. nevadensis* are similar cushion-forming species with white flowers and leaves that are semi-circular in outline. Further searching may reveal some special treasures such as:

Campanula herminii   *Ptilotrichium purpureum*
Dianthus subacaulis   *Ranunculus demissus*
Erigeron frigidus   *Sideritis glacialis*
Linaria aerugina   *Viola crassiuscula*

The campanula resembles our harebell. The dianthus is the sub-species *brachyanthus* and a rather beautiful alpine pink with 3-4cm flowering stems. *Erigeron frigidus* has large pink 'daisies' with a semi-double appearance. The linaria usually has brownish flowers but these can vary in colour. The ptilotrichium is like a neat form of *P. spinosum* with darker flowers. *Ranunculus demissus* is a tiny plant with relatively large 'buttercups'- it also grows in Greece and Asia Minor. *Viola crassiuscula* is a small pansy endemic to the Sierra Nevada with violet, pink to pale yellowish flowers

1-1.5cm across and with a short spur.

The summit of Trevenque 2079m is a limestone outcrop and here it is worth looking for:

Convolvolus boisseri   *Santolina elegans*
Helianthemum pannosum   *Scabiosa pulsatilloides*
Potentilla caulescens

The convolvolus is a particularly beautiful small plant with close tufts of small silvery leaves and pink flowers - a Spanish endemic. The potentilla, which is also found in the Alps, forms tufts and has white flowers. The helianthemum is also white-flowered.

7.10 **Las Alpujarras:** This is the name given to the region which covers the south slopes of the Sierra Nevada, the valley of the river Guadalfeo and the Sierra Lujar between it and the coast.

Travelling westwards from Almeria one can turn right at Pozuelo along C333 to cross the Sierra Lujar. The valley is cultivated and with extensive plastic greenhouses until one begins to climb after passing Albuñol. The hillsides are covered with almond orchards with a few crops of figs and vines but at about 1,000m one comes to rough scrub with *Genista umbellata, Lavandula stoechas, Stipa sp., Ulex parviflorus* and some other species such as *Phagnalon sordidum, Senecio linifolius, Ruta montana*. In places there are cork oak plantations. In autumn one may see the rather uncommon *Aster sedifolius* growing to about 40cm tall, with narrow leaves and capitulae having generally no more than seven purple, pointed ray florets. Two roadside weeds that one may meet with in summer include *Heliotropium europaeum* and *Mantisalca salamantica*. The second of these is a rather attractive mauve flowered cornflower-like perennial and a name that rolls off the tongue well. Descending then to Puerto Camacho one passes more vineyards and almond orchards which, with the addition of groups of poplar trees produce some magnificent autumn colours. The road from Puerto Camacho westwards to Velez Benaudella C332 has some interesting plants on the north-facing slope:

Antirrhinum hispanicum *Viola arborescens*
Dianthus lusitanicus *Trachaelium coeruleum*
Moricandia moricandioides

The antirrhinum is a neat small species with pink and white flowers. The dianthus resembles *D. monspessulanus* but has a basal tuft of leaves and the flowers have more pronounced hairs towards the centre of the petals. The moricandia is a more impressive species than *M. arvensis* which is so common in the region. It often grows in rock faces and has a rosette of sizeable leaves which look like those of a cultivated cabbage with a distinct bluish bloom. The inflorescences are taller and the flowers larger than those of *M. arvensis*. The viola is a neat sub-shrub and the flowers here were smaller and more rounded than the form commonly seen in the Algarve. The trachaelium is a fine species formerly much prized as a pot plant for its flat heads of blue-mauve flowers.

At Puerto Camacho one meets the main Motril-Granada road (N323). This valley, which is well irrigated by the drainage water from the Sierra Nevada, supports many crops including sugar beet, oranges and custard apples. By the roadside in late summer one can see the attractive introduced weeds *Mirabilis jalapa* and *Bidens aurea*. To get to the 'real' Alpujarras one should turn eastwards off the N323 at the old spa of Lanjaron to Orgiva. From here one can drive up a winding road to Trevelez. There is much spanish broom *Spartium junceum* by the roadside and one may find the short, pink-flowered *Antirrhinum hispanicum* and the much taller *Antirrhinum barrelierei, Digitalis obscura, Asperula hirsuta*. At the highest point of the road it is possible to see both *Gentiana alpina* and *G. brachyphylla* and in wet places *Pinguicula lusitanica* with tiny mauve-pink flowers. The high part of the Alpujarras has interesting small villages which have, so far, resisted the pressures of tourism and are well worth a visit for their own sake..

7.11 Addendum - Benidorm Region: Travelling north-eastwards along the coast from Almeria, one comes to the popular seaside resort of Benidorm. At first sight the high-rise hotels and provisions for discos and other forms of entertainment would suggest that it is not a suitable region for plant enthusiasts. However, there a many interesting species to be seen near here with the aid of a car and if it is deemed too noisy then the quieter resort of Calpé, some 20km to the north, may be more acceptable.

The region is nearly as dry as further south in the Cabo de Gata (7.8) and at Elche there are the most extensive commercial date plantations in Europe. These are worth a visit, especially as there is an interesting sub-tropical botanic garden nearby called El Huerto del Cura. Amongst vineyards and other crops, there are numerous orchards of loquat *Eriobotrya japonica* which are common as isolated specimens throughout the Mediterranean but rarely grown on such a scale. The fruit is delicious when fully ripe but is easily damaged and does not travel well consequently it is largely used for local consumption. In Spanish it is called 'Nispero de Japón' - literally Japanese Quince.

The most obvious place to botanise in the region is the Peñal d'Ifach, a huge rock off the coast from Calpé. It is a nature reserve with a well-appointed museum and research station easily reached by taking the road from the harbour; from there on a footpath goes to the top of the penyal. It covers some 35Hk with about 300 recorded plant species some of which are endemic to a small area around the coast here if not to the penyal itself. These include:

Centaurea boissieri spaehi   *Hippocrepis valentina*
Dianthus valentinus   *Scabiosa saxatilis*

The centaurea is a perennial with light mauve flower heads and pinnate leaves having narrow toothed leaflets. It can be seen low down near the harbour. The dianthus is a typical mat-forming pink that grows on the rocks. The hippocrepis is fairly widspread on the penyal and somewhat resembles the Balearic endemic *H. balearica* but has broader leaflets. The scabiosa is another evergreen mat-forming perennial of the rocks similar to *S.cretica* of Crete and Sicily. The common garrigue shrubs found here include:

Cistus albidus
C. monspeliensis
Ephedera distachya
Juniperus phoenicea
Lavandula dentata

Osyris quadripartita
Pistacia lentiscus
Rhamnus alaternus
Thymelaea hirsuta
Thymus vulgaris

The osyris closely resembles the widespread *O. alba* but has larger berries and leaves. The rhamnus here is sometimes claimed as the sub-species *myrtifolia*. Amongst these shrubby species grow:

Anthyllis cytisoides
Antirrhinum barrelieri
Arisarum vulgare
Asperula paui
Asteriscus maritimus
Coronilla juncea
Inula crithmoides
Lapiedra martinezii

Lavatera maritima
Lonicera implexa
Ononis natrix
Rubia peregrina
Smilax aspera
Sarcocapnos enneaphylla
Urginea maritima

The asperula is an uncommon species of limited distribution found also in the Balearics. It is a straggling plant with white flowers. Lapiedra is an interesting but rather unprepossessing member of the Amaryllidaceae (see 7.6). It grows mainly on the lower parts of the penyal and is not uncommon on the top of sea cliffs throughout the region.

Inland, to the west of Benidorm one has a good view of the mountains, mainly the Sierra de Bernia and the Sierra de Altana. These are picturesque and interesting though they only rise to about 1,500m and rarely carry snow. The highest has a road to the summit of Aitana 1558m which is out-of-bounds - unfortunately as *Saxifraga longifolia* grows there though it may be worth looking for it in other high parts of the region. It is well worth visiting Guadalest, Callosa and the Col de Rates 580m and plants one may encounter on the journey include:

Ballota hirsuta   *Globularia alypum*

Chamaerops humilis *Moricandia arvensis*
Cistus albidus *Rhamnus lycioides*
Daphne gnidium *Ulex parviflorus*
Erinacea anthyllis *Viola arborescens*
Erodium petraeum valentinum

    The erodium is a very attractive sub-species with carrot-like leaves and pink flowers. Another interesting plant which one may see usually at lower levels is *Leucanthemum coronopifolium* - one of the several species referred to as 'manzanilla' in Spain and used as camomile.

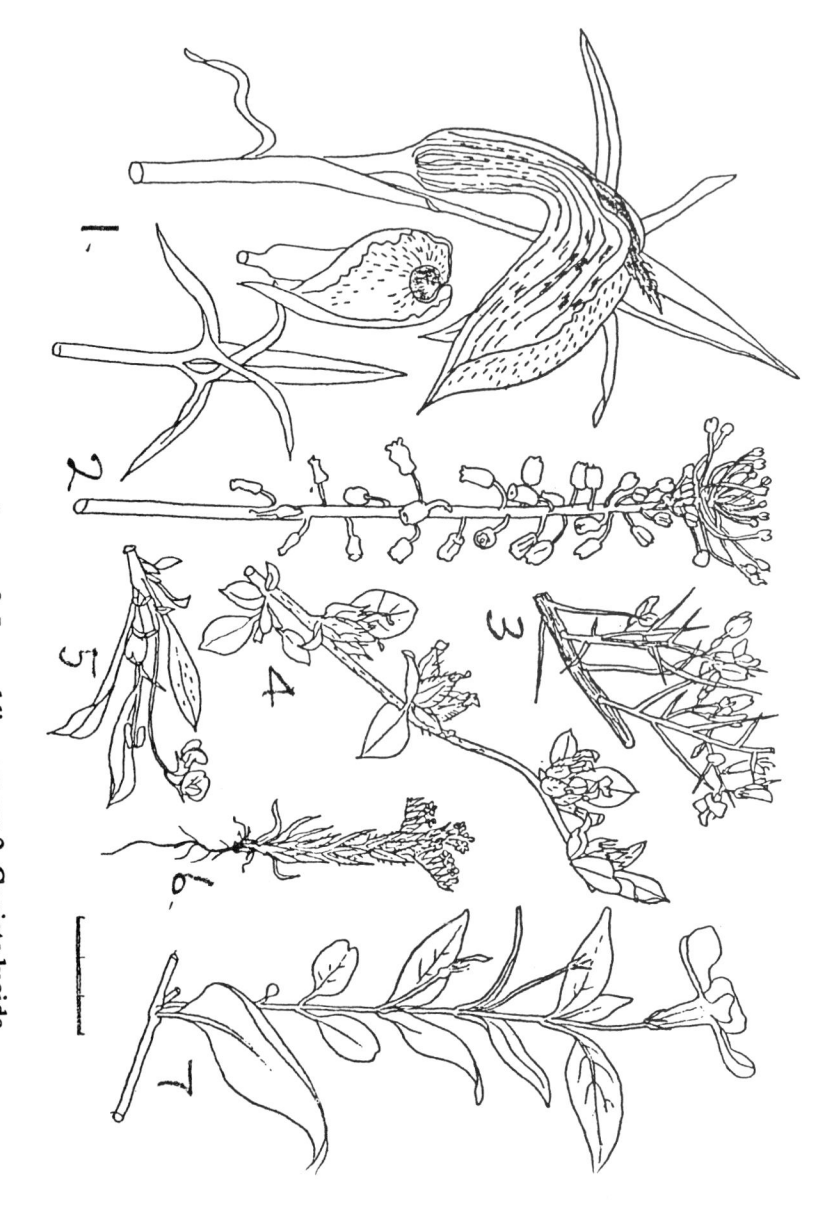

Majorca 1. Helicodiceros muscivorus 2. Leopoldia comosa 3. Genista lucida
4. Anthyllis tetraphylla 5. Scorpiurus muricatus 6. Neatostema apulum 7. Vinca difformis

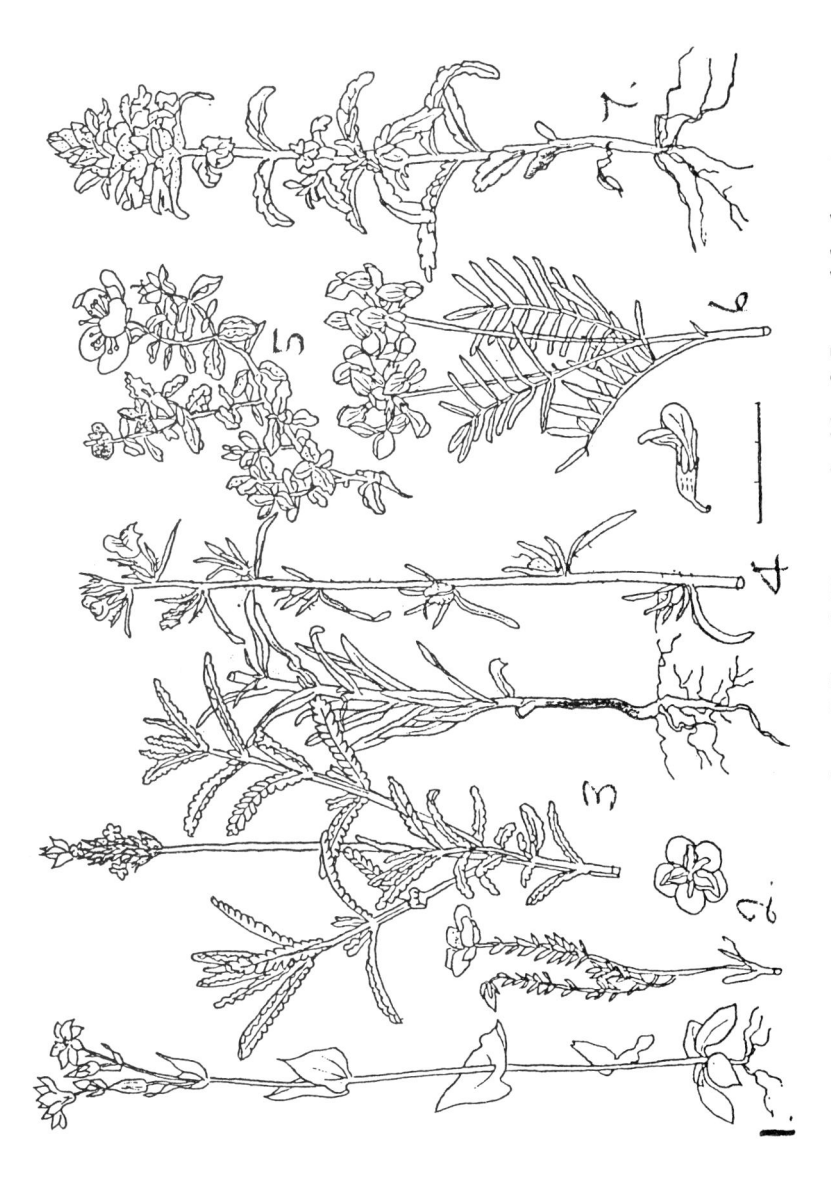

Majorca 1. Blackstonia perfoliata 2. Fumana ericoides 3. Lavandula dentata
4. Misopates orontium 5. Hypericum balearicum 6. Hippocrepis balearica 7. Bellardia trixago

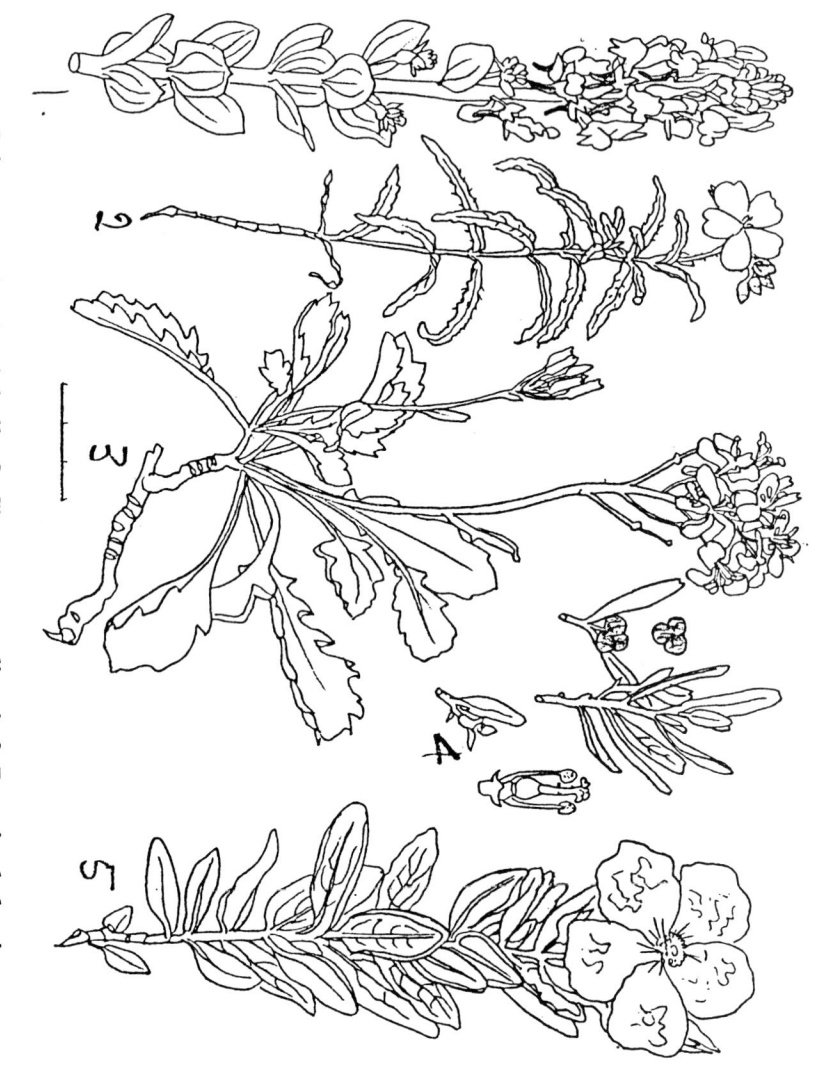

Majorca 1. Linaria triphylla 2. Cistus monspeliensis 3. Brassica balearica 4. Cneorum tricoccon 5. Cistus albidus

8. Majorca

8.1 Majorca - written Mallorca by its inhabitants - is more than just a holiday centre for sun seekers. It is the largest of the Balearic islands, has spectacular mountain scenery along its north-west coast, and is of special interest to plant enthusiasts and birdwatchers. By comparison with other Mediterranean islands of comparable size, the flora of Majorca comprises rather few species - about 1,300 - but these include some 50 endemics. Although it is Spanish territory, and Spanish is the official language, many of the inhabitants are proud of their own dialect of Catalan known as Mallorquin, which they speak amongst themselves.

There are plenty of holiday resorts on the island open throughout the year and any of them are suitable bases for exploring the flora. However, it is preferable to choose a site in the northern half, near to the mountains which are of very special interest for their endemic plants. Although much can be done on foot and travelling by public transport, the use of a car is a great help and car hire on the island is relatively cheap. A useful map is the Firestone T26 (scale 1:125,000)

From mid-April to mid-May is a good time to see most wild plants, and before the tourist season has reached its peak, but the autumn can be of special interest for some late-flowering species.

8.2 **The Terrain:** The most obvious physical feature of the island is the range of mountains that stretches some 80km parallel to the north-west coast and rises to the highest point at Puig Mayor 1445m. They fall steeply to the sea to give spectacular cliffs, especially between Sant Elm and Puerto de Soller. South east of this range is a plain where vegetable crops such as potatoes and broad beans are grown, centred on the towns of Inca and Sineu. In the south, the country is hilly in places and the coast has some good beaches. Most of the rocks are limestone, including those of the mountains.

There are salt pans in the north-east near the bay of Alcudia and in the south-west by Colonia. Both are special places for bird watchers and have a few interesting plants but the marshy land around those in the north-east, is gradually being drained. Fortunately, part of this area has recently been declared a Nature Reserve and it is to be hoped that this will delay or prevent further 'improvement'. On the east coast, near Porto Cristo, the caves of Drach are well worth visiting.

Prevailing north-westerly winds (the 'Tramontana') lose much of their moisture in the mountains where the rainfall may be as high as 1,500mm (60in) a year and low cloud and mists are commonplace in winter there but further south the weather is usually drier and sunnier.

8.3 Literature: The flora of Majorca is particularly well annotated. One of the earliest treatises on the topic is that by Knoche (1921-23) - a four-volume work that has recently been reprinted in Holland. Now this has, been superseded by another four-volume, up-to-date work by Bonafè (1977-80). This new work is in Catalan, most of which is not too difficult to understand if one can read Spanish, French or, preferably, both of these languages. Quite recently a very useful book by Beckett (1988) on the flora of the Balearics has become available. It is essentially a botanical key to the species found there and is very helpful to the serious enthusiast who does not want to purchase or carry around Boniafè's four volumes on a visit.

A particularly useful small book on the 'Plants of the Balearic Islands' by Bonner (1985) is available on Majorca and from some specialist booksellers in Britain. It was originally published in Catalan but now there is also an English version. An interesting book on the natural history of the island by Parrack (1973) has been out of print for some years and is, unfortunately, difficult to obtain. Recently a well-illustrated book in German on the flowers of Mallorca by Straka et al.(1987) has become available.

8.4 Roadsides: Walking along side roads is a good way to see many of the wild plants. Particularly noticeable are the two attractive thistles *Silybum marianum* and *Galactites tomentosa*.

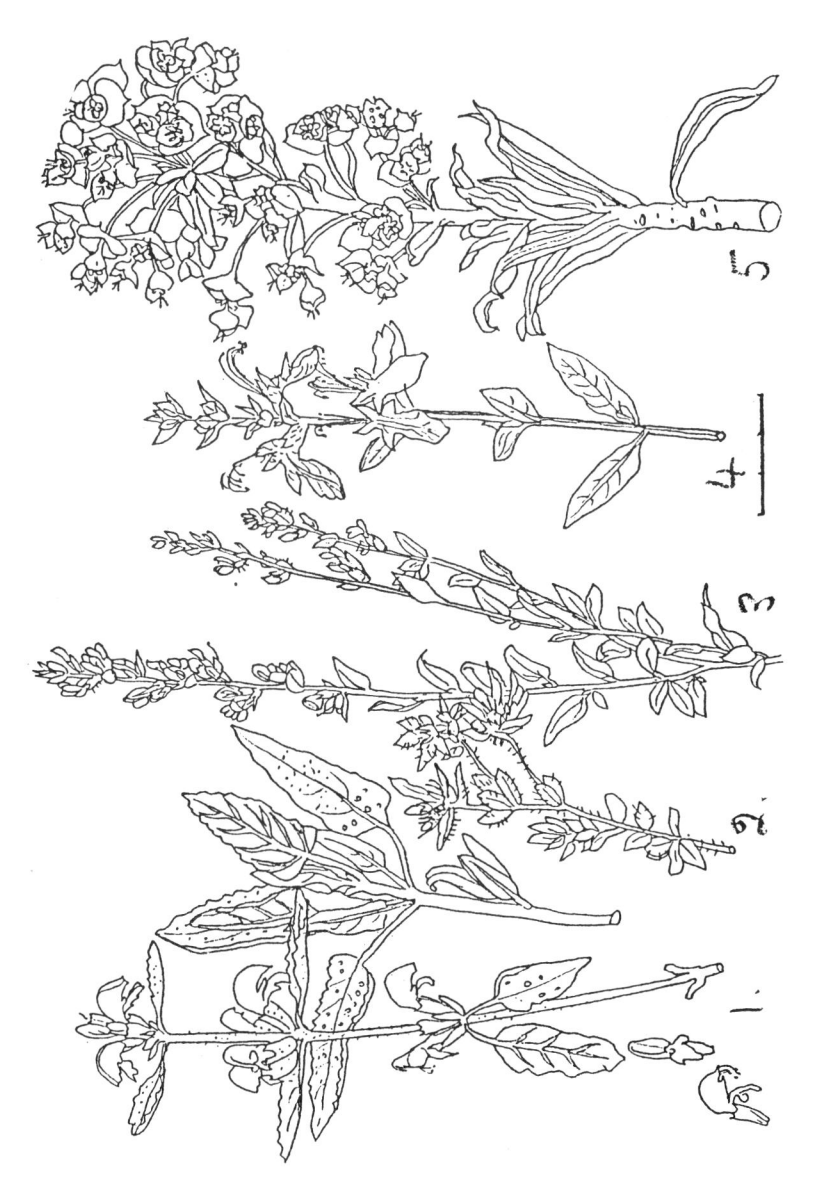

Majorca 1. Phlomis italica 2. Dorycnium hirsutum 3. Anthyllis cytisoides 4. Teucrium fruticans 5. Euphorbia characias

In some places an even taller thistle-like plant is to be seen. This is *Cynara cardunculus*, the wild form of the cardoon and perhaps the species from which the cultivated globe artichoke was derived. Other plants to look for include:

Allium roseum	*Ecballium elaterium*
Anthyllis tetraphylla	*Fagonia cretica*
Arisarum vulgare	*Foeniculum vulgare*
Barlia robertiana	*Linaria triphylla*
Bellardia trixago	*Misopates orontium*
Calendula arvensis	*Oxalis pes-caprae*
Centaurea calcitrapa	*Pallenis spinosa*
Cichorium intybus	*Solanum sodomaeum*
Chrysanthemum coronarium	*Urospermum dalechampii*
Convolvulus althaeoides	*Urtica pilulifera*

All of these are described in Polunin and Huxley's (1965) excellent small handbook on the Mediterranean flora though two of them are referred to there under different names: *Anthyllis tetraphylla* = *Physanthyllis tetraphylla* and *Misopates orontium* = *Antirrhinum orontium*. Some forms of *Allium roseum* on Majorca have relatively large pink flowers and are very attractive. The giant orchid *Barlia longibracteata* blooms quite early, sometimes starting in January, and its sombre-looking spikes of ash-grey flowers are so tall and conspicuous they can be identified when passing in a bus. Two species seen in moist or shady places by the roadside but not mentioned above are *Arum italicum* and *Vinca difformis* (= *V. media*). The first is a rather large cuckoo pint with a pale yellowish green spathe and the second the only wild periwinkle usually seen on the island.

Roads and tracks often pass through or along the edge of olive and carob orchards and the latter are somewhat of a special feature of the island. The carob *Ceratonia siliqua* is an unusual member of the Leguminosae (now Fabaceae) with trees that are either male or female and have small flowers without petals. A limited number of male trees are grown as pollinators in commercial crops so that the females produce their broad-bean like pods that dry to a dark brown or black colour. The pods have no coarse fibres and a sweet taste; they were once much used as winter fodder for cattle and sold to children in sweet shops as 'locust beans'. One of their main uses nowadays is for adding to chocolate. The seeds are of a rather uniform size and were once employed as a standard weight - hence the word 'carat'.

8.5 Garigue: This typical Mediterranean type of vegetation of small shrubs, which are grazed from time to time by sheep and goats, exists in patches throughout the island especially around the coasts and on the top of the smaller hills where cultivation is unprofitable. There are particularly good examples in the south-east of the island. The main shrubs to be found here are:

Anthyllis cytisoides *Myrtus communis*
Calicotome spinosa *Olea europea sylvestris*
Cistus albidus *Phillyrea angustifolia*
Cistus monspeliensis *Phillyrea latifolia*
Erica arborea *Pistacia lentiscus*
Fumana ericoides *Rosmarinus officinalis*
Lavandula dentata *Teucrium polium*

These species are commonly seen also as a constituent of the garigue on mainland Spain. The only comment needed about this list is that *Lavandula dentata* is the chief species of lavender found in the Balearics. The French lavender *L. stoechas* does occur, but less frequently. Amongst these one should be able to find the following herbaceous and bulbous plants:

Arisarum vulgare *Dorycnium pentaphyllum*
Asparagus acutifolius *Gladiolus illyricus*
Asparagus albus *Leopoldia comosa*
Asparagus stipularis *Neatostema apulum*
Asphodelus aestivus *Psoralea bituminosa*
Asphodelus fistulosus *Rubia peregrina*

Bellis annua *Ruscus aculeatus*
Blackstonia perfoliata

The rather charming small monk's cowl *Arisarum vulgare* - called 'frare bec' (friar's nose) in Mallorquin - is common. It flowers in early spring and produces its fruits in small clusters on stems coiled like those of a cyclamen. *Asparagus stipularis* is very prickly with sharp thorn-like cladodes 2-3cm long. In its early stage of growth the plant is soft and at that time it is eagerly sought after as excellent wild asparagus. *A. albus* has shorter 'thorns' and feathery leafy cladodes whereas *A. acutifolius* has no thorns at all. The psoralea smells strongly of bitumen when rubbed and the rubia is wild madder, which was once important for the production of a red dye. Amongst these grow a number of orchids:

Anacamptis pyramidalis *Ophrys tenthredinifera*
Ophrys bertolonii *Orchis tridentata*
Ophrys bombyliflora *Serapias lingua*
Ophrys fusca *Serapias parviflora*
Ophrys incubacea

These are in flower between March and early May except for the serapias which follows some two or three weeks later. The anacamptis ranges in colour from white through pale pink to purplish red, unlike populations of the pyramidal orchid in Britain, where the flowers are usually all of a bright pink colour. *Ophrys bertolonii* is a rather uncommon species but sometimes plentiful in Majorca; it is more or less absent from the mainland of Spain though there are a few records of it from Catalonia. Some specimens which seem to be *Orchis tridentata* have white flowers and it is possible that they are the species recently described as *O. conica*, though floras do not record it from the Balearics. Some of the less common species also found in the garigue are *Ophrys apifera, Op. ciliata, Orchis coriophora, O. italica, O. longicornu, O.mascula*. The first of these flower rather late - in May and June. The mirror orchid *Ophrys ciliata* is plentiful in some places and forms growing around the

S. France 1. Euphorbia biumbellata 2. Serapias neglecta 3. Iris lutescens 4. Globularia punctata 5. Polygala nicaensis

salinas near Colonia are sometimes tall and with a narrow lip and should probably be assigned to *Ophrys vernixia*. *Orchis longicornu* is an uncommon species that grows mainly in the coastal region of the south west around Llucmajor and in places in the mountains of the north of the island. *Orchis italica* which is quite common on the mainland of Spain is rare here but has been seen near the Bay of Alcudia.

In addition to the shrubs cited above, a number of less common species occur in the garigue including:

Anagyris foetida	*Genista lucida*
Cistus salvifolius	*Globularia alypum*
Cneorum tricoccon	*Juniperus oxycedrus*
Daphne gnidium	*Osyris alba*
Erica multiflora	*Thymelaea hirsuta*

Anagyris, the bean trefoil, grows to 2-3 metres tall, has trifoliate leaves and hanging yellow 'pea' flowers in winter. It is evil smelling and poisonous. Cneorum is an interesting shrub, rarely above 1m tall, and with small 3-petalled yellow flowers followed by distinctive tripartate fruits like three berries fused together. The ripening fruit is bright red but goes black at maturity. This species, the only one of the genus, has a very limited distribution and is found in the Balearics, a small area of Andalusia, South France (10.4), Sardinia and Yugoslavia. *Genista lucida* is an endemic of Majorca alone which somewhat resembles a calicotome but its leaves are not tripartite - it is, in fact, rather similar to the widespread *Genista acanthoclada*. Its distribution is localised in the central part of the island between Randa and Arta and in the west around Sant Elm and Santa Ponca. The globularia is a particularly attractive shrub with lilac-blue flower heads 1-2cm across. Osyris somewhat resembles *Phillyrea angustifolia* but the top end of the branches appear almost leafless. It has tiny yellow flowers and fleshy red fruits borne singly.

8.6 **Woods:** At one time the woodlands of Majorca were mainly of holm oak *Quercus ilex* which has now been extensively cut down and used to produce charcoal. Today most of the woods consist of Aleppo pine *Pinus halepensis* including a local form of this species with more upright-growing branches recently discovered near Lluchmajor and named the variety *cecilae*. The Aleppo pine gives light shade which is particularly pleasant on a Mediterranean summer afternoon and its woods harbour a number of interesting plants. The honeysucle *Lonicera implexa* is quite common and so is the sweet-scented *Clematis flammula* that flowers in midsummer. The evergreen winter-flowering *Clematis cirrhosa* may also be found in woods but it more often grows on the dry stone walls that are a feature of some parts of the island. It has hanging bell-shaped flowers that are white, or greenish, with small spots in the corolla. Selected forms of this species are becoming a popular garden plant in Britain. In autumn, the large daisy *Bellis sylvestris* can be seen in flower and in moist places the ground between the trees may be covered with the white flowers of *Allium tetraquetrum*. Several orchids grow in the woods, particularly *Ophrys incubacea* (O = *O. sphegodes atrata*) and *Anacamptis pyramidalis*.

In some woodlands, particularly those in the north east around Formentor, one can encounter *Chamaerops humilis*, the small European palm tree that grows to a height of 2m under favourable conditions. It is found in the north-eastern and western extremities of the island around, Formentor, Cala Ratjada and Sant Elm (sometimes spelt San Telm). respectively. It grows in garigue more frequently than in woods and there it forms low-growing clumps of spiny-stemmed foliage. At times the leaves are collected and woven into decorations that somewhat resemble our corn dollies - relicts of a more widespread use in the past and which may sometimes be purchased as inexpensive souvenirs in local open markets.

8.7 Sea shores: Although there are some good sandy beaches on the island, much of the coastline is rocky. Amongst the rocks, and sometimes in very exposed sites, one can expect to find:

| *Anthyllis fulgurans* | *Frankenia laevis* |
| *Arum pictum* | *Helichrysum stoechas* |

Asteriscus maritimus *Launaea cervicornis*
Centaurea balearica *Limonium spp.*
Crithmum maritimum *Phillyrea media*
Daphne rodriguezii *Reichardia picroides*
Daucus carota *Senecio rodriguezii*

The anthyllis is a small, spiny, leafless, endemic shrub with tiny pink or white 'pea' flowers. It grows along the north coast and at Formentor. The centaurea is a very rare Balearic endemic with yellow flowers. *Daphne rodriguezii* (= *D. vellaeoides*) is another rare endemic more likely to be found on Minorca (9.7). The launaea is a small, spiny plant with yellow composite flowers and somewhat resembles *L. spinosa* which is common on the coast of Almeria and other parts of southern Spain (7.4). *Reichardia picroides* is a dandelion-like plant. The senecio is a rather special endemic small 'composite' with pinkish-violet flowers and could be said to resemble a very dwarf, few-flowered greenhouse 'cineraria'. Growing in the sand just above high water mark, one finds a different plant association mainly comprised of:

Cakile maritima *Lobularia maritima*
Calystegia soldanella *Matthiola sinuata*
Cistus salvifolius *Medicago littoralis*
Crucianella maritima *Medicago marina*
Eryngium maritimum *Pancratium maritimum*
Euphorbia parialis

In contrast to the plants of rocky shores none of these is endemic - they are all seen fairly frequently in similar conditions throughout the western Mediterranean. The beautiful sea daffodil *Pancratium maritimum* flowers in late June or July and its ripe black seeds, which look like chips of charcoal, sometimes litter the shore in winter.

One other plant that deserves mention and is a common component of the flora of the Mediterranean is *Posidonia oceanica*. Though a flowering plant, it lives like a seaweed in salt water and is responsible for the accumulations of evil-smelling litter seen on some beaches. Its detached leaves and stems often get matted together into brown fibrous balls and are washed up on the beaches.

8.7 **Bahia de Alcudia:** This relatively large bay lies in the north east of the island just south of the interesting old Moorish town of Alcudia. Leaving Puerto de Alcudia southwards one comes to the lake Lago Esperanza, which is gradually being filled in, and then to the marshy area around salt pans at La Albufera. This is a good birdwatching place and one can nearly always count on seeing osprey, marsh harrier, stilts and other waders and an occasional flamingo that has lost its way. A track leads in from the Esperanza hotel and a short distance along here another track to the left goes to slightly higher ground and a picnic area. Here orchids are numerous and grow with *Leopoldia comosum* (= *Muscari comosum*) and *Muscari atlanticum* (= *M. neglectum*). On drier scrub land by the shore and south of Ca'n Picafort one should look for the endemic *Thymelaea myrtifolia*. It forms a silvery-looking bush with small, rounded, hairy leaves. *Orchis italica* has been recorded from this area but is rare in Majorca.

8.8 **The mountains:** To get to the mountains from Palma, take a road out in a north-westerly direction to Esporles and then to La Granja. Here is a museum of old farm implements and other curiosities - well worth a visit. It was originally a Moorish site and has a high, natural artesian fountain with large groups of aspidistra growing in the spray. From here one can turn south-westwards along a winding mountain road past Galilea and then return to Palma via Calviá. Instead of going to Galilea one may turn northwards at La Granja to join the winding, mountainous C710 road running roughly paralell to the north coast that continues all the way to Cape Formentor. It is possible to get to this road from Palma by going northwards through Valledemosa or by the C711 through Abunyola and Soller. A small scenic railway runs from Palma to Soller and from there a ride in an open tram takes one through orange orchards to Puerto de Soller on the coast - a very enjoyable trip but not the best way to see the plants.

The south-western part of the C710 passes near the sea at places and there are some magnificent views of the impressive cliffs, especially when the sun is setting. Along here, particularly at Mirador de ses Animes near Banyalbufar, grows the attractive shrubby *Lavatera maritima* with large pink petals that are dark at the base. Further north east along this road and past the join with the C711 is the Coll de Puig Major, at 1036m it is the highest part of the road. A few kilometers further on brings one to the Gorg Blau. This is a particularly good area for special plants. The shrubby endemic *Hippocrepis balearica* is fairly common growing on rocks here. Its Mallorquin name is 'Violeta de penyal' or cliff violet though its flowers are bright yellow but it does smell of violets. A true violet grows in shady places by here - the endemic *Viola jaubertiana*. Another plant that keeps mainly to the rocks is *Brassica balearica*, a rather dwarf species with somewhat fleshy leaves and yellow flowers, not unlike the more widely distributed *B. repanda*. Other interesting endemics include *Hypericum balearicum* - a low shrub with crinkled leathery leaves and typical small St. John's wort flowers - and *Pastinaca lucida*. The last of these is a kind of parsnip with umbels of yellow flowers and shiny foliage with a strong, objectionable smell. It is called the equivalent of 'stink weed' or 'devil's cabbage' by the locals. One should also look for *Helleborus lividus*, sometimes referred to as *H. lividus ssp. lividus* to distinguish it from the subspecies found in Corsica. It is a somewhat weaker growing plant than its Corsican relative and has pale pink flowers and leaves divided into three white-veined leaflets. Along with it grows *Helleborus foetidus var. balearicus* which can be distinguished by its green flowers and leaves divided into 7-10 leaflets. A prize to be found is the rare, beautiful, endemic *Paeonia cambessedesii* with large red flowers in late April or May and leaves with a reddish underside. It somewhat resembles *P. broteroi*.

Should one decide to ascend Puig Major 1445m or Tomir 1103m then more treasures are to be found. The early part of the ascent will take one through rough sheep-grazed pasture with wild olive and tufts of the 2m tall rough *Ampelodesmos mauritanica* which is sometimes used for roofing rustic buildings. Although it is a grass, it is one of the few species too tough to be eaten by sheep. Later one can expect to see the spiny endemic shrubs *Astragalus balearicus* and *Teucrium subspinosum*, and the mauve pink flowers of *Phlomis italica*. Some other special plants here are the cotton lavender of our gardens *Santolina chamaecyparissus*, the endemic foxglove *Digitalis dubia*, and in moist patches, the endemic *Ranunculus weyleri*. The last of these is a small, yellow flowered species. In autumn the local endemic *Crocus cambessedessii* can be found in flower with *Merendera filifolia*. The crocus is a dainty species with very narow leaves, flowers 2cm diameter, pale lilac, feathered outside with purple and conspicuous orange stigmas not unlike those of the saffron crocus which also grows in the island at lower levels. Finally, when in this region, one should look up occasionally for black vultures. These huge birds which soar magnificently are more common in this part of Majorca than any other part of Europe.

Back on the C710 road near the reservoir at Gorg Blau there is a very winding road that leads northwards to the coast at Sa Calobra. From here a track goes down to near sea level at the Torrent de Pareis. In the moist river bed grows the shrub *Vitex agnus-castus* which has buddleia-like inflorescences in summer. It is sometimes called the chaste tree and has the reputation of calming 'natural inclinations' but at times has been used also as an aphrodisiac! The summer snowflake *Leucoium aestivum* produces its hanging white snowdrop like flowers in March between the vitex bushes - it is the sub-species *pulchellum* that flowers earlier than the type. On leaving this area one may notice from a height that the sea water looks bright blue where it mixes with the fresh, limy water from the torrente.

The north end of the C710 takes one to the lighthouse at Cabo Formentor. In the spring *Cyclamen balearicum* produces its dainty white flowers; it has mottled leaves that are reddish on the underside. At one time it was thought to be endemic to the Balearics but has since been found in the South of France. Another interesting plant is *Helicodisceros muscivorus* sometimes included in the genera *Arum* or *Dracunculus*. This curious aroid has unusually-shaped leaves with long thin sections pointing in different directions and a spathe and spadix with rather long hairs. The inflorescence is held more or less horizontally or downward tilted and the name 'dragon's mouth' is an apt description. It is rare outside the Balearics, though it is to be found in Corsica (11.10),but is quite plentiful in places on Majorca, especially in stony areas around the coast though it does not often produce flowers. An endemic found here and at other places along the north coast is *Aristolochia bianorii*. It has stalked leaves and solitary flowers which are brownish yellow with brown stripes. Whilst one is in the Cabo Formentor area one should look for Eleanora's falcons. These graceful birds nest in colonies near the lighthouse and are most likely to be seen in late August and early September. They are fast enough to be able to fly down migrating swallows.

Senecio rodriguezii Fornells, Menorca. (p.118)

Asclepias curassavica Barbate de Franco, Spain. (p.44)

Iberis candolleana Mont Ventoux, S. France. (p.141)

Iris pseudopumila Lentini, Sicily. (p.201)

9. MINORCA

9.1 The inhabitants call their island Menorca - Minorca is the anglicized version. It is about one sixth the size of Majorca and has a reputation for providing 'quieter' holidays. Its climate is somewhat more humid than that of Majorca, so substantial grassland can be sustained and dairy products are an important export. For this reason also the growth of wild plants is more lush than on Majorca, including that of the orchids which often abound here. Fewer species of wild plants occur than on the larger island but Minorca can boast three endemics that are found nowhere else; *Lysimachia minoricensis, Viola bifoliolata* and *Daphne rodriguezii*. The lysimachia probably has disappeared from the wild but is sometimes grown in gardens on the island and at least one British seed firm has offered it in their catalogue. It has marbled leaves and very small pink flowers - an interesting, though not showy, species for the gardener who likes uncommon plants which are reasonably hardy and not difficult to grow. The viola was at one time thought to have become extinct but it has recently been rediscovered in the north east of the island. The daphne is described later (9.7).

Most plants are to be seen in flower on Minorca from the end of March to the beginning of May but, as in most parts of the Mediterranean, there are some interesting autumn-flowering species such as *Merendera filifolia* and *Narcissus serotinus*.

Minorca is only approximately 45 x 15km, so travelling around is not arduous. Buses are often a convenient way of getting to interesting sites, especially those that serve the main C-721 road which traverses the length of the island from Mahon in the east to the old Moorish capital of Ciudadela in the west. There are good Firestone maps available on the island itself and some others which are produced locally. Occasionally the place names on different maps, and in older literature, do not always agree due to indecision over the use of the local Catalan dialect or Spanish. 'Cala' means a bay and 'son' of or appertaining to.

9.2 **The terrain:** A geological fault runs very roughly along the line of the C-721 road and has given rise to the fine natural harbour of Mahon. South of this line the rock is predominantly limestone and the soils are fertile and it is here that the best cattle pastures are situated. It is in this area also that one finds most of the fine stone which is so suitable for building and has made possible the construction of the extraordinary number of megalithic monuments.

North of the fault there are areas of schists, hard rocks and less fertile acid soils carrying holly oak, heathers and plantations of Aleppo pine. It is north of this line, also, that the main hills are to be found though the tallest Monte Toro, (a few kilometers due east of Mercadel that lies on on the main C721 road) only rises to 357m. The costal regions, especially where there are limestone cliffs, are good places to look for plants.

9.3 **Literature:** The most useful booklet on the subject for most visitors is undoubtedly 'Plants of the Balearic Islands' by Bonner (1985) but for those who wish to make a somewhat fuller study the key book for the identification of species is that by Elspeth Beckett (1988). One of the oldest works on the flora of the Balearics is that by Marès and Vigineix (1880) and a copy can be consulted by arragement at the museum in Mahon. It is in French and gives specific references to sites for plants in Minorca but on up-to-date maps it is often difficult to locate the places it quotes. Other useful works are mentioned under 8.3.

9.4 **Around Mahon:** Near Mahon harbour the newcomer to the Mediterranean flora will see many of the common species that flower in spring:

Anagallis arvensis   *Convolvulus althaeoides*
Asphodelus fistulosus   *Galactites tomentosa*
A. aestivus   *Malcomia maritima*
Bellis annua   *Oxalis pes-caprae*
Calendula arvensis   *Silybum marianum*

It is a short walk from Mahon to Cala de San Esteban. Once out of the built-up area one soon enters what might be described as typical Minorcan countryside with dry stone walls, small patches of bushes, rock outcrops and intervening areas of short grass. Here and there a typical local water well is covered with a whitewashed canopy. The overall effect is somewhat like a carefully designed garden scheme and perhaps could give ideas for home garden planning. Groups of the prickly pear cactus *Opuntia ficus-indica* are established and produce edible fruit sometimes referred to as 'figues de Barbarie'. The thornless variety is frequently cultivated here and used as cattle fodder when the grass supply is limited in the height of summer. The prickly *Smilax aspera* scrambles along the walls with scented white flowers in late summer and clusters of shiny red fruits in spring. Another small evergreen climber here is *Clematis cirrhosa* which has speckled, cream-coloured, bell-shaped flowers in winter and is becoming popular as a garden plant in Britain. In shelter, at the base of walls grow the showy periwinkle *Vinca difformis*, the greyish-leaved *Artemesia gallica* (a sub-species of *A. caerulescens*) and the charming small *Arisarum vulgare*, Friar's cowl or 'Frare bec' (friar's nose) in the local Catalan language. Other bushes which fit in with the scheme include the shrubby horsetail-like *Ephedera fragilis*, sometimes with red berries and the Sodom apple *Solanum sodomeum*. Plants of the spiny asparagus species *A. stipularis*, *A. acutifolius* occur here and there and their young, soft, shoots are collected by the inhabitants on week end outings as a much-prized vegetable, especially those of the first of these species.

Arriving at Cala de San Esteban (St. Stephen's Bay), one comes to garigue comprised typically of *Pistacia lentiscus, Cistus monspeliensis, C. albidus*. The last of these, which has mauve-red flowers, is confined to the western Mediterranean and not found east of Italy. Orchids are fairly common amongst the bushes, especially *Ophrys bombyliflora, O. speculum, O.lutea, O. tenthredinifera*. In the drier parts the charming small mirror orchid is very plentiful. In the local Catalan dialect it is called 'Sabeta del bon Jesus' (Jesus' sandal). A plant that grows alongside it in the short turf by the sea is *Allium ampeloprasum* - the wild form of our cultivated leek. Another bulbous plant in this kind of environment is *Urginea maritima* with huge, half-buried bulbs and tall racemes of numerous pinkish-white flowers in late summer.

9.5 **The South:** Nearly all the interesting megalithic monuments and most of the holiday beaches lie in this limestone area. At the land side of the sandy beaches one may see many typical Mediterranean shore plants:

Cakile maritima	*Launaea cervicornis*
Crithmum maritimum	*Lobularia maritima*
Eryngium maritimum	*Matthiola sinuata*
Glaucium flavum	*Pancratium maritimum*
Inula crithmoides	*Verbascum sinuatum*

The launaea is a low spiny composite with small yellow flower heads. It is endemic to the Balearics but similar to *L. spinosa* found in Andalusia (7.5). The verbascum forms large attractive rosettes of felted leaves with wavy margins; its branched metre-high flowering stems bearing yellow flowers are produced in June and July.

Behind the shore there is often an area of garigue which is an especially good place to search for plants. The popular sites of Cala de Santa Galdana and Cala'n Porter with the nearby Calas Coves are well worth a visit. The main shrubs here are:

Cistus albidus	*Olea europea sylvestris*
C. monspeliensis	*Phillyrea angustifolia*
Juniperus phoenicea	*Pistacia lentiscus*
Myrtus communis	*Rosmarinus officinalis*

Cistus salvifolius is another member of the genus which occurs on Minorca, though it is less common there than the other species. It has particularly beautiful saucer-shaped white flowers. Numerous orchids are to be found and often they are especially lush specimens here:

Anacamptis pyramidalis *Ophrys ciliata*
Ophrys bertolonii *O.tenthredinifera*
O.bombyliflora *Orchis tridentata*
O.fusca *Serapias vomeracea*
O.lutea

In Minorca the anacamptis flowers are usually much paler in colour than those of the typical British form of pyramidal orchid and not infrequently pure white. It flowers later than most other species listed above except for the serapias. *Ophrys bertolonii* is, in general, a rather uncommon species but it is fairly widespread here. Its dark violet lip has a large 'mirror' and when the flower is observed closely in side view it can be seen to deserve its Italian folk name 'L'ucellino che si taglia allo specchio' - little bird dashing itself against a mirror. *Ophrys ciliata* is usually to be found under the name of *O. speculum* in older books. The form of *O. tridentata* here is often light coloured and tends towards the newly-described species *O. conica*. One may even be fortunate to come across a specimen of the rare *Orchis longicornu*. Furthermore there is always the chance of finding hybrids between ophrys species such as *Ophrys x kallista* (*O. bertolonii x tenthredinifera*) seen at Calas Coves. Other plant species to be found growing amongst the shrubs of the garrigue include:

Allium roseum *Leopoldia comosa*
Bellardia trixago *Merendera filifolia*
Evax pygmaea *Orobanche ramosa*
Helichrysum stoechas *Romulea columnae*

The merendera is an autumn-flowering bulbous plant with mauve flowers, similar to a colchicum and the romulea has tiny whitish star-shaped flowers in spring. In addition to the above species it is worth looking for *Cyclamen balearicum*, especially in damp places amongst rocks. It has white flowers in spring and leaves that are mottled rather than patterned.

9.6 Monte Toro: At 357m this is the highest part of the island and is situated roughly in its centre so the view from the top on a fine day is very impressive. One approaches it from Mercadel which lies on the main C721 and in the ditches around this small town *Narcissus tazetta* provides a colourful display during the winter. Although it is native to the island it has been cultivated there and many of the plants may be 'escapees'. It is acompanied in places by *Leucoium aestivum* - the summer snowflake that flowers in May and June. The plants in the Balearics are the sub-species *pulchellum* which has smaller flowers and blooms later than the type.

A winding road takes one to the summit of Monte Toro where there is a monastery. Amongst the boulders near the summit grow the purple-flowered *Phlomis italica* and *Euphorbia characias*.

9.7 The North: Much of the area north of the main C721 road has an acid soil with plantations of Aleppo pine and high garigue or maquis comprised mainly of *Erica arborea* which grows to 2m and has white or pale pink flowers in summer, *E. umbellata* that only reaches 1.5m and produces bright pink flowers during the winter, together with the strawbery tree (*Arbutus unedo*) and rosemary (*Rosmarinus officinalis*). Amongst the woods and scrub grows the delicate *Crocus cambessedesii* which flowers in winter and is generally over by March. It is rather well described by Bowles (1924) as "One of the daintiest of all, this tiny species with leaves like fine grass and flowers less than an inch in length looks as though the Fairy Queen had tried to make a crocus for a doll's house." It is endemic to Mallorca and Menorca.

Some of the costal regions, especially in the north and northwest, have some interesting plants. Down by the sea near Fornells one may see *Senecio rodrieguezii* like a small, pink, florists 'cineraria'. It is endemic to the Balearic Islands and is sometimes described as a sub-species of *S. leucanthemifolius*. Further east near Es Grau grows the rare endemic *Daphne rodreguezii* resembling a dwarf form of the common Mediterranean species *D. gnideum* but with hairy leaf margins and purple flowers. Its main stronghold is the Isla Colom, a short distance off-shore from Es Grau, and many of the sites where it was found on the mainland have been lost to 'development'. In other areas, and especially around the cliffs at Cala Mesquida, there are patches of the small prickly endemic *Teucrium subspinosum* which grows with romuleas and merendera. Near here one may have the good fortune to see *Helicodiseros muscivorus* (= *Dracunculus muscivorus*) and *Digitalis dubia*. These are more common on Mallorca and the first is not often seen in flower.

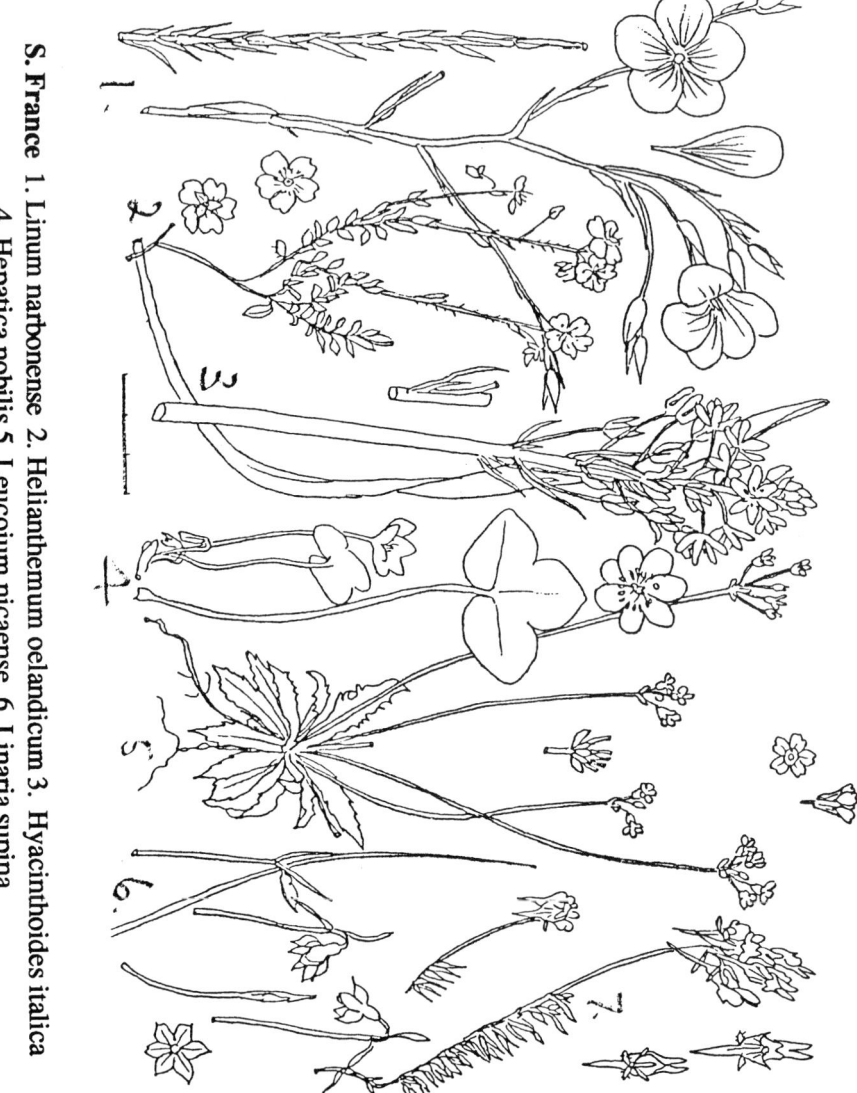

S. France 1. Linum narbonense 2. Helianthemum oelandicum 3. Hyacinthoides italica 4. Hepatica nobilis 5. Leucoium nicaense 6. Linaria supina

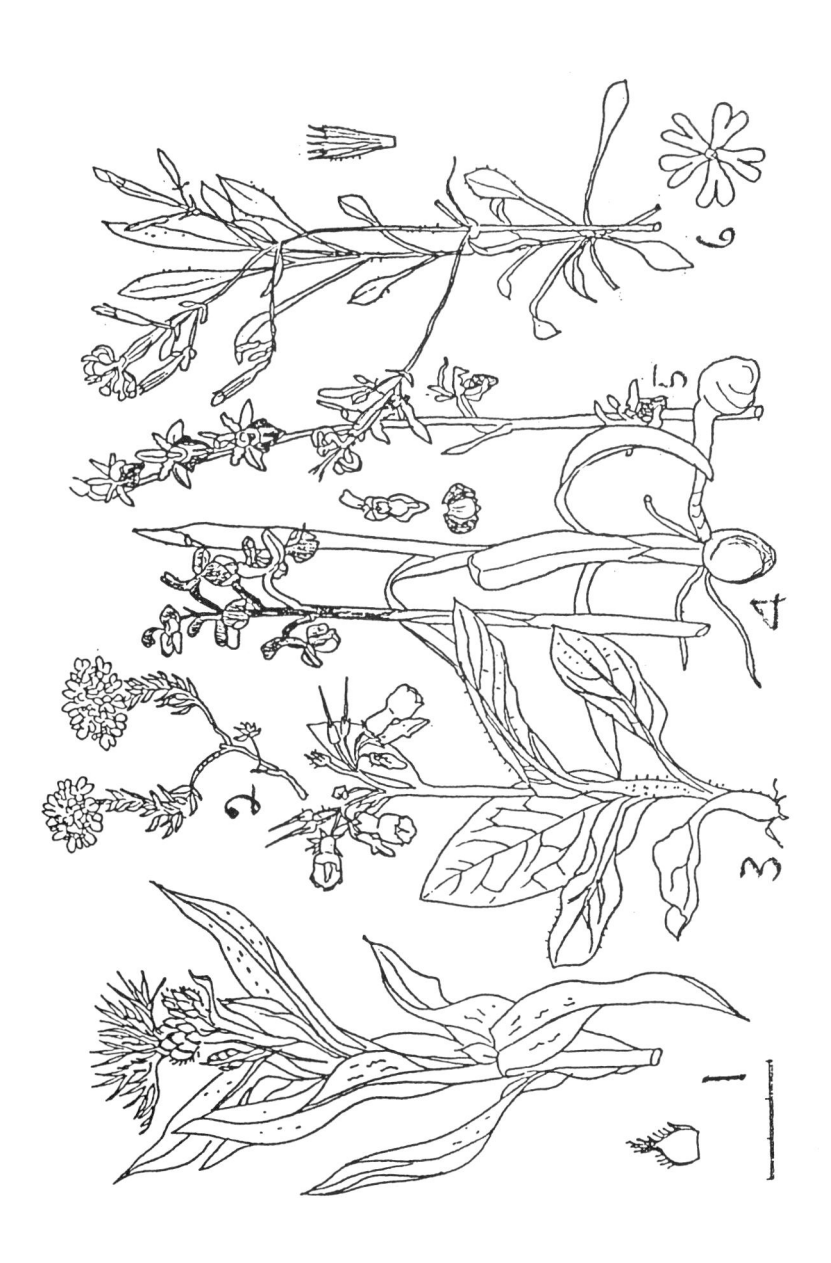

S. France 1. Centaurea triumfetti 2. Iberis candolleana 3. Symphytum tuberosum 4. Orchis champagneuxii 5. Ophrys provincialis 6. Silene italica

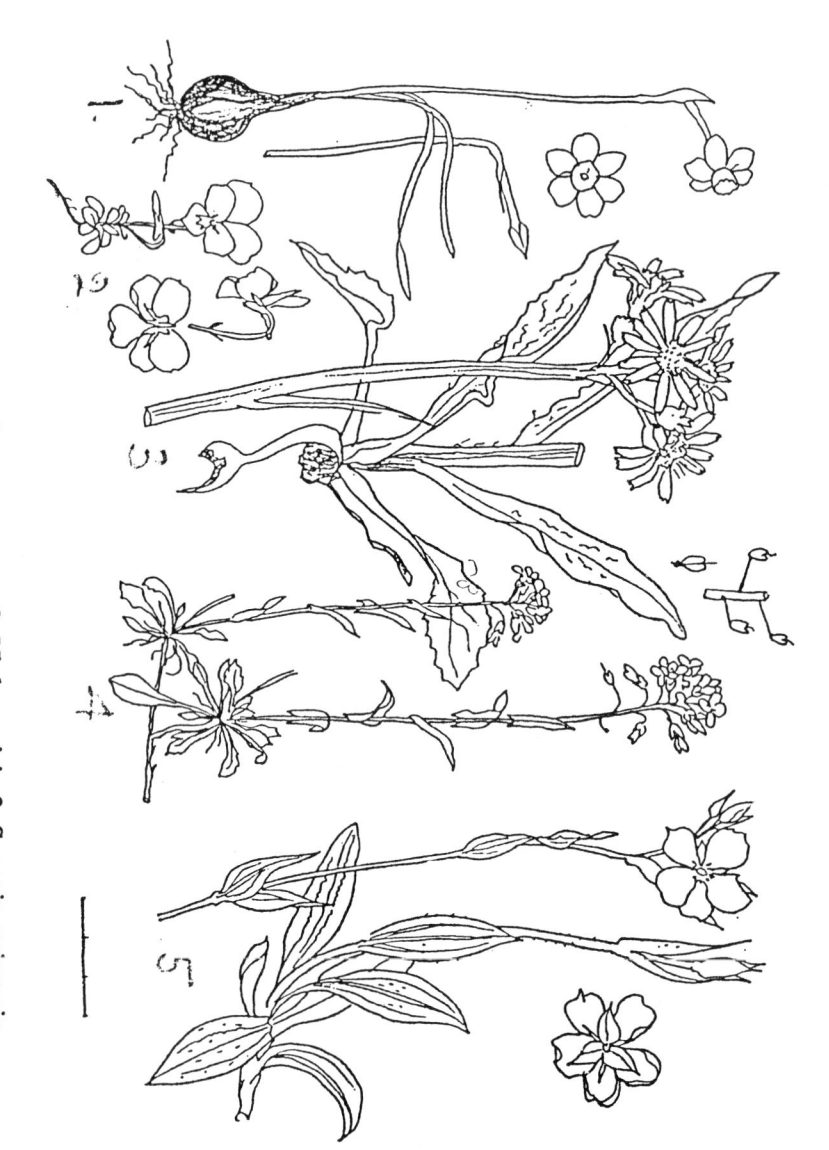

S. France 1. Narcissus assoanus 2. Viola cenisia 3. Senecio ovirensis 4. Thlaspi praecox 5. Tuberaria lignosa

10. SOUTH OF FRANCE

10.1 The French Riviera or Côte d'Azur was the first region of the Mediterranean to be developed for large scale tourism. Even before the turn of the century the well-to-do British flocked there to escape the northern winter and it is not by chance that the main coastal thoroughfare of Nice is called the 'Promenade des Anglais'. Nowadays, most British holiday makers prefer to go to Spain, Portugal or Greece though the Côte dAzur is naturally still very popular with the French themselves and nationals of neighbouring countries on the European mainland. The whole of the southernmost part of France is of considerable interest to the plant enthusiast and well worth visiting in spring when most species are in flower and before the summer tourist traffic has reached its peak.

The region to the west of the Rhône valley is referred to as Languedoc and to the east as Provence, and although there is a considerable similarity in the flora between them, each has enclaves of less common and endemic species. The truly Mediterranean vegetation is confined to a rather narrow strip along the coast and much of this has been lost to urbanisation. Furthermore, recent exceptionally cold winter weather has damaged or killed some of the more tender cultivated and wild plants there. The special feature of the flora of southern France is a mingling of plants from northern Europe with Mediterranean species via the Massif Central, the Alps and the Pyrenees.

Package holidays are offered for resorts on the Côte d'Azur and these make quite good bases for studying the interesting flora of the Alpes Maritimes and the Var but the best way to see the area is to travel independently by road. If not by car, then a bicycle would do and not be out of place on French roads, though most cyclists there are more intent on getting from one place to another as quickly as possible than stopping to observe the countryside. They may, however, appreciate it sufficiently to shout in passing to the plant hunter "C'est beau, la nature!" For road travel, the two Michelin 1/200000 maps 240 and 245 are almost essential and there is an even larger scale (1/100000 No. 115) for the Alpes Maritimes

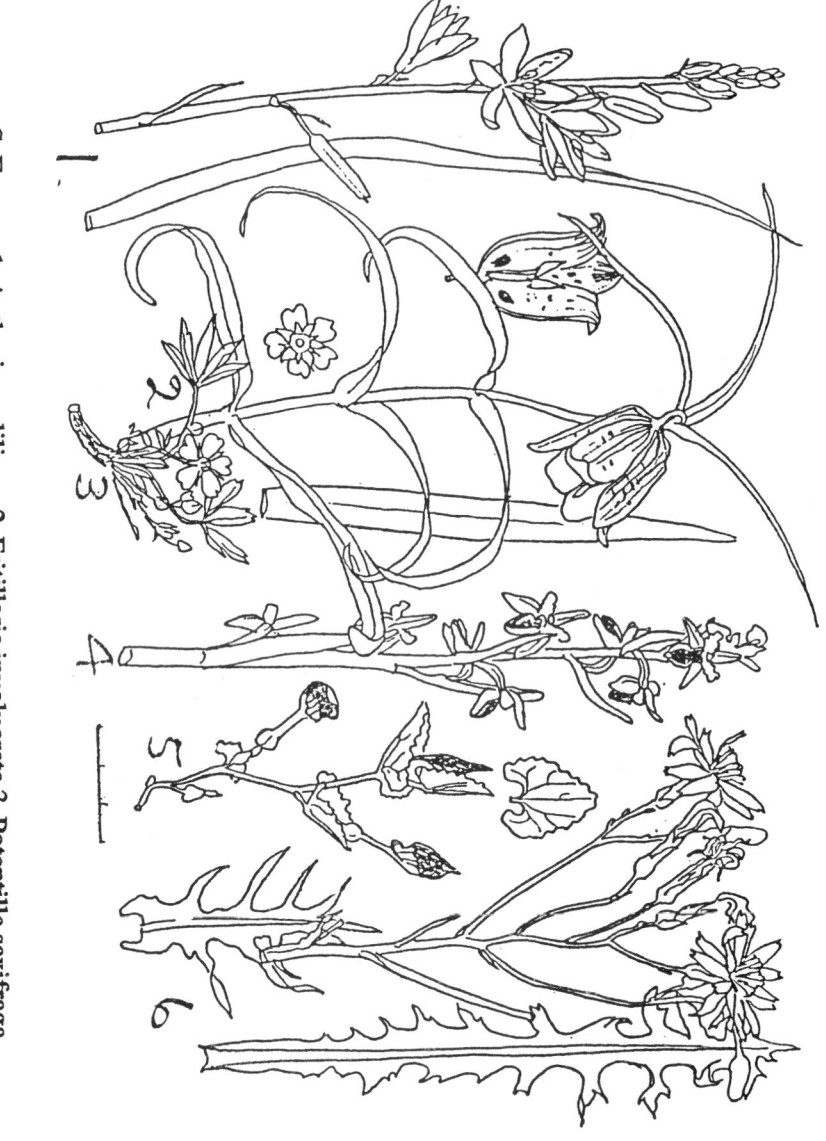

S. France 1. Anthericum liliago 2. Fritillaria involucrata 3. Potentilla saxifraga 4. Ophrys araneola 5. Aristolochia pistolochia 6. Lactuca perennis

Department. With these it is possible to plan ones route along well-surfaced smaller roads and enjoy some of the beautiful parts of rural France. It is a real pleasure to stop by the wayside for a midday snack of baguette, camembert and bottle of local wine to the accompaniment of the cuckoo and nightingale - yes, he sings fervently both night and day!

Two important establishments for the study of the flora of the region are the Institut Méditerrnaéen d'Ecologie et de Paleoecologie at Marseilles and the Conservatoire Botanique National de Porquerolles on the small island of Porquerolles, more or less south of Hyères.

10.2 The Terrain: The region is effectively divided in two by the lower reaches of the river Rhône which terminates in an extensive delta of salt marshes and land-locked lagoons - generally called the Camargue. To the west of this, in Languedoc, the coastal area between Montpellier and Narbonne is rather low lying but becomes more hilly southwards towards Perpignan and the Spanish border. The hills and soil of this region are essentially based on limestone but with extensive alluvial deposits around the river Aude in Bas Languedoc. Much of the lower-lying region is devoted to vines for wine production. By contrast, the Pyrenees, which form the boundary of the region to the south are mainly composed of granite. So also are the higher hills to the north of the region -The Cevennes which form the south west extremity of the Massif Central and culminate in the Montagne Noire, roughly 20km north of Carcassonne.

Around 200 million years ago the sea broke through this barrier of granite and volcanic rocks of the Cevennes, somewhere north of where Beziers now lies, and laid down a flat deposit of calcareous material similar to that of the costal region. This has now been tilted and fissured by faults to give an exposed high limestone region known as the Les Grands Causses. The limestone, subject to erosion from carbon dioxide dissolved in rainwater, has produced some spectacular gorges of which the Gorges du Tarn, just north of the region described here, are perhaps the best known.

To the east of the Rhône, in Provence the land generally rises more steeply from the sea than in Languedoc; indeed, very steeply towards the eastern border with Italy where some 40km north of Nice there are 3000m peaks of the Alpes Maritimes. Most of the region, including the Alpes Maritimes and the Alpes de Provence in the north, are predominately limestone and there are several chains of limestone hills running roughly east to west in the eastern section that are well worth special visits for botanising. They include: Mt. Ventoux (east of Orange) which continues eastwards in the Montagne de Lure; Montagne du Lubéron (roughly east of Arles) which extends westwards to Les Baux; Massif de la Sainte Baume (east of Marseille). By contrast, the Massif des Maures and Massif de l'Esterel (near the coast between Toulon and Cannes) are of cristalline acidic rocks, more resembling the geology of Corsica with which they were once joined. As in Languedoc, the limestone has developed several deep gorges, notably the Grand Canyon du Verdon which is the largest and perhaps the most spectacular in the whole of France.

10.3 **Literature:** Considering that this was the first region of the Mediterranean to be examined in detail by plant enthusiasts, there is surprisingly little up-to-date literature in English specifically about its flora. A useful, well-composed book on Provence is 'Flowering Plants of the Riviera' (Thompson 1914) which, in spite of its age, may still be obtained second-hand from some specialists. It is, as one might expect, slightly old-fashioned in its nomenclature. There are several French Departmental floras for the region but most are difficult to obtain or consult, however the recent 'Flore du Département de Vaucluse' (Girerd 1990) is available and very useful. 'Flowers of South-West Europe (Polunin & Smythies 1973) describes and illustrates many species generally found in the south of France and deals in a small way with the Grands Causses area of Languedoc and with the Camargue. The usual general books on Mediterranean flora (Blamey & Grey-Wilson 1993), Schönfelder & Schönfelder 1990) are helpful but they do not mention all the interesting endemic species of the region. 'The Alpine Flowers of

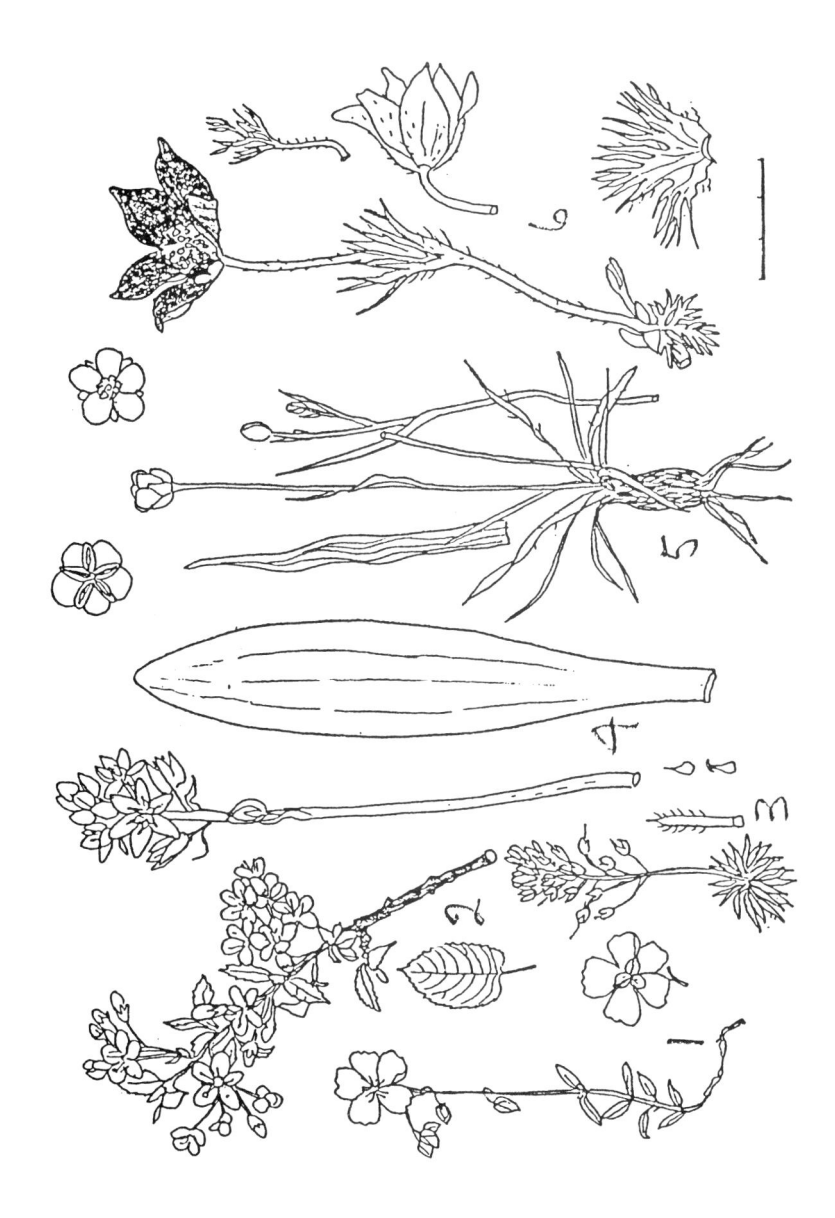

S. France 1. Helianthemum apenninum 2. Prunus mahaleb 3. Draba aizoides 4. Scilla liliohyacinthus 5. Ranunculus gramineus 6. Pulsatilla rubra

Britain and Europe' (Grey-Wilson & Blamey 1979) is also helpful for some areas as it includes species from the Pyrenees, Cevennes, Alpes de Provence and Maritime Alps.

An authorative work on the flora of France as a whole is the 3-volume 'Flore descriptive et illustrée de France, de la Corse et des contrées limitrophes' (Coste, H. 1999-1906). It has many helpful line drawings but is not easily available and somewhat out of date.' Flore de France' (Guinochet, H. & Vilmorin R 1973-1984) is more up-to-date but comprises five expensive and bulky volumes. Flora Europaea is obviously useful though it tends to be a 'lumper' and ignores a few of the especially interesting endemic varieties and sub-species described here and is due for revision.

10.4 **Les Corbières:** This region in the south west of Mediterranean France, comprising part of the Departments of Pyrénées Orientales and Aude, is famous for its wine. It is also known touristically as Le Pays Cathares after the proto-Protestant sect which was quelled in the name of the Catholic Church as 'heretical' during the 13th century. They built fortified hill towns of which Carcassonne is the best restored and most famous. Several of these ruins are worth visiting for their flora as well as their historic interest.

Out of Perpignan the D117 road travels westwards and after some 40km reaches the village of St. Paul de Fenouillet which is a convenient starting place for a days tour of a typical small gorge and one of the finest ruins of a Cathar fortress. Near the village one may see *Tulipa australis* growing on the edge of fields and the strange, poisonous shrub *Coriaria myrtifolia*. Northwards the road rises steeply to the Gorges de Galamus and, on the rocks and by the roadside here, one may encounter the following shrubs:

Amelanchier ovalis   *Helichrysum stoechas*
Bupleurum fruticosum   *Rhamnus alaternus*
Buxus sempervirens   *Thymus vulgaris*
Cneorum tricoccon   *Viburnum tinus*
Coronilla emerus

Most of these are common throughout the region though the cneorum is not seen in all parts. It is a low evergreen shrub growing to about 30cm with yellow flowers in spring followed by small fruits resembling three tiny oranges fused together which start green then change to yellow, red and black as they ripen. It occurs in Spain and the Balearics (8.5)

Non-shrubby species growing with the above include:

Asparagus acutifolius *Linaria supina*
Cheiranthus cheiri *Muscari neglectum*
Euphorbia characias *Ruscus aculeatus*
Helianthemum apenninum *Ruta angustifolia*
H. oelandicum *Scabiosa atropurpurea*
Hesperis laciniata

The asparagus is one of the 'feathery' types similar to the cultivated plant; the prickly species do not grow in France. The cheiranthus is the wild wallflower, now known sometimes as *Erysimum cheiri*. The hesperis is a species of dame's violet which is rather special to the south of France though it may be seen also in the south of Spain as at the Serrania de Ronda (6.6). The linaria resembles *L. alpina* but has plain yellow flowers - it grows along the wall by the road that skirts the short gorge with its spectacular views.

Coming out of the gorge northwards, the rocky hillsides are covered with bushes of box and the buckthorn *Rhamnus alaternus*. Here and there one may see a rosaceous shrub or small tree with white flowers in spring. It may be one of four species which are widespread throughout the south of France:

1. *Prunus spinosa* The blackthorn, is spiny and has black bark.

2. *Prunus mahaleb* St. Lucie's cherry is not spiny and has heart-shaped small leaves. It is used commercially as a stock for grafting cherries.

3. *Pyrus amygdaliformis* is a stocky, somewhat spiny small

tree with elongated leaves that are white with hairs when they emerge but turn green later. It is particularly common in parts of Greece and produces small apple-shaped fruits.

4. *Amelanchier ovalis* (= *A. vulgaris*, *Aronia rotundifolia*) The snowy mespilus, also has whitish young leaves but is distinguished by its elongated petals - those of the pyrus are rounded.

The common hawthorn also occurs and has similar white flowers but may be recognised by its pinnatifid leaves.

Further along towards Soultage dense short maquis in places is comprised of box with some *Juniperus communis* and *Erica arborea*. This is a domain of the wild boar with warning signs saying "Chasse au grand gibier". Along the roadside one may expect to see, *Cistus albidus, Euphorbia sp, Helianthemum apenninum, Helleborus foetidus*, as well as a boar!. The road to the ruins of the Cathar stronghold of Peyrepertuse leaves from Duilhac and many orchids grow near here including *Aceras anthropophorum, Ophrys lutea, Ophrys sphegodes provincialis, Orchis mascula and Orchis purpurea*. Other plants near the road to the car park are *Helianthemum oelandicum, Helleborus foetidus, Muscari neglectum* (= *M. atlanticum*) and *Polygala vulgaris*. The last of these is generally the beautiful blue form and may possibly be *P. calcarea* - it is a difficult genus. From the car park a steep footpath leads to the ruins and there are many interesting plants as the track turns round the north side near the top. Amongst the box scrub with a few bushes of *Lonicera xylosteum* grows a small-flowered form of *Arabis alpina, Chelidonium majus, Lilium martagon* and *Saxifraga granulata*. The most exciting plant here is a fritillary which occurs in considerable numbers. Unfortunately I was unable to identify it for it was not yet in flower during early May but it may be *F. pyrenaica*.

Continuing along the road that returns to St. Paul de Fenouillet one crosses a pass near another Cathar ruin - Quéribus. Here is scrub composed of:

Cneorum tricoccon   *Pistacia lentiscus*

Genista hispanica *Rhamnus alaternus*
Jasminum fruticans *Quercus coccifera*
Phillyrea angustifolia

Amongst these grow *Ophrys fusca, O. lutea, O. sphegodes provincialis* with more muscari and a reddish-flowered form of *Anthyllis vulneraria*.

10.5 **Montagne Noire:** This region in the north west of the Corbières lies some 20km north of Carcassonne and a good base from which to examine it is the small town of Mazamet. Along the picturesque road approaching it from the south the fine blue flowers of *Buglossoides purpureocaerulea* (O = *Lithospermum purpureocaerulea*) and the yellow *Chelidonium majus* can be seen in spring. Both are uncommon British natives, the chelidonium being often a relict from old gardens where it was grown for the yellow juice reputed to be a cure for warts.

The 'montagne' is composed of granite and metamorphic rocks as distinct from the limestone hills further south. The highest point is the Pic de Nore 1210m which is easily reached along a good road and in spring provides an excellent botanical outing. One leaves Mazamet southwards by the D54 through the Gorges de l'Arnette past a series of evil-smelling, small, old factories that process wool and leather. The D87 road turns off to the east and one is quickly in unspoiled deciduous woodland, mainly of oaks but including some other species such as *Acer campestre*. The lush undergrowth has a distinctly 'northern', rather than a Mediterranean, composition and includes:

Alliaria petiolata *Lamium maculatum*
Cruciata laevipes *Mercurialis perennis*
Glechoma hederacea *Orchis mascula*
Lamiastrum galeobdolon *Stellaria holostea*

All of these grow wild in Britain. The lamium here is in great variety with pale pink to purple flowers and differing leaf variegations; some clones would make good additions to the named varieties grown in gardens. In addition to the above, there are plants of *Senecio ovirensis* - a variable species and the plants here may be the sub-species *gaudinii* (= *Senecio gaudinii*). It is a rather fine ragwort growing to 20-30cm and with a kind of umbel of 4-8 fine ragwort-like flower heads. The leaves and stems are sparsely covered with somewhat woolly hairs.

The road climbs out of the wooded valley into more open ground with plantations of conifers and beech woodland having a surprising, thin undergrowth of holly bushes. Along the roadside here grow *Anemone nemorosa, Corydalis solida* and a Soloman's seal, possibly *Polygonatum multiflorum*. Higher still, are dense beech woods carpeted with *Scilla lilio-hyacinthus* making a fine show in early May. The flowers are star-shaped, mid-blue in colour and the bulb consists of loose scales similar to those of a lilium. It is sometimes called the Pyrenean squill but it also grows in other parts of central and southern France.

Approaching the summit, at Col del Tap, there are plantations of stunted pines and plants one might expect to see in the Scottish highlands; bilberries, bracken, wild raspberries and heather. In the rough grass round about are the dark-blue stars of *Scilla bifolia*, much *Erythronium dens-canis* and an occasional plant of *Narcissus pseudo-narcissus* possibly the sub-species *pallidiflorus*. A dark-blue pansy can also be found here - *Viola lutea ssp. sudetica*. The summit is treeless with an almost complete cover of heather *Calluna vulgaris* amongst which grow the erythronium and *Scilla bifolia*. A large radio-transmitting station marks the top and from here on a clear day one has a magnificent view of the distant Pyrenees. In spring they may appear as a disembodied white line high up on the horizon.

10.6 **Causse du Larzac:** 'Causse' is a Provencal word describing a calcareous plateau and derives from Latin meaning chalk. The Causse du Larzac is the largest and most southerly of the Grands Causses which lie north of Bédarieux and Lodève; further north are the Causse Noir, Causse Méjean, Causses de Sévèrac and

S. France 1. Iberis amara 2. Aristolochia clematitis 3. Smilax aspera

S. France 1. Daphne laureola 2. Corydalis solida 3. Antirrhinum latifolium

Sauveterre. The area is of a somewhat inhospitable environment for humans and plants alike. The stony calcareous land has little soil, is mostly unsuitable for cultivation and supports few trees to break the wind. Winter is cold there with frequent snow squalls and in summer the ground is baked dry. Its only real use to man is as rough pasture and thousands of sheep regularly nibble the plants. In spite of these disadvantages it has a surprisingly varied and interesting plant life.

The small town of Lodève (important for its uranium mines), on the N9-E11 road between Clermont l'Hérault and Millau, is a convenient base from which to examine the region. From there, leaving northwards one can take the D25 eastwards to St. Pierre de la Fage and then either north-east along D25 or north-west along D9. Here one is on the causse where the stony grassland has scattered bushes of box with a few blackthorns and junipers. The ground carries a surprisingly rich collection of smaller species, including:

Aphyllanthes monspeliensis	*Muscari neglectum*
Asphodelus albus	*Narcissus assoanus*
Astragalus monspeliensis	*Orchis mascula*
Carlina acanthifolia	*Plantago afra*
Chaenorrhinum rubrifolium	*Platanthera bifolia*
Eryngium campestre	*Ranunculus gramineus*
Euphorbia cyparissias	*Teucrium polium*
Geum sylvaticum	*Valeriana tuberosa*
Globularia vulgaris	

Not all of these are equally distributed. The narcissus, an attractive small jonquill with thin, rush-like leaves, occurs in patches and may colour the ground yellow in April. It has suffered a number of name changes and was formally called *N. juncifolius* and later *N. requenii* but its most up-to-date, valid name is *N. assoanus*. Primarily a Spanish species, it is found in several parts of southern France on either side of the Rhône valley. The carlina is the impressive, large carline thistle, scattered throughout the causses and also in the Pyrenees. Its dry flowering heads are collected and nailed on house doors for it is prized locally as a rustic weather forecaster and called the 'Baromètre de Berger'. The bracts round the flower head move inwards, according to the humidity. The globularia is another attractive plant common in rocky places throughout southern France. It is a variable species and the plant here could be the same as *G. punctata*. The valeriana is fairly widely distributed in rocky calcareous regions of Portugal, Spain, France and Italy. It generally grows to 10-15cm tall with compact heads of pink, or occasionally white, flowers and has a short cylindrical tuberous root.

The D25, previously mentioned, leads to St. Maurice-Navacelles and thence along D130 to the impressive Cirque de Navacelles. The road workers here, as in other places, have clipped the wild box bushes short top and sides to give them a neat appearance. The verges are home to large patches of *Iris lutescens* (= *I. chamaeiris*) with both yellow and purple forms and some intermediates. This is one of the most spectacular and delightful of all the plants found in France. It shares the pitch here with *Alyssum spinosum, Astragalus monspeliensis, Ferula communis* and *Narcissus assoanus*.

The D9 from St. Pierre de la Fage takes one northwards to Le Caylar and from there on to the well-restored Templars' fortified town of La Couvertoirade. In addition to the species already mentioned one may come across a thrift (possibly *Armeria arenaria*) and large patches of *Pulsatilla rubra* flowering in April. It differs from *P. vulgaris* by having darker, reddish purple flowers that tend to hang their heads. The usual cowslips and *Helleborus foetidus* (which merits the French name of 'Pied de griffon' - griffon's foot - rather than the English 'Stinking hellebore'!), and *Hepatica nobilis* (= *H. triloba*) lurks under the shelter of the few bushes to be seen. Surprisingly, one may also encounter *Draba aizoides* growing on rocks nearby.

A number of orchids, in addition to *Orchis mascula* and *Platanthera bifolia* already mentioned, can be found in the Causses, including:

Aceras anthropophorum   *Orchis militaris*

Ophrys apifera *Orchis morio*
Op. aymonii *O. palustris*
Op. sphegodes *O. purpurea*
Op. insectifera *O. ustulata*

The most interesting of these is *Ophrys aymonii* that was first found in 1959 and is apparently endemic to the Causses region. It resembles *Ophrys insectifera* and differs from it by having a broader lip which is brownish and with a distinct yellow border. It appears to be intermediate between *Op. insectifera* and *Op. lutea* and possibly may be a hybrid between these two species. It flowers in May and June.

Further north the impressive 50km long Gorges du Tarn lie between the Causse de Sauveterre and the Causse Méjean and are well worth a visit though they hardly fall within the Mediterranean zone. The interesting plants to be seen here include *Aquilegia hirsutissima, Campanula persicifolia, Erinus alpinus* and many others. Another place to the south which is well worth a visit is the charming old Cevennes village of St. Guilhem-le-Désert that lies some 15km east of Lodève. Here many of the houses have a *Carlina acanthifolia* and a wild boar's tail nailed to the door. In the stony countryside one may encounter *Aristolochia clematitis, Hieracium stelligerum, Iberis amara, Malva moschata, Plumbago europea,* and *Smilax aspera*. The plumbago is an unusual plant which is distinct but puzzling to identify if one has not seen it before. It is more common to the west of the Rhône than the east and produces its violet or pink flowers in late summer when most species have finished flowering.

10.7 **La Bambouseraie:** This is not a site for wild plants but unique European collection of bamboos that will interest any plant enthusiast. It is situated at Anduze which on the N107 about 15km south-west of Alès. It was founded by Eugène Mazel in 1856 and has large groves of the giant bamboo *Phyllostachys pubescens* and many other species together with an 'Asiatic village' of houses built to demonstrate the use of the material. Here one can relax,

S. France 1. Dianthus sylvestris 2. Althea officinalis 3. Plumbago europaea 4. Heliotropium europaeum

have a meal and buy plants for the garden.

10.8 **The Camargue:** This region of the Rhône delta, correctly called 'Plaine de la Camargue' is well known to tourists for its unique style of cattle ranching and its birdlife. It is a 'Réserve Naturelle, Zoologique et Botanique' but does not merit a special visit to see the plants unless one is especially interested in maritime species. In many respects the flora is like that of the Coto de Doñana of Spain (5.7) though less rich in unusual species but it is easier to gain access.

There are effectively three types of habitat, seashore, sand dunes and salt marshes called 'sansouires'. The seashore plants are similar to those one might encounter along the sandy shores of Languedoc and where these exist in Provence and include:

Cakile maritima *Matthiola sinuata*
Calystegia soldanella *Medicago marina*
Crucianella maritima *Otanthus maritimus*
Eryngium maritimum *Pancratium maritimum*
Euphorbia parialis

The sand dunes in places have pine trees and here one may find many species that appreciate a sandy soil and are able to stand drought such as:

Alkanna tinctoria *Hypercoum procumbens*
Bellardia trixago *Ononis natrix*
Coris monspeliensis *Tribulis terrestris*

The coris is an attractive species with violet blue flowers from March until June; fairly widespread in the south of France and also found in Spain and Italy growing not only in sand dunes but sometimes in garigue. Here it is not far from the town which gave it its name - Montpellier.

The salt marshes are colonised by plants which are fairly widespread in the Mediterranean and likely to be of interest only to those who specialise in the difficult task of naming them correctly. However, one relatively uncommon species occurs here - *Heliotropium curassavicum*. It is a somewhat insignificant species with a woody base and more or less procumbent branches bearing small white flowers. It hails from North and South America and the West Indies and has established itself in Sardinia and Sicily as well as the Camargue

10.9 **Mont Ventoux:** This mountain in Provence, about 30km east of Orange, lies more or less at the northern limit of the Mediterranean influence. It is at the west end of a limestone ridge that stretches eastwards in the Montagne de Lure of Haute Provence towards Digne. The summit reaches 1909m and is often snow covered in winter. It has an interesting, though limited, flora and, as with the other east-west ranges here, the vegetation on the southern face often differs markedly from that of the north.

The summit can be reached by taking the D974 road eastwards out of Bedoin and from there the D974 goes westwards along the northern face to Malacène though this road is sometimes blocked by snow as late as April. The lower slopes are mainly covered with deciduous oak, beech and a few cedars (*Cedrus atlantica*) which have been introduced and thrive especially well so that they produce numerous self-sown seedlings and threaten to become dominant as eucalyptus trees have done in other parts of the Mediterranean. In places the *Quercus pubescens* oaks are short dense and mixed with box and hazel to form a maquis-like vegetation. *Hepatica nobilis* and *Helleborus foetidus* are common. Other species one may encounter are the attractive perennial cornflower *Centaurea triumphettii* and *Saponaria ocymoides*.

Higher up, the trees are very stunted and associated with some pines and junipers. Approaching Le Chatelet, where there is a winter sports centre, one reaches the tree line and is faced with an apparently bare summit of white limestone scree and here it can be very windy, cold and uncomfortable - it well deserves its name as 'The windy mountain'. In this apparently inhospitable climate there are a number of scattered low-growing plants including:

Aconitum anthora
Androsace vitaliana
Eryngium spina-alba
Euphorbia cyparissias
Helleborus foetidus
Hypericum hyssopifolium
Iberis candolleana
Lavandula angustifolia
Papaver aurantiacum
Saxifraga oppositifolia
Viola cenisia

The aconitum produces its rather beautiful yellow, or rarely blue-flowered, 'monkshoods' in August and is found also in the Pyrenees and Alps. The androsace (= *Vitaliana primulaeflora*) is the subspecies *cinerea* which differs from most other forms by preferring limestone. The eryngium is a stocky, low-growing species with heads of bluish white flowers in July, sometimes referred to as the silver eryngo. The iberis is an especially interesting, low-growing candytuft with white or pale pink flower heads in April. It is endemic to Mt. Ventoux and a few sites in the Department of Drôme to the north. Its nearest relatives are *Iberis aurosica* and *I. pruitti* from the Pyrenees. The papaver and saxifraga are two well known species which grow as far north as Greenland and were left behind here during the last ice age.

Mont Ventoux may also be approached from the east along the D164 road leaving from the small town of Sault that is a centre for lavender production, using *Lavandula angustifolia*. Lower down the hybrid, between this and *L. latifolia*, known as 'le lavandin', is generally grown for this purpose. Another route to the summit is from Malaucène eastwards along D974 which skirts the north side of the ridge with a distinctly cooler summer climate than that of the south facing sites. Here one might find *Epipactis atrorubens, Lilium martagon* and *Orchis pallens*.

10.10 **Plateau de Vaucluse and Montagne du Luberon:** Although the first of these is called a plateau, it falls away southwards and is, in effect, an east/west range of hills lying parallel to and south of Mt. Ventoux. The Luberon is another but higher similarly orientated range still further south stretching some 60km between Cavaillon and Manosque. Both are composed of limestone and are wooded on northern slopes, even in the highest parts.

Coming south from Mt. Ventoux, it may be worth taking the road from Villes sur Auzon along the upper reaches of the Gorges da la Nesque. Not far from this village at a hairpin bend there is a GR91 footpath to Combe de l'Hermitage and somewhere along here grows the endemic *Leucoium fabrei* (Girerd 1991). However, it is small and has a short flowering period at the beginning of May so one may well be disappointed. It is very similar to *L. nicaense* (= *L. hiemale*) which is found further east in the Alpes Maritimes (10.4) but has slightly larger flowers. Even without seeing it, one's visit may not have been in vain, for this is a particularly good area to find the fine yellow snapdragon *Antirrhinum latifolium*. Other plants growing near here include *Aristolochia pistalochia* and *Linaria supina*.

The north facing side of the Plateau de Vaucluse has considerable woodland on the higher slopes, mainly of deciduous oaks in the west but some beechwoods to the east. Smaller plants include:

Centaurea triumphetti	*Ornithogalum umbellatum*
Hepatica nobilis	*Polygonatum odoratum*
Lathraea clandestina	*Satureia grandiflora*
Lithospermum purpureo-caeruleum	*Valeriana tuberosum*

The lathraea is the large-flowered, and rather rare, parasitic toothwort. The saturea (= *Calamintha grandiflora*) is an attractive species with relatively large pink flowers. One may see also *Narcissus assoanus*, sometimes in considerable quantities as at the Col de Murs 627m on the D4 road that runs south-east out of Carpentras. Two other plants of special interest here are *Erysimum burnati* and *Thlaspi praecox*. The first is a treacle-mustard with rather large yellow flowers and the second a white-flowered pennycress that has a limited distribution in the south of France but is plentiful at the Col de Murs. Several orchids may be seen, especially *Orchis purpurea* and *Ophrys araneola* (= *O. sphegodes litigiosa*)- a rather tall species with up to 12 rather small flowers that have a yellowish edge to the lip. It sometimes grows in large clumps, suggesting that it regularly propagates vegetatively.

Coming down the south-facing slopes of the Plateau de Vaucluse it is obvious that the region is drier and has a distinct Mediterranean feel with species such as *Spartium junceum* and rosemary. The N100 road that leaves Cavaillon eastwards runs more or less parallel to the Montagne de Luberon and just north of it, at the west end, lies the interesting village of Roussillon where there are considerable deposits of ochres that are mined as colouring matters. These materials consist of sedimentary deposits containing variable amounts of iron and are acidic and as a result a small region here carries plants that will not readily grow on the predominant limestone of the surrounding area such as:

Cistus laurifolius	*Moenchia erecta*
Linaria arvensis	*Senecio lividus*
Lupinus angustifolius	*Spergula arvensis*

None of these plants is very rare or interesting, though the cistus is a fine species sometimes grown in gardens in Britain and possibly the hardiest of the genus. The linaria somewhat resembles *L. supina* but has purple flowers and the senecio is a typical groundsel. Around here, and also on hilly exposed regions, one may encounter the grass *Stipa pennata* with attractive feathery seeding heads in June. It is frequently collected for decoration and a vase-full aptly describes its local name of 'plumet de Vaucluse'. It certainly resembles the plume used to add panache to some ceremonial army headware.

The Luberon was a Protestant stronghold of the Vaudois, whose beliefs resembled those of Calvinism - did they give their name to the Canton of Vaud in Switzerland?. The western part is referred to as the Petit Lubéron and boasts one of the first and most flourishing cedar (*Cedrus atlantica*) plantations in France. The eastern part is higher and its flora is probably more interesting. It is crossed by three good roads of which the D33 from Granbois to Céreste is probably the most interesting. The northern slope is almost

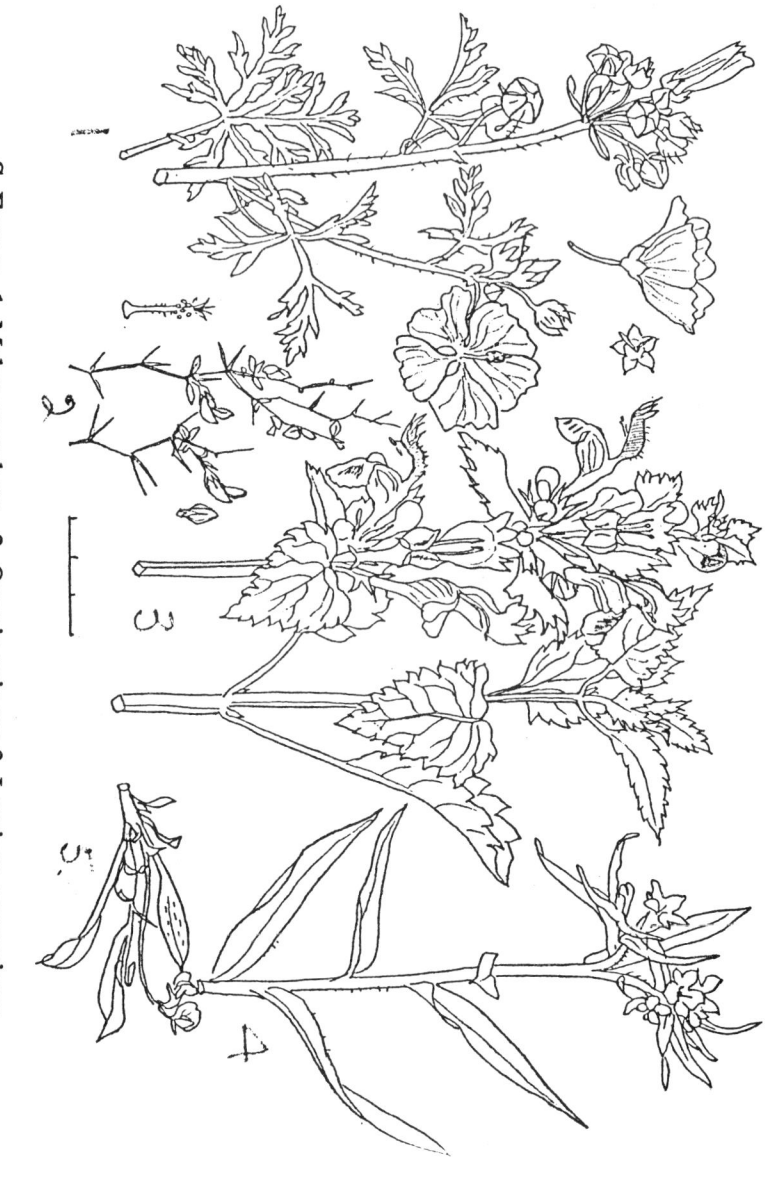

S. France 1. Malva moschata 2. Ononis spinosa 3. Lamium garganicum 4. Buglossoides purpurocaerulea 5. Scopiurus muricatus

entirely covered by deciduous woodland comprised mainly of *Quercus pubescens* with an undergrowth of box, but in places there are stands of beech and it is likely that these were at one time more extensive. Under the trees and along roadsides here one may see:

Astragalus monspeliensis	*Hippocrepis emerus*
Bellis sylvestris	*Hypericum montanum*
Cephalanthera longifolia	*Lilium martagon*
Colchicum autumnale	*Narcissus poeticus*
Dictamnus albus	*N. assoanus*
Helianthemum oelandicum	*Ranunculus gramineus*
Hepatica nobilis	*Saponaria calabrica*

Most of these are widespread but the dictamnus and lilium are certainly not common and found mainly in the western part. Plants of *Narcissus poeticus* are dotted around under the trees but in open ground towards the N100 near Céreste they grow in great numbers so that the fields look white in April.

The southern slopes are drier, hotter and have fewer trees. A selection of plants that one might see here and as far south as the river Durance includes the following:

Antirrhinum latifolium	*Lathyrus setifolius*
Centranthus ruber	*Leopoldia comosa*
Cistus albidus	*Linum campanulatum*
Euphorbia spinosa	*Lithodora fruticosa*
Jasminum fruticans	*Tragopogon porrifolius*
Lactuca perennis	

The euphorbia is a prickly 'cushion' plant usually growing on rocks and takes the place of *E. acanthothamnus* which one sees so often in Greece. The lathyrus is an annual with small, brilliant dark red flowers. The linum is a very attractive woody-based perennial with rather large yellow flowers found in east Spain, southern France and north west Italy. The lithodora (O = *Lithospermum fruticosum*) is a small shrub with beautiful blue flowers. It is not very plentiful here but also occurs in Spain (7.5). The tragopogon is salsify - a variable plant - and the form here is the subspecies *australis* in which the pointed bracts are longer than the flower heads.

10.11 **Massif de la Sainte Baume:** This is another limestone ridge running approximately east to west, about 12km long and some 16km east of Marseille. It slopes fairly steeply towards the south but the north side drops away dramatically with a impressive limestone cliff over 300m high in places. There is an exciting footpath GR98 along the top that passes through the highest point of St. Pilon 994m. 'Baume' is a Provençal word meaning a cave and refers to a specific site of pilgrimage in the north face below the summit where there is a chapel to Mary Magdalene, who is said to buried at the Basilica in the charming small town of St. Maximin, 15km to the north.

This massif, of special botanical interest, is best approached from the east along the D95 road that leads from Mazaugues along the north side. One passes at first through light woodland of pines with some *Pinus halepensis* and deciduous oaks and occasional crab apple trees *Malus sylvestris* with beautiful pink blossom in April. In places there is a fairly dense covering of *Erica scoparia* or possibly *E. multiflora*. There are a few scattered bushes of *Juniperus communis*, some thyme and *Helichrysum stoechas*. As one approaches the Hôtellerie, more interesting plants may be seen:

Cynoglossum officinale	*Ranunculus gramineus*
Genista hispanica	*Saxifraga granulata*
Genista lobelii	*Symphytum tuberosum*
Geum sylvaticum	*Tulipa australis*
Iris lutescens	*Valeriana tuberosa*
Juniperus phoenicea	*Vincetoxicum hirundinaria*
Phlomis herba-venti	

The iris covers large areas and is a real spectacle in spring. It is generally the yellow-flowered form, but purple and variegated specimens may also be found. *Genista lobelii* is a small, dense, prickly shrub superficially resembling *G. acanthoclada* which is common in the eastern Mediterranean. The particular juniper is not common here as in other parts of the Mediterranean. The phlomis is an attractive species that produces its mauve flowers in June. The symphytum is a rather low-growing comfrey with yellow flowers and, as its name indicates, has small root tubers around the base of the stem.

Several orchids may be seen, including *Barlia longibracteata, Cephalanthera longifolia, Dactylorhiza sambucina* and forms of *Ophrys scolopax*. The most outstanding of these is the dactylorhiza that covers the ground in places, both under the trees and on open ground. The plants are mostly the yellow-flowered form but purple specimens may be found. It is interesting that Thomson (1914) comments that it has been seen in the Forêt de Sainte Baume though he, of course, calls it *Orchis sambucina*. It is encouraging to hear that an orchid has made surprising progress during the last eighty years.

The feature of major botanical interest is the deciduous woodland which shelters immediately under the north facing cliffs of the massif. It is protected from the heat, drought and bright light of the Mediterranean and has a northern European character. It was considered to be sacred by the Celts and has remained more or less untouched by human activity except for the removal of dead trees. One may walk through it on well defined pathways on the way to the cave where Mary Magdalene is said to have spent her last days and which has been an important route for Christian pilgrims over hundreds of years. The path leaves from the Hostellerie and the forest is quiet, cool and dark, inspiring a feeling of timelessness and mystery. It is a carefully guarded nature reserve and one may not stray from the path, but it is easy to see the ground cover which includes *Daphne laureola, Lilium martagon* and *Narcisus poeticus*. *Paris quadrifolia* is also said to grow here.

The road continues westwards past the Hostellerie to the village of Plan d'Aups where in spring the roadside is lined with woad *Isatis tinctoria* that is common in parts of the south of France but is primarily an Asiatic plant and could be an escape from cultivation. At one time it was important for the purple dye produced by grinding the leaves and fermenting them. The route then continues southwards through a dry rocky landscape with much *Genista hispanica* and *Spartium junceum*. On the rocks one may see the wild wallflower and many typical Mediterranean species.

10.12 **Massif des Maures:** This rather extensive region near the coast, stretching from Hyères to Fréjus and up to 25km wide, is of special interest botanically for it is composed mainly of acid rocks such as granites and schists rather than the limestone of most of Provence. Because of this it carries considerable woods of cork oak and sweet chestnut, with much *Erica arborea* in places. The main local products from here are cork, chestnuts and briar pipes. The word briar comes from the French 'bruyère', meaning heather. It is sometimes thought that the region is given its name by reference to the Moors who occupied part of it at one time but the word 'Maures' is Provençal, meaning dark and refers to the appearance of the woodland there. Much of the land is covered by trees and it is here that some of the worse forest fires occur during the summer.

As a starting point to examine the region one may stay at Le Lavandou on the south coast or Bormes les Mimosas, though unfortunately the mimosas once grown to produce perfume have been seriously damaged by recent winter frosts, though they are now beginning to recover. Coming up the winding road northwards over the Col de Gratteloup one passes through an area of the typical flora of southern Provence including such species as:

Acanthus mollis	*Lathryus articulatus*
Althea officinalis	*Lavatera arborea*
Calicotome spinosa	*L. maritima*
Cistus albidus	*Rosmarinus officinalis*
C. salvifolius	*Ruta angustifolia*
Convolvulus althaeoides	*Pistacia lentiscus*
Galactites tomentosa	*P. terebinthus*

Helichrysum stoechas *Thymus vulgaris*
Lavandula stoechas *Tragopogon porrifolius*

These are all plants common to many parts of the Mediterranean and are described in most general books on the subject. One comment, however, is that the helichrysum and lavender have the specific name of 'stoechas' which is derived from the old Greek name for the Iles d'Hyères which lie off the coast south of Le Lavandou and can be seen from here. Both of them are, of course, common on the islands.

After crossing the N98 one comes to the winding D41 leading to Collobrières and here is a choice of going westwards to Pierrefeu or eastwards to Grimaud and the second of these is probably the most interesting for its flora. There are also several smaller roads that lead northwards from here and these are also worth following, especially the one that goes to Nôtre Dame des Anges. In the more open areas several of the species listed above occur, though *Cistus albidus* is decidedly less common here and *C. monspeliensis* is found in places. Other species here include:

Allium roseum *Silene italica*
A. triquetrum *Tuberaria lignosa*
Arum italicum *Urospermum dalechampii*
Lactuca perennis *Verbascum blattaria*
Lathyrus latifolius

The alliums are mainly found in wet ditches. The silene resembles white campion but the petals are narrower and tend to curl inwards in dry weather; it is rather a common feature of the roadsides in eastern Provence. The lathyrus is the broad-leaved everlasting pea but here it is usually the sub-species *angustifolius* - which seems a contradiction in terms. The tuberaria (O = *Helianthemum tuberaria*) is an uncommon and attractive plant with a rosette of leaves and stems to about 30cm carrying 5-10 yellow flowers which, unlike the more widespread *Tuberaria guttata*, have no dark blotch at the base of the petals. The verbascum here frequently has white flowers.

Surrounding woodland includes sweet chestnut and cork oak and is often rather dense. *Symphytum tuberosum* is a common component of the undergrowth but there are other species including bracken. The area is a good place to find orchids such as:

Cephalanthera longifolia   *Orchis champagneuxii*
Epipactis helleborine   *O. provincialis*
Limodorum abortivum   *Platanthera bifolia*
Neotinea maculata

The most interesting of these is *O. champagneuxii* which usually grows in the woods but is sometimes found on open ground. It is often classed as a sub-species of *O. morio* but, in many respects, it more closely resembles *O. longicornu* with its long, strait spur and more or less unspotted lip. Its main distinguishing feature from both of these species is that its new tuber is separated from the older one on a long stalk.

One can cross the region also from Cogolin to Grimaud and then along the D48 to Vidauban. The flora along the southern part of this route is similar to that just described but it changes as one approaches Vidauban. The well-watered or marshy area north from La Miquelette until one crosses over the river Aille is a very productive botanising site. On the drier ground there is woodland of well-spaced *Pinus pinea* under which one may find: *Anemone hortensis, Ophrys sphegodes, Op. scolopax, Ranunculus flabellatus* and *Serapias vomeracea*. In the lower-lying parts, the marshy area with small streams has a very exciting flora:

Anemone hortensis   *Linum narbonense*
Anthericum liliago   *Orchis sp.*
Aristolochia pistolochia   *Ranunculus gramineus*
Bellis annua   *Serapias neglecta*
Gladiolus italicus   *Tuberaria guttata*
Iris lutescens   *Tulipa sylvestris*

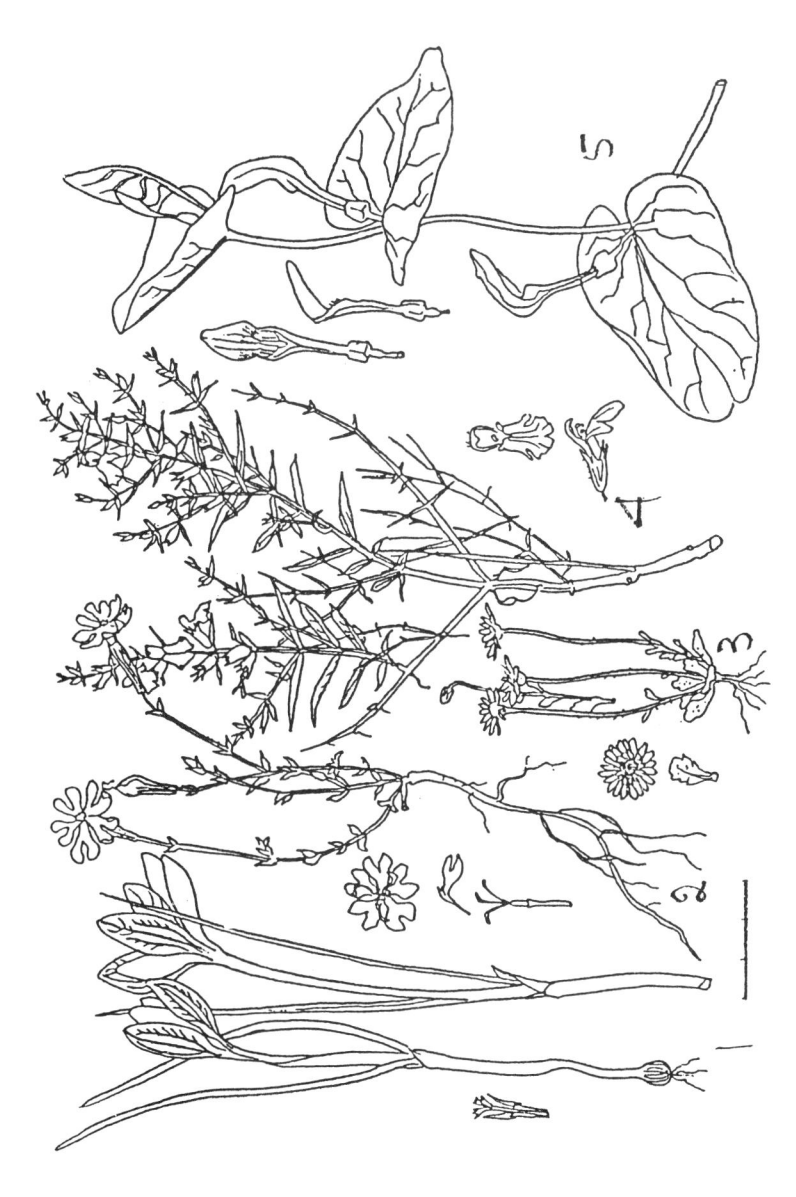

Corsica 1. Crocus corsicus 2. Silene sericea 3. Bellis annua 4. Stachys glutinosa 5. Aristolochia rotunda

Corsica 1. Lotus ornithopodioides 2. Euphorbia pithyusa 3. Halimium halimifolium 4. Genista corsica 5. Saxifraga pedemontana cervicornis

The anthericum is not uncommon in Provence but usually seen as individual plants; here it grows in great profusion with its relatively large white flowers in April. The orchis species is probably *O. champaneuxii* but this needs checking as it is near to *O. morio picta* and growing in conditions that might suggest that it is *O. laxiflora* - a question for some orchid expert. The serapias is an uncommon species, also found on Elba, Corsica and Sardinia. It flowers from March to May, is short-growing and has the largest flowers of the genus; they vary from ochre-coloured to pink and occasionally dark brown-crimson.

The Massif des Maures certainly merits several days botanising and the Massif de l'Esterel which lies to the east of it is similar and well worth a visit.

10.13 **Around Draguignan:** This town lies north of the Massif des Maures in a relatively unurbanized limestone region bordering on Haute-Provence. Much of the area to the south is cultivated, primarily with vines, but pleasant and with plenty of roadside and woodland plants of species already described. To the north is the Grand Plan des Canjuers which rises to over 1000m and is mainly covered with garigue and maquis in places. It is mostly used for army exercises and is largely out-of-bounds. To the north west of Draguignan lies the celebrated Grand Canyon du Verdon which is spectacular, populated by tourists and takes about a day to drive round. Few special plants are to be found there though *Iris lutescens* is frequent in places and one may see *Fritillaria involucrata* (10.14).

10.14 **North of Nice and Cannes:** Much of the costal region is now urbanized but there is still evidence of the wild Mediterranean flora with such plants as *Aphyllanthes monspeliensis, Asphodelus fistulosus, Cistus incanus* and *Lavandula stoechas* by the roadside. However, north of Vence and Grasse one comes to higher limestone hills which form the lower part of the Alpes-Maritimes. A particular route that should not be missed is the D2 northwards out of Vence to Coursegoules and Gréolières. The road rises steeply and high up, with a view over the sea and round about the Col de Vence at 970m, is a rocky area well worth exploring. The plant to look for here is the diminutive *Leucoium nicaense* that produces its tiny white snowdrop-like flowers in late April and May. Because of its size it may well be overlooked. There is a rich flora here and at the Col de Vence itself which includes:

Barlia longibracteata   *Orchis olbiensis*
Chrysanthemum cineraria   *O. tridentata*
Euphorbia spinosa   *Senecio cineraria*
Genista hispanica   *Thymus vulgaris*
Globularia vulgaris   *Urospermum dalechampii*
Helianthemum oelandicum

Orchis olbiensis is generally classed as a sub-species of *O. mascula* and is very common here. The senecio (O = *S. maritima*) is the wild form of the grey-leaved plant so common in park bedding schemes and it may be the sub-species *bicolor* (14.6). There are scattered bushes of blackthorn and *Pyrus amygdaliformis*.

Just north of here is an enclosure where horses are kept and a little further on a wooded area called La Garussière which should not be missed by any plant enthusiast. The woods are composed primarily of hornbeam and *Acer opulus*. Under the trees is a particularly rich flora of:

Arum italicum   *Lavandula angustifolia*
Corydalis solida   *Lilium martagon*
Crocus versicolor   *Muscari neglectum*
Cruciata laevipes   *Orchis olbiensis*
Daphne laureola   *Paeonia peregrina*
Erythronium dens-canis   *Polygonatum odoratum*
Fritillaria involucrata *Primula veris*
Helleborus foetidus   *Stellaria holostea*
Lathyrus latifolius   *Saxifraga granulata*
Lamium maculatum   *Viola odorata*

The crocus is widespread and the only member of the genus

to be found in the region. It produces its pale violet flowers in February and early March. The fritillaria is endemic to the south of France and north-west Italy. In March and April its stems produce one, or sometimes more, pale green flowers chequered with mauve brown, the degree of marking varying greatly between individuals. The paeony, which is sometimes classed as the sub-species or variety *paradoxa*, is not plentiful here and is more usually found in mountain pastures. It is a handsome plant with large rose-coloured flowers in May and June and leaves which are somewhat more divided than those of *P. officinalis* and with a greyish overtone due to the hairs.

Travelling northwards, the woodland gives way to garigue of box, kermes oak and juniper with some bushes of *Pyrus amygdaliformis* and under vegetation including the fritillaria. In woodland, just before the garigue, grows *Hyacinthoides italica* (O = *Scilla italica*), with conical-shaped inflorescences of typical blue flowers in April.

Continuing along the D2 road one comes to Coursegoules, Gréolières and to Logis-du-Pin. This is not a particularly interesting route for plants but one may see the orchids *Listera ovata, Platanthera bifolia* and *Dactylorhiza majalis* with broad, lavishly spotted leaves. *Colchicum autumnale* is plentiful here and the small, white-flowered *Potentilla saxifraga* also occurs in the turf. On the hills in summer the striking *Lilium pomponium* produces its beautiful scarlet but evil-smelling flowers.

The Gorges du Loup are not far from Grasse and worth a visit for the scenery. The road along the eastern side gives the best views and in the woods here grow *Asparagus acutifolius* and *Cephalanthera longifolia*. The road round the western side of the gorge passes through interesting woods composed almost entirely of the hop hornbeam *Ostrya carpinifolia*. The D12 road, near the attractive Provencal hill village of Gourdon, climbs steeply to the Col de l'Ecre 1120m where one can see *Erythronium dens-canis, Fritillaria involucrata* and *Narcissus poeticus*. On the way there are large areas covered with broom *Cytisus scoparius* (= *Sarothamnus scoparius*), which is rather short-growing and compact (cf 6.5) and seems to be distinct from our British form. Although it is plentiful near here it is by no means common in the rest of the south of France. Near the broom grow *Hyacinthoides italica* and *Orchis olbiensis* in exposed places.

North of Nice the ground rises steeply and one is soon into the Alps with true alpine flora that is obviously extremely interesting but will not be discussed in this book.

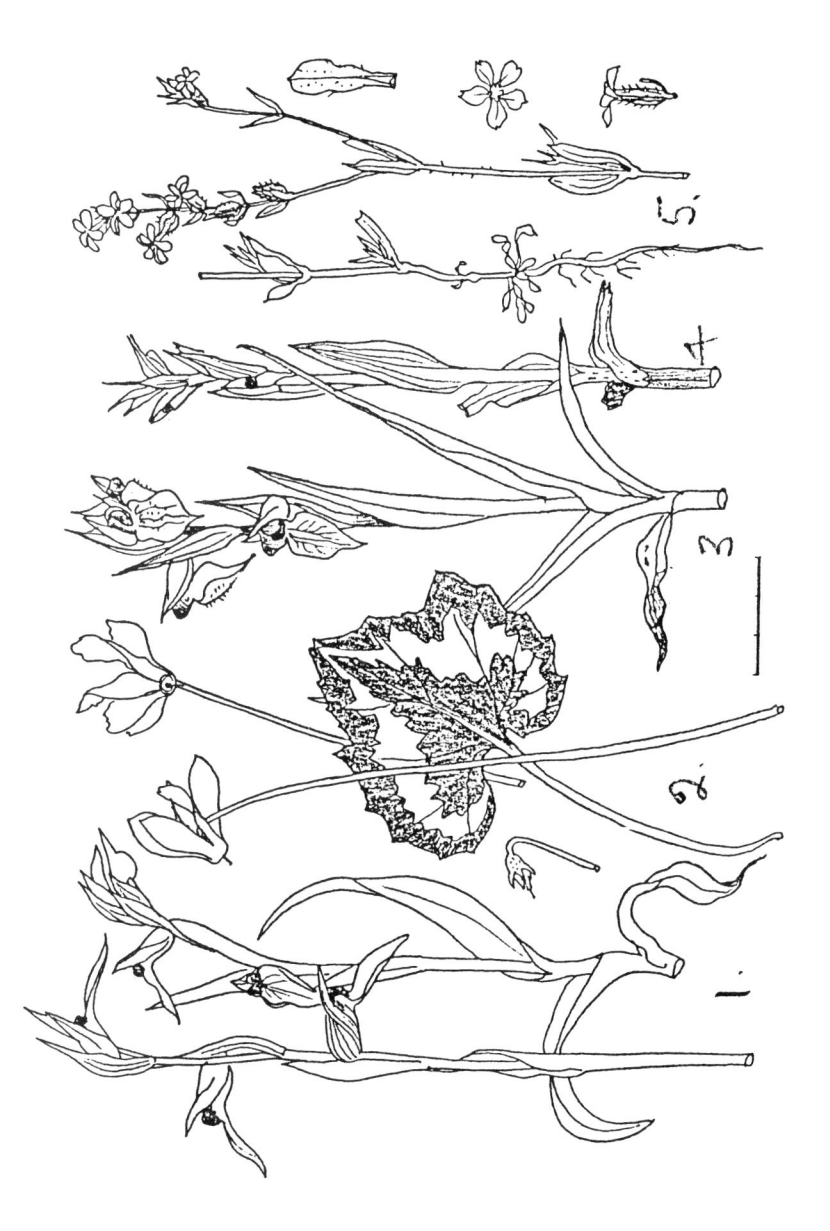

Corsica 1. Serapias lingua 2. Cyclamen repandum 3. Serapias cordigera 4. Serapias parviflora 5. Silene quinquevulneraria

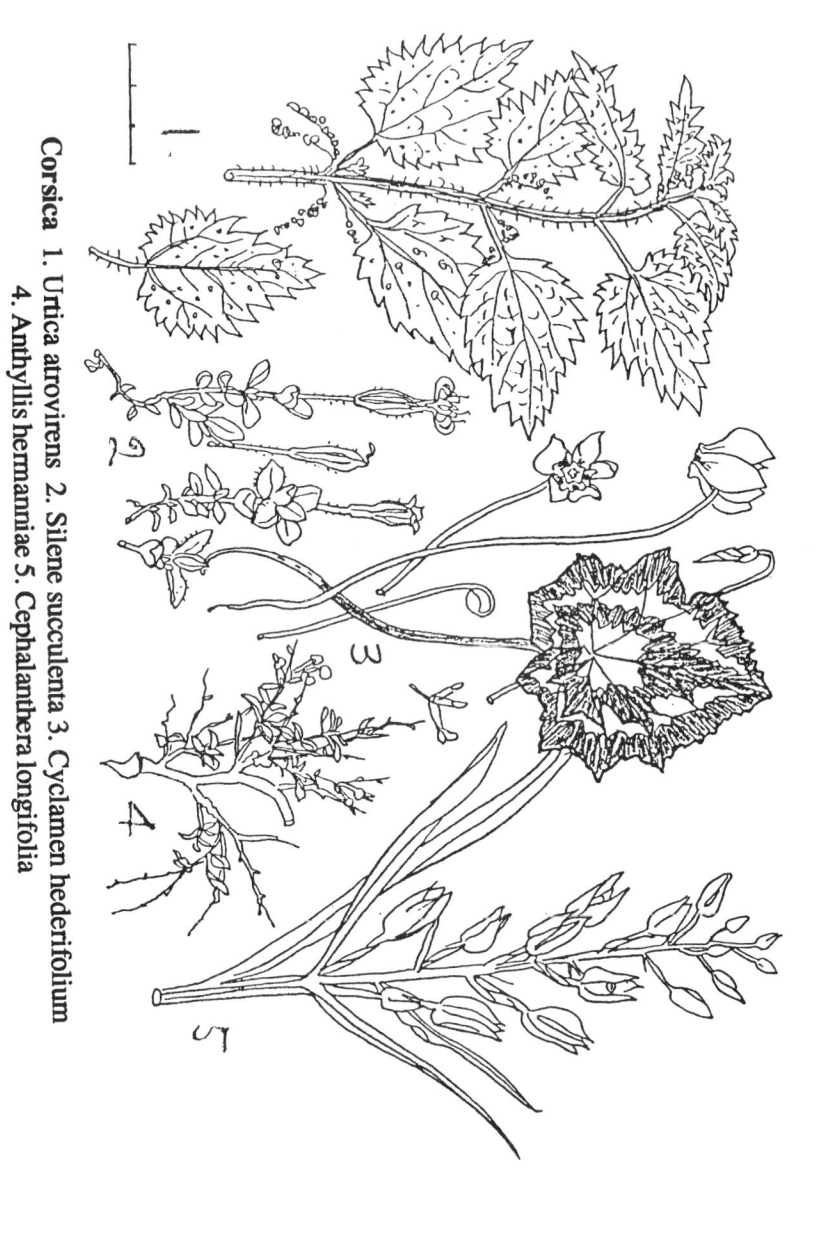

Corsica 1. Urtica atrovirens 2. Silene succulenta 3. Cyclamen hederifolium
4. Anthyllis hermanniae 5. Cephalanthera longifolia

11. CORSICA

11.1 Corsica is sometimes referred to as 'The Mountain in the Sea'. It is the fourth largest island of the Mediterranean and has spectacular scenery with jagged peaks of pink granite mountains that are snow-capped for much of the year. A few readers may associate it with romantic rumours that it is still a wild land of bandits but this is far from the truth - it is a sophisticated 'département' of France. However, the Corsicans have their own language, which resembles Italian, though most of them also speak French. A few Corsicans seek separation from France, but the only sign of their protests against the establishment likely to affect visitors is the defacement of signposts - so if you are motoring, be sure you carry a good, up-to-date map, such as the Michelin 90.

Several tour operators arrange package holidays to Corsica but most of these are not available until May which is somewhat late in the year to see the best of the lowland flowers. However, they can be helpful to the plant enthusiast wishing to study the rather special mountain flora that is at its peak in June and July. For those who prefer to make their own way to the island there are regular air and sea services from France and Italy and accommodation and car hire is easy without prior booking if one goes there in spring. Although it is quite possible to drive from one end of the island to the other in a day it is, nevertheless, difficult to make a reasonable study of the flora by staying at a single base.

Corsica offers a great deal to the plant enthusiast. It has over 2100 plant species and an astounding 11.8% of all taxons (species, sub-species and varieties) there are endemic (Gamisans 1991). Furthermore, one can enjoy botanising in comfort and with gastronomic style for in places the food is gourmet standard. Wild boar paté and a liqueur called 'myrthe' (flavoured with *Myrtus communis*) are two of the specialities.

11.2 **The Terrain:** The island is divided roughly into two geologically different regions. South-west of a line from L'Isle Rousse to Solenzara is mostly comprised of granite and it is here

Gargano 1. Doronicum orientale 2. Lamium bifidum 3. Aristolochia pallida 4. Ranunculus millefoliatus

that the tallest mountains occur. There are twenty peaks over 2,000m - the highest being Monte Cinto at 2,710m. Snow generally starts to fall on high ground in this region during late October and some patches remain on the peaks until the end of August. Nearly all of this mountainous area is included in 'Le Parc Naturel Régional de la Corse' which stretches from coast to coast between Porto-Vecchio and Calvi. Plants and animals are protected in the park but there are no visible boundaries and visitors are free to travel there as they wish. A footpath GR 20, some 220km long and marked by red-painted stones, traverses the high peaks in a south-west direction from Calenzana in the north to Conca in the south.

In the other region, north-east of the line from L'Isle Rousse to Solenzara, the underlying rock consists mainly of schists, but even here the land is hilly and rises to 1,767m at Monte San Petrone in the Castagniccia and to 1,307m at Monte Stello in Cap Corse. There are some interesting, though relatively small, areas of calcareous rocks east of St. Florent in the north, near the east coast and in the extreme south around Bonifacio, where they occur as rugged chalk sea cliffs. Alluvial regions are found around La Canonica in the north-east and the Plain of Aléria near the central east coast.

Both the granites and the schists give rise to acid soils and in this respect Corsica differs from most other Mediterranean regions which have predominately alkaline base. This acidity is the main reason for the presence of thick, dense maquis in Corsica; some of the component species such as *Arbutus unedo* and *Erica arborea* only grow well in acid conditions. It is significant that 'maquis' is a Corsican word for this type of vegetation.

Although the climate of Corsica is typically mediterranean, the island attracts an appreciable rainfall on account of the mountains and thunderstorms are a typical feature of summer afternoons. The island's precipitation maintains some sizable rivers such as the Tavignano that flows into the sea near Aléria. However, in places the ground can become very dry and forest fires are a constant hazard; some started by lightning and some by human carelessness or arson.

11.3 **Literature:** The plant species of Corsica are well-covered by French floras but the most useful up-to-date treatise is the 'Flora d'Italia' by Pignatti (1982) - a three volume work in Italian which describes all the plants of mainland Italy and its islands, together with those of Corsica and Malta. Nearly all species are illustrated by line drawings and distribution maps. The most complete flora of the island alone is that by Briquet (1910-55) which comprises three volumes in French but some of its nomenclature is already out of date. A very useful modern check list, based on Briquet, is that by Gamisans (1985).

A project for a series of publications through the Botanic Garden of Geneva, Switzerland, entitled ' Complements au Prodrome de la Flore Corse' is under way. So far monograms on six plant families have been produced and an 'annexe' entitled 'La Vegetation de la Corse' (Gamisans 1991). This last work is especially useful for it describes the plant communities in detail and gives some 400 line drawings which include many of the endemic plants; it is indeed a 'must' for the enthusiast who can understand French.

On the island one can purchase a popular, illustrated, booklet called 'Parc Naturel Regional de la Corse - Plantes et Fleurs Rencontrées' by Conrad (1976) and another book in French by Brun *et al* (1975) entitled 'La Nature en France, Corse', which deals with plants, animals and geology of the island is well worth having if one can locate a copy. As a start, the Guide Michelin is a good general introduction to the island and gives some limited information on the plants and agriculture of the region.

11.4 **Cap Corse:** This, the most northerly peninsula of the island, is quite hilly; the land rises steeply from the sea with impressive cliffs on the western side but slopes more gently in the east. There is a good road (D80) round the region with some side roads that make incursions inland. Approaching Patrimonio in the south west in spring one may see many of the roadside flowers which are common and abundant throughout the island including; *Cyclamen repandum* with mauve red flowers, *Allium triquetrum, Anemone stellata*, and the attractive dandelion-like *Urospermum dalechampii*.

At Patrimonio, where the land is chalky, a number of orchids occur in fair quantities: *Aceras anthropophorum, Orchis morio picta, Orchis papilionacea, Ophrys bombyliflora, Ophrys sphecodes* and *Serapias cordigera.*

Along the west side of Cap Corse, the road hugs the hillside and there are relatively few places where one can park a car safely or even get off the road. However, it is easy to see inland the typical tall maquis which can reach a height of 2-3 metres and is often so dense in places as to be inpenetrable to human beings. This is the home of the wild boar - a serious agricultural pest and in parts of the island grants are given for the clearing of maquis so that it may be kept under control near to habitation. The main plant species of the maquis include:

Arbutus unedo	*Lavandula stoechas*
Calicotome villosa	*Myrtus communis*
Cistus incanus	*Olea europea sylvestris*
C. monspeliensis	*Phillyrea angustifolia*
C. salvifolius	*Pistacia lentiscus*
Erica arborea	*Rosmarinus officinalis*

It is said that Napoleon claimed that he could smell his island before he could see it when approaching from the sea. This was no doubt due to the mixture of scents from the calicotome, cistus, lavender, rosemary and other plants of the maquis.

By the side of this road grow scattered plants of *Pancratium illyricum* which produces its head of beautiful starry, white flowers in April. It is endemic to Corsica, Sardinia and the surrounding small islands but there are unconfirmed reports that it has been seen in parts of southern Spain. It resembles the sea daffodil *Pancratium maritimum* which favours sandy areas by the shore throughout the Mediterranean including Corsica, and flowers in mid-summer with more trumpet-shaped flowers. *Pancratium illyricum* is found in many parts of Corsica often quite high up some of the alpine valleys but rarely on sandy coasts. When not in flower it may be distinguished from *P. maritimum* by its broader and less twisted leaves.

In fields by the coast road near Macinaggio grows *Oxalis pes-caprae* - a beautiful but pernicious weed that is less well established here than in most other parts of the western mediterranean. Two other roadside plants which occur near Macinaggio are, the 1.5-2 metre tall shrub with silvery leaves and yellow flowers *Anthyllis barba-jovis* and the neat small, shrubby *Euphorbia pithyusa*. Some seven kilometers south of Macinaggio a road crosses the penisula via Luri through lush green lanes where there are the ferns *Asplenium onopteris* and *A. adiantum-nigrum* and the creeping, moss-like *Selaginella denticulata*. Amongst the cyclamen and anemones here grows a particularly tall, robust figwort with dull purple flowers *Scrophularia trifoliata* - a species endemic to Corsica, Sardinia and Elba.

Further south down the D80 a road leads inland up the Sisco valley which is lush and well-watered. A speciality here, close by the stream, is the giant fern *Woodwardia radicans* with fronds 2m long that root at the tips to produce new plants when they touch the ground. On the rocks grows the tiny creeping *Soleirolia soleirolii* (O = *Helxine soleirolii*) a member of the nettle family endemic to Corsica and Sardinia and once popular in Victorian conservatories as 'mind your own business'. It is sometimes grown in gardens in Britain where it can become a troublesome weed. Another small plant with a localised distribution found here is the lobelia-like *Solenopsis minuta* (O = *Laurentia bivonae* = *Lobelia tenella*) that produces its yellowish white flowers bordered with violet in summer. If one progress right to the top of the road and climbs to the peak of Monte Stello one might see the rare *Morisia monanthos* mentioned in 11.9.

11.5 **Les Agriates:** This area, known also as 'Le Désert des Agriates', is a rather inhospitable, sparsely populated region in the north of the island. It has been, and still is to some extent, used as a winter pasture for flocks of sheep which are taken to graze in the high Asco valley during summer. The weather can become very hot here during July and August and in winter high winds and driving rain are frequent. A road (D81) crosses the area from St. Florent to Lozari. Approaching the region over the Col de San Stefano one passes degraded maquis and hundreds of huge, ancient specimens of *Helleborus lividus* ssp.*corsicus* (N = *H. argutifolius*) known in our gardens simply as *Helleborus corsicus*. It is common throughout the island where it grows in a wide range of habitats but is endemic to Corsica and Sardinia. The type species *H. lividus*, is a weaker growing plant, found only on Majorca and is thought to have come to Corsica from there when the two islands were joined as part of the Tyrrhenian continent sixty million years ago.

At the small port of St Florent on the south east edge of Les Agriates there are marshy areas with much *Iris pseudacorus* and on moist waste land one may find the bristly-fruited silkweed *Gomphocarpus fruticosus* - a rather uncommon species from S. Africa which was introduced as a possible substitute for cotton, but no longer cultivated. Travelling west from St. Florent, the road passes through stony pastures with scattered arbutus, cistus, phillyrea and pistacio bushes. Two superficially similar, very prickly small bushes here are the endemic *Stachys glutinosa* and *Genista corsica*. The former has small white flowers and a strong sage-like smell and the latter small yellow flowers. Although they are both especially evident in this region they occur in many other places on the island. Asphodels, which are avoided by sheep and often a sign of overgrazing, are a prominent feature here and their flowers pervade a strong catty smell. They may be either *Asphodelus aestivus* or *A. ramosus* for the species hybridise readily. The Corsican name for asphodel is 'luminelli' or 'candellu' which alludes to the fact that the dried inflorescences will burn like torches and were once used in processions on all saints day - an important feature of a local culture which pays so much attention to family tombs!.

Around the village of Casta, in moist grassland one can see much *Bellis annua* accompanied by cyclamen, anemones and the orchids: *Orchis morio picta, O. papilionacea, Dactylorhiza sambucina*. Careful searching may reveal the annual blue stonecrop *Sedum caeruleum* growing in cracks amongst the rocks.

11.6 **Etang de Biguglia:** This lies to the east of the section of the N198 road between Bastia and Borgo and the sea. It is the largest of several such partly saline lakes along the east coast of the island and is, separated from the sea by a narrow spit of sand. Eels are harvested from here and there are several special kinds of fish.

Driving eastwards from Borgo past the airport and interesting ruins of La Canonica the roadsides are sometimes lined with large specimens of the milk thistle *Silybum marianum*. Arriving at the seashore by Plage de Pinetto one will see the ubiquitous *Cakile maritima* which is as much at home in Scotland, Norway and North America as it is in the Mediterranean. With it grows *Medicago marina*, common on Mediterranean coasts but not extending further north than Brittany. A more exciting plant here is the attractive *Centaurea pullata* with pink, white-centered cornflowers; it grows on sandy road verges rather than by the seashore. Other species to look out for in this area include the prickly teasel *Dipsacus ferox* and the pink-flowered Silene sericea.

The road turns northwards at La Canonica along the sandy spit between the sea and the etang. Here are extensive areas covered with *Cistus salvifolius* and its taller relative *Halimium halimifolium* which has yellow flowers with a dark patch at the base of the petals. Both species may be parasitised by the curious flowering plant *Cytinus hypocistis*. Around the lake there are bracken, blackberries and some giant reed and here and there large patches of the yellow bartsia *Parentucellia viscosa*. In February and March *Narcissus tazetta* and the rare *Romulea rollii* with white or lilac flowers are to be found in bloom. The romulea is usually treated as a sub-species of *R. columnae*.

11.7 **Around Porto-Vecchio:** The N198 which leads down the east coast of the island to Porto-Vecchio is a fast road and there are few places to stop. However, if one can do so in spring one will see the roadsides are bright with the flowers of such species as *Bellardia trixago, Hyoseris radiata, Lotus ornithopodioides, Lupinus angustifolius, Urospermum dalechampii* and a few orchids here and there. The hyoseris is a dandelion-like plant in which the leaves tend to be held upright rather than in a flat rosette. The lotus is a taller growing species than our own bird's foot trefoil and has tiny yellow flowers in fours which are followed by groups of pods that closely resemble birds' feet. Near Porto Vecchio are cork-oak and mixed woods with much *Quercus ilex*. Orchids abound here in clearings and field corners. Serapias are something of a speciality of Corsica and cover the ground in enormous numbers in places, especially *S. lingua* and *S.cordigera. Orchis papilionacea* is nearly as plentiful and *Serapias neglecta* and *Orchis laxiflora* are not uncommon. One should look carefully at the plants thought to be *Serapias neglecta* for they may be confused with *S. nurrica*, which was first described in 1974. In clearings the ground is sometimes white with thousands of flowers of the tiny *Bellis annua*.

11.8 **Bonifacio:** Approaching Bonifacio from Porto-Vecchio along the N198 it is worth turning left after some 14km through Suartone on D158 and past the village towards the coast. Here one comes to a marshy area the Marais des Tre Padule, where one can see *Narcissus serotinus* in the autumn and the interesting *Ambrosinia bassii* which flowers from December until March. The last of these is a small aroid having a tiny boat-shaped spathe with a distinct projection at the top and carried close to the soil. The local French name for it is Sabot de Noël (Christmas slipper). It is a rare species in south-west Europe but common in parts of North Africa (25.8).

The attractive old Genoese walled town of Bonifacio is situated on a peninsula of pink-coloured chalk cliffs. The whole area is of particular interest to botanists and a walk from here to the lighthouse is especially rewarding. Patches of garigue are largely made up of the following shrubby species:

Astragalus massiliensis	*Juniperus phoenicea*
Cistus monspeliensis	*Pistacia lentiscus*
C. salvifolius	*Rosmarinus officinalis*
Helichrysum italicum	*Thymelaea hirsuta*

Within this and in grassy patches between the shrubs grow:

Artemisia vulgaris	*Lobularia maritima*
Asphodelus microcarpus	*Matthiola sinuata*
Chrysanthemum coronarium	*Matthiola tricuspidata*
Crithmum maritimum	*Paronychia argentea*
Eryngium maritimum	*Smilax aspera*
Evax pygmaea	*Trifolium stellatum*
Leopoldia comosa	

Orchids here include *Gennaria diphylla, Ophrys lutea, Op. sphegodes, Orchis papilionacea* and *Serapias parviflora*. Both *Orchis longicornu* and *O. pauciflora* have been found near Bonifacio, though not around the lighthouse area. Other rare plants to look out for include *Scilla obtusifolia, Urginea fugax, Urginea undulata*, all of which flower in August and September. Another prize to search for is the small crucifer *Morisia monanthos* (= *M. hypogaea*) known only from here, on Monte Stello, Cap Corse and in Sardinia. It is a relict species from North Africa and has a neat rosette of jagged leaves and small yellow flowers from March to May followed by fruits that bury themselves in the soil.

11.9 The mountains: Here one can find an extraordinary number of interesting endemic species in June and July. The main valleys and peaks to which they lead are from north to south:
1. Asco (Monte Cinto 2710m) Road D147
2. Golo (Monte Cinto 2710m, Tozzo 2007m) Road D84
3. Restonica (Monte Rotondo 2622m, Cardo 2453) Road D 623
4. Vizzavona (Monte d'Oro 2389, Renoso 2032) Road N193
5. Solenzara (Monte Incudine 2136m) Road D268

The first four of these valleys are usually approached from the N197 and N193 roads which pass through Ponte Leccia and Corte. If one is in this region in April or May it is worth making a diversion to find the endemic wild cabbage *Brassica insularis*. It has relatively large white or very pale yellow flowers with a scent

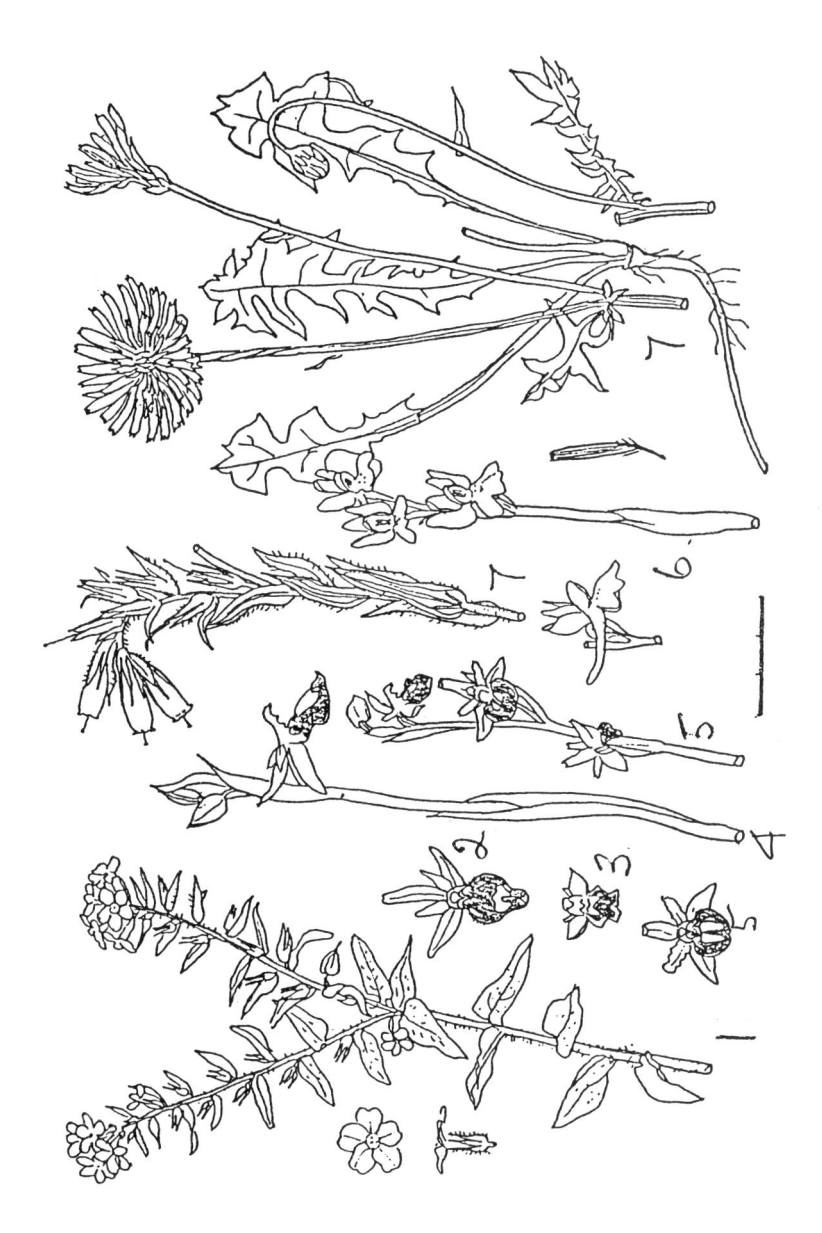

Gargano 1. Alkanna tinctoria 2. Ophrys bertoloniformis 3. Ophrys lutea melana
4. Ophrys bertolonis 5. Ophrys garganica 6. Orchis pauciflora 7. Crepis rubra

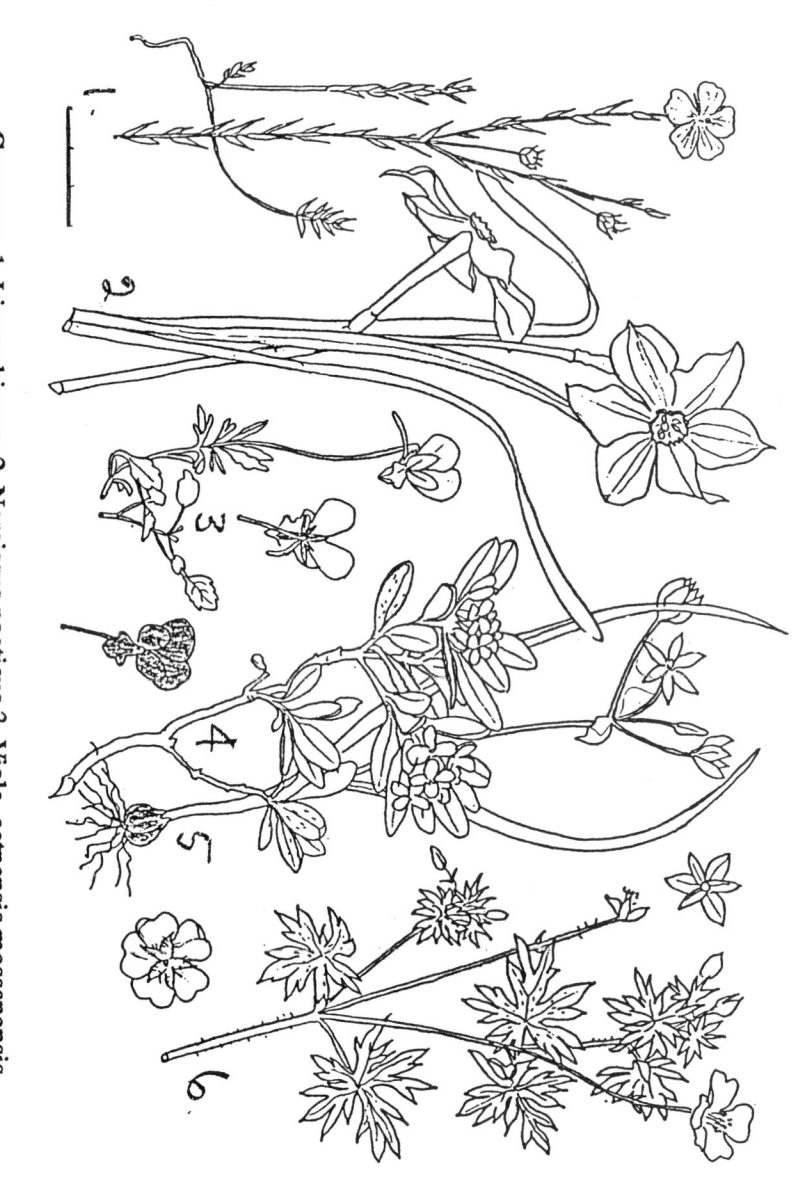

Gargano 1. Linum bienne 2. Narcissus poeticus 3. Viola aetnensis messanensis 4. Daphne sericea 5. Allium triquetrum 6. Geranium sanguineum

said to resemble that of orange blossom. At one time it was widespread on the island but has now become restricted. However, it can be seen on the railway embankment near Omessa, some 10km north of Corte, and in the Défile de l'Inzecca by Vivario about 15km south of Corte.

The Col de Vizzavona is of special interest as one can climb to 2,000m from here in a reasonably short time without the aid of a car by taking the train to Vizzavona station; the circuit on foot from there to the summit of Monte d'Oro takes some 4½ hours for anybody who is reasonably fit. The area around the col itself is clothed with very dense natural beech woods and the trees have trunks of smaller diameter than we are accustomed to at home. Very little grows under their thick canopy but careful searching in clearings can reveal *Cyclamen hederifolium, Cephalanthera rubra* and *C. longifolia*. The last of these grows in other parts of the island and is picked and sold in bunches near Bastia on the 30th of April when it is called 'Muguet' and used as a substitute for the true lily of the valley *Convallaria majalis*. By the roadside at the Col de Vizzavona a dead- nettle with white, pink-flushed flowers may be seen - the endemic *Lamium corsicum*.

The Solenzara valley leads to the Col de Larone which is of special interest in April for it is one of the few places on the island where *Anemone apennina* grows in profusion. In the woods both mauve and white forms occur and present a beautiful sight around Zonza when flowering with *Cyclamen repandum*. At the col itself *Crocus corsicus* is plentiful in spring though it not confined to this area alone and grows in most of the alpine valleys. The mauve flowers are attractively 'feathered' on the outside. The other endemic Corsican species *C. minimus* has smaller flowers, blooms somewhat later, and is generally found at lower levels.

On the way up the mountain valleys the feathery white flowers of the manna ash *Fraxinus ornus* are a feature in March and April. Other trees include *Acer monspessulanum, A. obtusatum, Alnus cordata*. Under the trees and in river valleys one should look for:

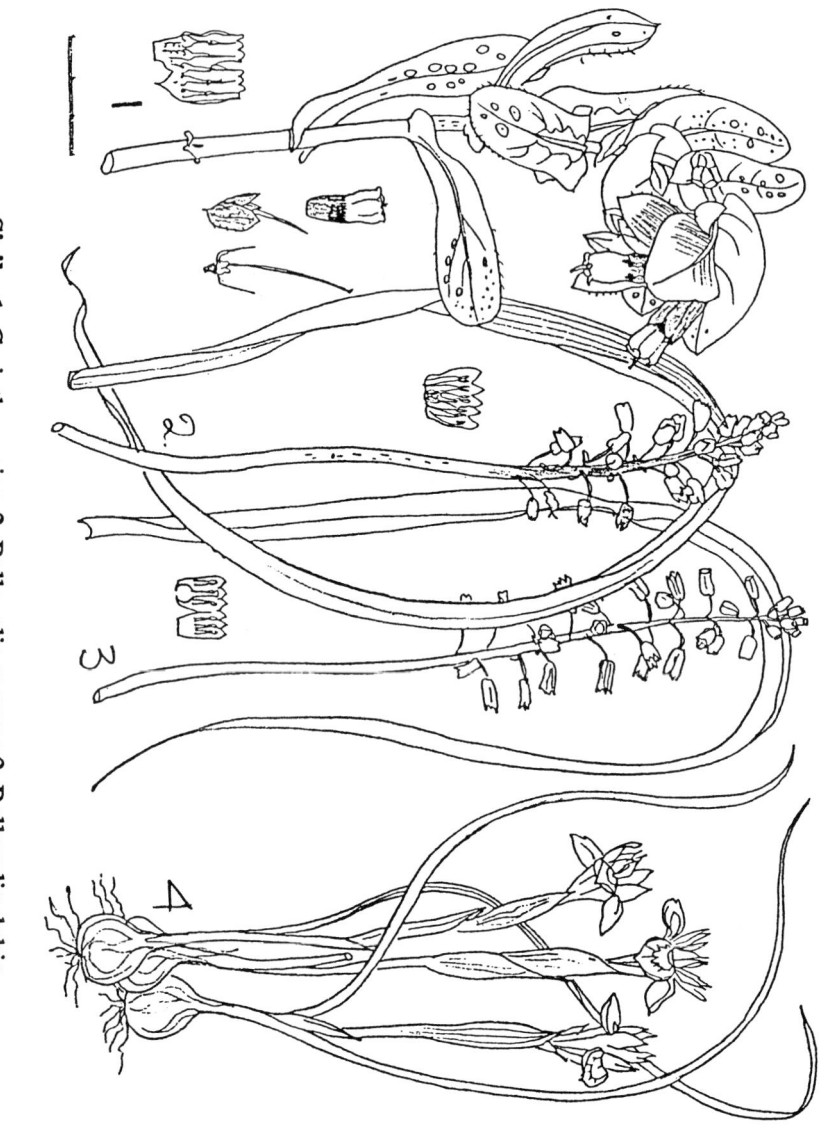

Sicily 1. Cerinthe major 2. Bellevalia romana 3. Bellevalia dubia 4. Gynandriris sisyrinchium

Cerastium soleirolii
Dianthus godronianus
Gentiana asclepiadea
Mentha requienii
Myosotis corsicana
Myosotis soleirolii
Saxifraga stellaris
Thymus herba-barona
Urtica atrovirens
Viola biflora

With the exception of the gentian, saxifrage and viola, all of these are Corsican endemics or of limited distribution. The dianthus is similar to *D. sylvestris*. The mentha is the smallest of all mint species - a creeping plant with pinkish flowers and a characteristic peppermint smell. Neither of the myosotis is a very distinguished looking forget-me-not, both have typical blue flowers. The uncommon nettle *Urtica atrovirens* has dark purplish green leaves. It is rare but also found in Tuscany and Sardinia.

Higher up the mountains one meets the real alpines:

Acinos corsicus
Aquilegia bernardii
Aquilegia litardierei
Arabis alpina
Armeria leucocephala
Armeria multiceps
Astragalus sirinicus
Bupleurum stellatum
Cymbalaria hepaticifolia
Doronicum corsicum
Doronicum grandiflorum
Draba dubia
Draba loiseleurii
Geum montanum
Helichrysum frigidum
Leucanthemopsis tomentosa
Leucanthemum corsicum
Phyteuma serratum
Potentilla crassinervia
Saxifraga cervicornis
Saxifraga stellaris
Sorbus aucuparia praemorsa
Viola nummulariifolia

Again, all of these, except the arabis, geum, *Doronicum grandiflorum* and *Saxifraga stellaris*, are endemics, and most of them are illustrated by Gamisans (1991). The acinos (= *Calamintha corsica*) is a small plant with lilac flowers. *Aquilegia litardierei* is a charming short-growing species with blue flowers in which the petals of the 'cup' are flattened as if to give a 'double' look. It is more or less confined to the Monte Incudine area whereas the taller

A. bernardii is similar to *A. alpina* and more widespread on the island; it has especially large flowers of a soft blue colour and has consequently suffered severely from over-picking by tourists in recent years. Of the armerias, *A. leucocephala* has white and *A. multiceps* pink, flowers. The astragalus is an endemic sub-species *genargenteus* which resembles the very prickly *A. massiliensis*. *Cymbalaria hepaticifolia* is a small creeping plant not unlike the ivy-leaved toadflax. *Doronicum corsicum* has rather ragged-looking flowers and is not as attractive as *D. grandiflorum* which is also found in the Alps. *Draba dubia* has white, whereas *D. loiseleurii* has yellow, flowers and leaves margined with bristly hairs. The helichrysum is a very attractive cushion-forming plant with silvery white capitulae 1-1.5 cm diameter greatly praised by Farrer (1928) but said by him to be nearly impossible to keep in cultivation. The leucanthemopsis is, perhaps the most charming of all the Corsican endemic alpine species. It is short-growing and has carmine-pink or white daisies whereas the leucanthemum more resembles an ox-eye daisy and occurs in two forms, one with much divided leaves. The potentilla has yellow flowers and 5-lobed leaves covered with sticky down. *Saxifraga cervicornis* is a refined 'mossy' type with white flowers and is sometimes referred to as a sub-species of *S. pedemontana*. The sorbus is a dwarf-growing form of the mountain ash.

A special feature of the Corsican mountains is the 'pozzines' - impervious rocky beds of old glaciers surfaced with thin layers of acid, peaty soil that is covered with short turf and dotted with numerous puddles and small lakes. Here one finds a special flora comprised mainly of the following species:

Bellis bernardii	*Pinguicula corsica*
Bellium bellidioides	*Plantago subulata insularis*
Narthecium reverchonii	*Ranunculus marschlinsii*

Surprisingly the bellium, which resembles a small garden daisy that propagates vegetatively by runners, is found also growing in stony places by the sea in southern Spain. The other species are confined to Corsica. *Bellis bernardii* is another small daisy, somewhat like *B. annua*. The narthecium closely resembles our bog asphodel *N. ossifragum* and is the only species of this genus on the island. The pinguicula, with violet flowers, cannot be confused with any other as it is also the only one of the genus to be found in Corsica. The plantago is a very dwarf species with narrow grass-like leaves and the ranunculus is also tiny with reniform basal leaves and very small yellow flowers.

A feature around the pozzines and in the higher valleys is the extensive stands of the dwarf alder *Alnus viridis*. It is common also in parts of the Alps but the sub-species special to Corsica is *suaveolens*. It grows to 1-3m tall, often in dense thickets, and the bare stems in winter give a purplish haze to the landscape. The common alder *A. glutinosa* also occurs but one should look for the taller *A. cordata* growing in mountain valleys. It is a handsome species of southern Italy, Corsica and Sardinia which grows to 4-8m, has heart-shaped leaves and larger fruits than those of other members of the genus. In Britain it is sometimes planted as a roadside tree and quickly grows into a handsome conical specimen. It will tolerate drought and grow in limey or acid soils.

11.10 The West Coast: Here is rugged scenery not unlike the north west coast of Scotland and there are interesting plants to be seen. Shortly after leaving Bonifacio on the N196 there is a turning to the left that leads to the Ermitage de la Trinité. Here grows the endemic *Centranthus trinervis* which is similar to our red valerian *C. ruber* but has narrower, more lanceolate leaves. Continuing until one comes near to the coast again at Roccapina, careful searching in moist places may reveal *Ambrosinia bassii* (11.8). At Propriano, on the sandy coastline one can see a number of typical shore plants such as:

Briza maxima	*Medicago marina*
Cakile maritima	*Otanthus maritimus*
Calendula arvensis	*Raphanus raphanistrum*
Crithmum maritimum	*Silene sericea*
Euphorbia parialis	

After Propriano the road bears inland again and climbs towards Petreto-Bicchisano. In moist places here *Dactylorhiza insularis* may be seen - it is a rare sub-species of *D. sambucina* which grows also in southern Spain, Elba, Sardinia and possibly Sicily. It has broader leaves than the type and always yellow flowers with one or more distinct orange-red markings on the lip. It is also recorded from Punta Pozzo di Borgo west of Ajaccio.

At Ajaccio itself the attractive *Ornithogalum arabicum* is reputed to grow on the walls of the old city but it is not the only plant of special interest in the region. To the west lie the islands called Isles Sanguinaires which can be reached by boat from the port of Ajaccio. Here grow the interesting species:

Helicodiceros muscivorus   *Nananthea perpusilla*
Leucoium roseum   *Tribulus terrestris*

The helicodiceros (= *Arum muscivorum*) is the dragon arum, fairly common on Majorca but rare elsewhere. It flowers in spring and early summer and its weird spathes have stiff hairs on the inner surface. Corsicans call it Orecchia di Porcu (pigs' ears) which aptly describes the inflorescences. The leucoium is an endemic species and a real gem with minute flowers, and is said to cover the ground pink when it blooms in the autumn. But the pink is so pale that one could be forgiven for calling it white until one has turned up the flower and looked inside. The nananthea a distinct endemic with very limited distribution - a very minute chamomile-like plant with white flowers. The tribulus belongs to the family Zygophyllaceae; it is a small, prostrate-growing plant with pinnate leaves and small yellow flowers reminiscent of a potentilla. Its fruits have thorny protuberances that are so hard that they can damage the feet of animals and, it is said, puncture bicycle tyres. With careful searching some of these plants can also be seen around Ajaccio itself and by the shore near Porticcio just to the south. Three additional species that grow nearby are *Calystegia soldanella, Silene succulenta* in the coastal sand and, further inland, the attractive autumn-flowering *Arum pictum* which has dark mauve-black spathes. The silene is an uncommon plant and here the sub-species *corsica*. It somewhat resembles the widespread *S. colorata* but has pale yellow flowers and rather succulent leaves.

After Ajaccio the main road bears inland and to keep to the coast one must get on to the D81 through Cargèse. Les Calanche is a good plant hunting area where one can find:

Dianthus furcatus	*Limonium dictyocladum*
Leucoium longifolium	*Silene nodulosa*
Lilium croceum	*Stachys glutinosa*

The dianthus is a sub-species *gyspergerae* and has typical pink flowers in May and June. The leucoium is a spring flowering species with white flowers similar to the more familiar *L. trichophyllum* of Spain. The limonium is an endemic with much branched leafless stems and bears mauve flowers in June to August. *Silene nodulosa* is another endemic which has narrow basal leaves and 20-30cm tall inflorescences that carry one to five pink flowers in June and July. The lilium is the orange lily of the Alps and is most likely to be found here near the village of Piana.

The maquis in this region is especially fine and contains several aqdditional species to those noted earlier (11.4) such as; *Bupleurum fruticosum, Lonicera etrusca, Viburnum tinus*. The bupleurum is not a typical hare's ear but a small evergreen shrub with heads of greenish flowers.

Past Les Calanche one may turn inland eastwards along D84 through Evisa, the Forêt de Aitone, the Col de Vergio 1477m, the Gorges de La Restonica and on to the Golo valley and Corte. Rising up to Evisa one may encounter the small yellow-flowered leguminous shrub *Anthyllis hermanniae* which occurs in the Balearics and eastwards to Crete and *Euphorbia Lathyris* the caper spurge. The Forêt de Aitone has some magnificent but geriatric and diseased old sweet chestnut trees and after that one rises to the Col. Here one is above the tree line and in windswept scrub with *Berberis aetnensis, Genista lobelii var lobeloides, Juniperus communis ssp. alpina, Plantago sarda var sarda, Thymus herba-barona* and a small viola, possibly *V. nummularifolia*.

Continuing along the coastline past Porto at Plage de Bussaglia grows *Erodium corsicum*, an endemic to Corsica and Sicily with pink, violet-veined 2cm diameter flowers in summer. Continuing north along the coast there are still more interesting plants. If one turns inland from the coast at Galéria along the Fango valley to Pirio one may find *Leucoium roseum* and *Ranunculus bullatus* flowering together there in quantity during the autumn. The leucoium may also be seen further north at La Revellata near Calvi.

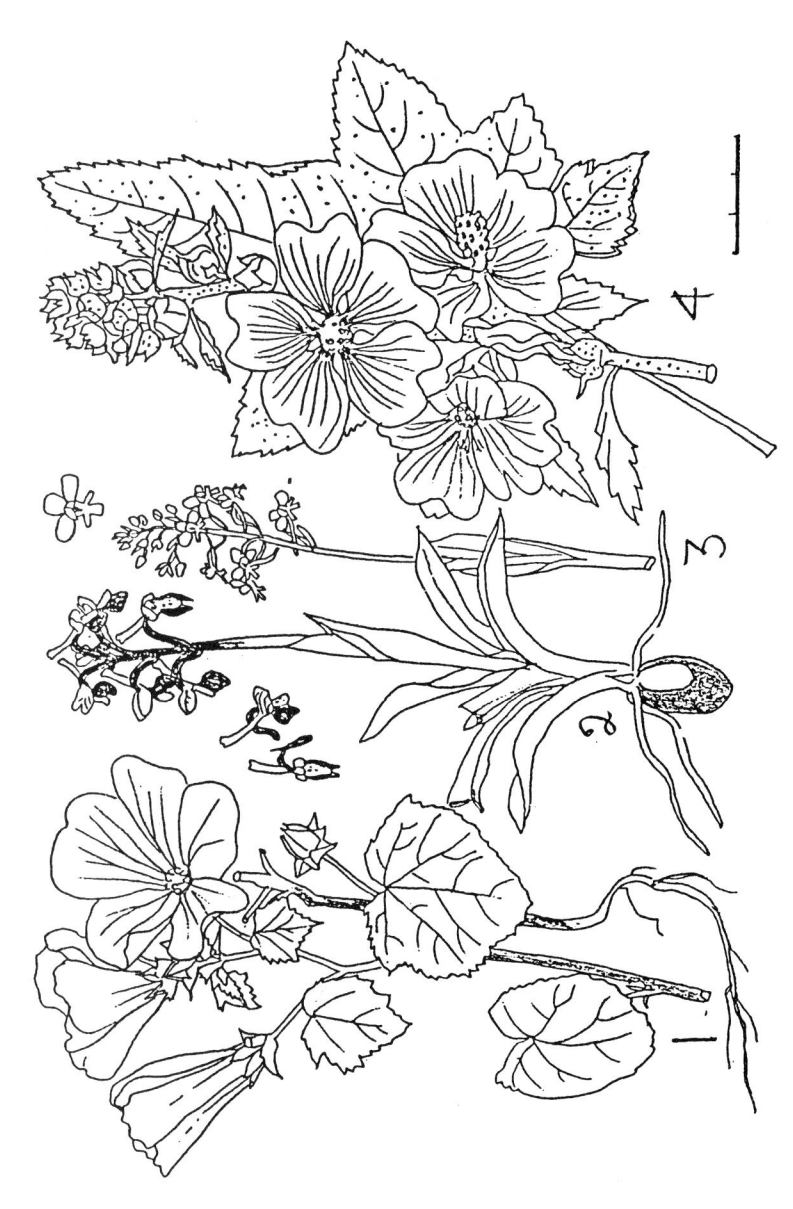

Sicily 1. Lavatera trimestris 2. Orchis longicornu 3. Orchis brancifortii 4. Lavatera olbia

Sicily 1. Hermodactylis tuberosus 2. Iris planifolia 3. Iris pseudopumila

12. The Gargano

12.1 The Promentorio del Gargano is a rocky peninsula, approximately 45 x 45km, which juts into the Adriatic Sea almost due east of Rome. It is a largely unspoiled area of limestone hills many of which rise to 600 - 1,000m. There are summer holiday resorts around the coast, but these are hardly known to British holiday makers. One can travel there by train from Rome to Rodi Garganico in the north or fly to the airports of Bari or Foggia. When visiting the region in the spring, and the flora is at its best, it is not difficult to find accommodation without booking in advance.

The Gargano may be described as a kind of Mecca for orchid enthusiasts, with over sixty species and often found in larger populations than in any other part of Europe. Anybody kneeling on the ground there will probably be a German, Swiss or Dutch national paying homage to the orchids by taking close-up photographs. However, orchids are not the only plants of interest to be found on the rock-strewn grassland and the charming woodland reserve of Foresta Umbra. Although most species here are typical of the Italian mainland, there is a slight Balkan influence. The northern part of the Adriatic Sea is relatively shallow and evidence suggests that the Gargano was once joined to what is now northern Greece and Albania and it seems that there may have been an exchange of plant species between these regions in geological times.

Excellent maps of the Gargano are readily available, including a tourist version of 1:100,000, which even names the farms (they are referred to here as 'Mas' - as in Provençal). It is easy to get around by car and one can eat well in the region, especially at some of the fish restaurants near the coast.

12.2 **The terrain:** Manfredonia, with a modern port and some industry, is the main town of the region. It has an interesting thirteenth century castle built by its founder, Manfred, king of Sicily. The other towns and villages are relatively small. Most of the terrain consists of rocky pastures which are often surprisingly green, for this area of the Mediterranean is not so dry as south east Spain

Sicily 1. Lamium flexuosum 2. Erysimum bonannianum 3. Symphytum gussonei 4. Bellis margaritaefolia 5. Linum punctatum 6. Cymbalaria pubescens

or further east in Greece. There are sea cliffs near the small coastal towns of Mattinata and Vieste. Somewhat east of the centre of the peninsular lies the Foresta Umbra which is a nature reserve with picnic sites and a zoo of local wild animals. It is a popular week-end holiday place for local people but the area is so large that one can still enjoy solitude there. Two large, brackish lakes, Largo di Lesina and Largo di Varano, lie along the north coast cut off from the sea by a narrow sandy strip planted with trees.

12.3 Literature: This part of Italy is, of course, well covered by Pignatti's excellent Flora d'Italia (1982). The region itself is dealt with by a series of articles in Italian by Fenaroli (1966-) but these are not readily available to the general public. An extremely useful publication for orchid enthusiasts is that by Lorenz and Gembardt (1987) which lists, describes, gives detailed distribution maps and some illustrations of the orchid species of the Gargano. It is in German and unfortunately not readily available. If one cannot obtain this then the new field guide by Buttler (1991) is essential for any visitor looking for orchids there and probably more up-to-date in the changeable area of orchid nomenclature.

12.4 Monte San Angelo: The majority of visitors will arrive in the Gargano at Manfredonia and it is only a short distance from here to the interesting small town of Monte San Angelo. About 5km along the road 89 north-east out of Manfredonia there is winding road which climbs to the town standing on a rocky bluff. It is a place of pilgrimage and the fine church commemorates a visitation of the Angel Gabriel. In spring the cliffs around the town are colourful with flowers including:

Arabis alpina　　　　　　*Doronicum orientale*
Aubretia columnae　　　*Saxifraga granulata*

The arabis (= *A. albida*) is the sub-species *caucasica* and the aubretia the sub-species *italica*.
In grassy fields near here one may see the following in spring:

Asphodeline lutea *Helianthemum jonium*
Anemone hortensis *Hermodactylus tuberosus*
Arabis verna *Lamium garganicum*
Bellis sylvestris *Saxifraga granulata*
Helianthemum apenninum *Viola aetnensis*

Some of these need qualification. The bellis flowers here in spring, whereas the typical species is autumn-flowering. Pignatti (1982) states that some forms flower in spring and hybrids *Bellis sylvestris x perennis* are not uncommon and often spring-flowering. *Helianthemum jonium* is a rare, endemic but unprepossessing plant with heads of 3-7 hanging, small yellow flowers that appear to remain partly closed. It can not be confused with *H.apenninum*, which is a typical rock-rose with white flowers. The lamium is on its home territory here, though it is widespread throughout the Mediterranean and rather variable with several sub-species. The viola is endemic to Puglia and Sicily and is found on Mount Etna as its name indicates but here the form may be sub-species *messanensis*. It occurs throughout the Gargano and is very similar to another long-spurred pansy found over the water in Albania and described in the Albanian flora (Demiri 1981) as V. speciosa (15.7). This situation needs further investigation. Of course there are orchids here, especially:

Ophrys bombyliflora *Ophrys lutea*
Ophrys fusca *Orchis italica*
Ophrys incubacea *Orchis papilionacea*

Ophrys incubacea is the up-to-date name for what used to be called *O. sphegodes ssp atrata*.

Departing from Monte San Angelo in a westerly direction one comes to road 272 that crosses the western end of the peninsular via San Giovanni. The terrain is mostly of exposed, rocky pastures over 600m. Here are large groups of *Iris pseudopumila* - a dwarf rhizatomous species of the *I. germanica* group and distinguished from the more widespread *I. lutescens* (= *I. chamaeiris*) by having a long (twice the length of the ovary) perianth tube and retaining its leaves during the winter. It has a limited distribution but is common in Sicily (13.9) and found in parts of what used to be called Yugoslavia. Another similar, but slightly taller and larger-flowered iris grows with it in places. This is *I. revoluta* which is endemic to this part of Italy and has dark purple flowers whereas those of *Iris pseudopumila* are usually yellow in the Gargano. *Asphodeline lutea* and *Asphodelus aestivus* grow in fair quantity with the irises and in places there are groups of *Narcisus tazetta*. *Narcissus serotinus* may also be found here, flowering in the autumn. Several of the ophrys species special to the area can be found in this region, though they do not always grow by the wayside and have to be searched for. Furthermore they are not always easy to identify with certainty and some enthusiasts may find it difficult to agree with the nomenclature. They include:

Ophrys apulica = *O. holoserica ssp apulica*
O. bertoloniformis = *ssp of O. bertolonii*
O. garganica = *O. sphegodes ssp garganica*
O. promontorii = *O. bertoloniiformis*
O. pseudobertolonii ssp bertoloniiformis
O. sipontensis = *O. sphegodes ssp sipontensis*

Fortunately all of them are described and illustrated in Buttler (1991). Another rather unusual orchid to be looked for here is *Serapias orientalis* (= *S. vomeracea ssp orientalis*). It is a rather impressive species somewhat resembling *S. neglecta* (*S. cordigera ssp neglecta*) with relatively large flowers that are dark brownish-red or sometimes ochre-coloured. It is primarily an eastern mediterranean species which is common in north-east Cyprus but also found in Corsica and parts of Sicily.

A more direct way to reach the 272 road from Manfredonia is to drive out almost due north up the winding route to Campolato. Where the hairpin bends take one up Scaloria there are rock faces and one may see *Onosma echioides*. Like most other species of the genus, it is variable, and here the flowers are nearly white. Beside the road at this point grows *Asphodelus tenuifolius* which looks like a minature *A. fistulosus* and is, indeed, not distinguished from that species in Flora Europaea. It is, however, an annual rather than perennial and has a limited distribution in Italy but is relatively common in parts of south-east Spain (7.4).

12.5 Manfredonia to Vieste: By the roadside leading out of Manfredonia one may see a number of species typical of the Mediterranean and also northern Europe such as:

Asparagus acutifolius   *Euphorbia lathyris*
Calendula arvensis   *Isatis tinctoria*
Centranthus ruber   *Orlaya grandiflora*

After passing through Mattinata the 89 road north from Manfredonia climbs inland through charming countryside on its way to Vieste on the coast at the eastern extremity of the peninsular. The dry cliffs by Mattinata support large groups of the shrubby *Euphorbia dendroides*. In the greener area above, one may be surprised by the profusion of orchids in spring, including:

Aceras anthropophorum   *Ophrys tenthredinifera*
Anacamptis pyramidalis   *Orchis coriophora*
Barlia robertiana   *O. italica*
Himantoglossum hircinum   *O. morio*
Ophrys arachnitiformis   *O. papilionacea*
Op. bertolonii   *O. pauciflora*
Op. bertoloniiformis   *O. provincialis*
Op. biscutella   *O. purpurea*
Op. bombyliflora   *O. quadripunctata*
Op. garganica   *O. tridentata*
Op. incubacea   *O. ustulata*
Op. lutea   *Serapias vomeracea laxiflora*
Op. lutea melena   *Serapias lingua*
Op. sphegodes

For details of these one needs, once again, to consult Buttler (1991). However, it is worth drawing attention to *Op. garganica*, named after the region, which is widespread and common here but easily confused with *Op. incubacea* (= *O. sphegodes ssp atrata*) that is also widespread here. The latter has more pronounced and hairy side lobes to the lip. With these orchids one may see *Anthyllis vulneraria, Psoralea bituminosa* and *Ranunculus millefoliatus*. The anthyllis is a form with fine dark red flowers - a complicated genus which splits up ino a number of species in Italy. The ranunculus is a buttercup with finely divided leaves resembling those of a carrot.

There are a number of scattered shrubs here including *Coronilla emerus, Pistacia lentiscus, Quercus coccineus, Rosmarinus officinalis* and *Spartium junceum*. All are typical of most parts of the Mediterranean though the coronilla grows wild as far north as southern Sweden and is sometimes cultivated in gardens. Its English name of scorpion senna derives from the pods which do look like a scorpion's tail, sting and all.

About the highest point of the road, at some 670m near Valico del Lupo (presumably they did have wolves here in the past) - there are patches of mixed woodland. The main trees include the holly oak, acers, hornbeam and the hop hornbeam *Ostrya carpinifolia* - a small tree which is an important component of the forests in the Balkans on the other side of the Adriatic. Under the trees in spring there is a carpet of *Anemone apennina* with both mauve and white forms. Some look very like the common wood anemone A. nemorosa which is also found in the region. To confuse matters, mixed in with them are groups of *Anemone hortensis* (O = *A. stellata*). Other plants here include:

Aristolochia pallida	*Doronicum orientale*
Crocus sp.	*Lamium bifidum*
Cyclamen hederifolium	*Ruscus aculeatus*
Daphne laureola	

The lamium is an interesting small dead nettle that sometimes has variegated leaves and bears white flowers in which the 'hood' has two projections, like fingers making the 'V' sign. The crocus are autumn- flowering species and may be *Crocus longiflorus* or *C. thomasii*, both with a limited distribution.

Continuing along the route to Vieste one may notice *Daphne sericea* beside the road. It has shiny evergreen leaves and strongly scented, pink flowers in March and April that turn brown when they fade. It is primarily an eastern Mediterranean species found also in Crete and parts of mainland Greece. Other roadside shrubs include *Euphorbia characias, Spartium junceum* and *Juniperus communis*. Amongst them grow *Anchusa arvensis, Linum bienne* and a fine form of the variable honeywort *Cerinthe major* with almost entirely yellow flowers.

12.6 **Foresta Umbra:** This impressive deciduous forest lies slightly east of the centre of the peninsula. It can be reached without difficulty from Vieste by taking the 89 road and turning southwest at Segheria il Madrione, or it can be approached from Monte San Angelo. The edges of the forest include the species seen near Valico del Lupo (12.5) but towards the centre these give way to magnificent beech trees. In the ground carpet of leaves grow *Anemone apeninna* and *Cardamine bulbifera* (O = *Dentaria bulbifera*). The latter has white, rather than mauve flowers, and some leaves with 3 instead of the usual 5 leaflets and is the local variety *garganica*. Other plants occurring round the edges of the forest and in clearings are:

Alliaria petiolata   *Doronicum orientale*
Allium triquetrum   *Euphorbia amygdaloides*
Arum italicum   *Ranunculus ficaria*
Asperula odorata   *Sanicula europaea*
Cyclamen repandum

The alliaria (*O = Alliaria officinalis = Sisymbrium alliaria*) is our familiar garlic mustard or jack-by-the-hedge, which has an objectionable smell. 'Jack' is an old rustic word for a lavatory. The other species here well known to British natural historians are the woodruff (asperula), the lesser celandine (ranunculus) and the sanicle (sanicula). A number of orchids grow in and around the Foresta Umbra including:

Cephalanthera damasonium   *Epipactis microphylla*
C. longifolia   *Limodorum abortivum*
C. rubra   *Neottia nidus-avis*
Dactylorhiza romana   *Platanthera chlorantha*
Epipactis helleborine

Of these, only the neottia and dactylorhiza are common and the latter is almost invariably the yellow-flowered form. One will need to search hard to find the others. However, in grassland around the forest there is a good chance of seeing the orchis species; *Orchis coriophora, collina, italica, morio, papilionacea, pauciflora, provincialis, purpurea.*

After leaving Monte San Angelo to approach the Foresta Umbra from a westerly direction, and negotiating two hairpin bends, a road leads off in a north-east direction and at the farmhouse Mas Lombardo and Chiancatapilone (all marked on 1:100,000 map) there is light woodland in which grows *Narcissus poeticus* that looks truly beautiful in this wild setting. It is at its best in April and in grassland around here and other parts of the Gargano *Narcissus tazetta* also can be seen in quantity but flowering somewhat earlier - in March.

12.7 Vieste to Lago di Varano: The coast road from Vieste to Peschici passes many holiday camp sites and rocky outcrops with several caves. There are various cistus bushes along here, especially at Grotta dell'Acqua, including *C. salvifolius, C. incanus* (= *C. villosus*), *C. monspeliensis* and the rather uncommon *C. clusii*. Some of the bushes are difficult to identify with certainty and it seems that the last two mentioned occasionally hybridise. Other shrubs with them include *Calicotome villosa, Myrtus communis* and *Pistacia lentiscus*. There is also an inland route

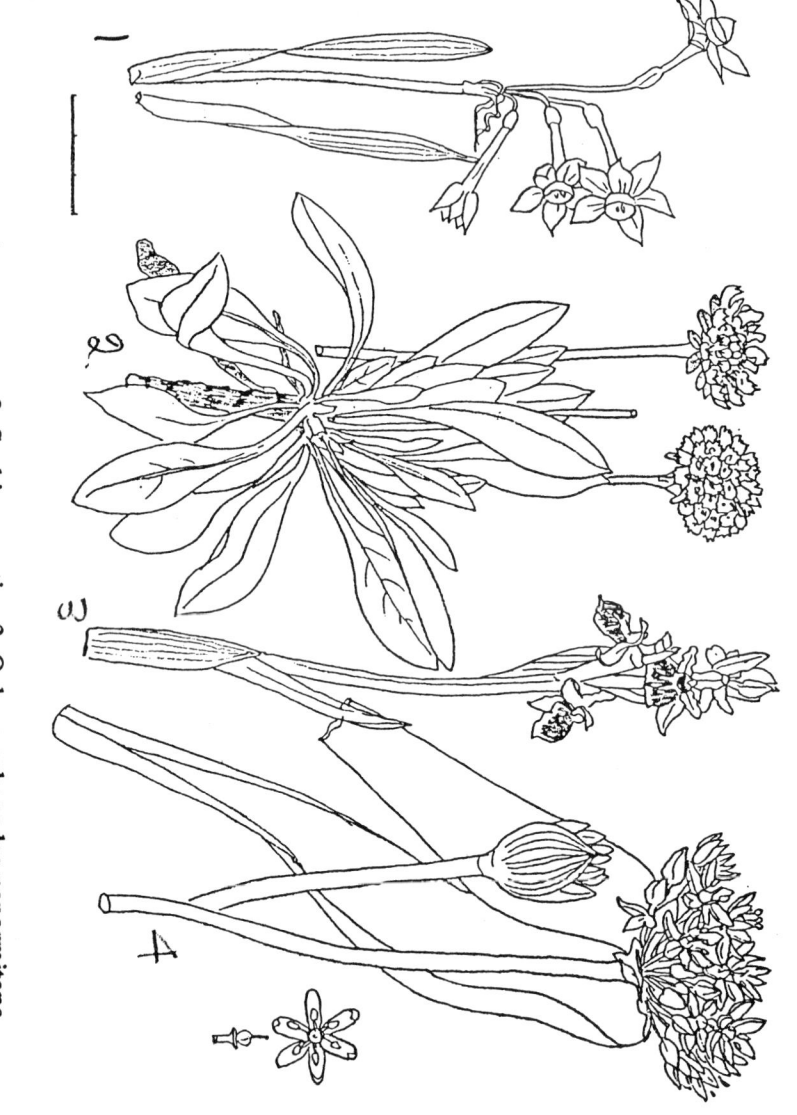

Sicily 1. Narcissus tazetta 2. Scabiosa cretica 3. Ophrys sphegodes panormitana 4. Allium nigrum

from Vieste to Peschici via Segheria il Mandrione. By the roadside one may see such widespread species as *Anagallis foemina, Anthyllis tetraphylla, Cerinthe major* and *Cynoglossum creticum.* In places there are large groups of the sophisticated dandelion-like *Urospermum dalechampii* with large sulphur-yellow flower-heads that often appear to have a small reddish spot in the centre. A somewhat similar plant here, but with flesh-pink flower heads, is *Crepis rubra.* It rather resembles the so-called Greek dandelion *C. incana* but has larger heads that hang downwards when 'in bud', furthermore it is an annual and not perennial like the Greek version. It is suficiently attractive to have caught the eye of seedsmen and is now offered amongst the annuals as 'pink hawksbeard'.

The road from Peschici continues along the coast to Rodi Garganico where there are more holiday campsites and then along the sand spit that separates the lake from the sea. The spit is planted in places with pines and eucalyptus and there are few species growing below them. However, large patches of the beautiful blue dyer's alkanet *Alkanna tinctoria* occur. Under the pines in places grow carpets of an asparagus, probably *Asparagus maritimus* which is a sub-species of *A. officinalis*, the wild form of our cultivated vegetable. The only other plants that seem to thrive here are wild madder *Rubia peregrina* and *Smilax aspera.*

One may wish to return from this area via Carpino and Monte San Angelo. Several of the orchids and other plants already mentioned can be seen near the road, and additional species such as *Geranium sanguineum.* Bushes by the roadside include the common hawthorn *Crataegus monogyna*, though it is a variable species and *Pyrus amygdaliformis* which has white flowers in spring and small, apple-shaped fruits in September. The last of these is not very common in Italy but grows in quantity on some Greek islands, especially Crete.

12.8 **Addendum:** Most botanising of the Gargano has been done in the eastern half, but the high remote parts of the western section would be well worth exploring. It is also not far from here to the Abruzzi mountains which are part of the Apennines and have an interesting flora including, as a taste, *Gentiana cruciata*, *Iris graminea*, *Ornithogalum pyrenaicum*, *Streptopus amplexifolius*, *Traunsteinera globosa*.

Ophrys garganica Mattinata, Gargano, Italy. (p.187)

Pancratium illyricum Cap Corse, Corsica. (p.163)

Cyclamen repandum var. *rhodense* Mt. Profitis Ilias, Rhodes. (p.369)

Sternbergia sicula Sellia Gorge, Crete. (p.351)

13. SICILY

13.1 Sicily is the largest of the Mediterranean islands and merits several visits to study its interesting flora of some 3,000 taxons, about 10% of which are endemics. It is not one of the most popular with holiday-makers, perhaps because it has relatively few extensive beaches. However, at present British, German and Scandinavian companies offer package holidays to Taormina and places near there on the east coast, whereas French and Italian companies specialise in Cefalù on the north coast. The independent traveller can conveniently fly to Palermo or Catania airports and make his or her own way by car, though those who are not familiar with Italian driving techniques may find the traffic in towns somewhat frenetic!

In addition to the flora there are interesting buildings and archaeological sites to be seen, for the island has been ruled by many different peoples and at one time supported an important classical Greek culture. 'Round the island tours' are frequently available for those who are interested in this aspect. However, ornithologists will usually be disappointed for the shooting of birds is an especially popular pastime for the Sicilians. Nevertheless some bird watchers may be prepared to share the view of migrating birds with the hunters at Monte Ciccia near Messina between mid-April to mid-May.

Several maps are available, especially the 1:350,000 scale route map by Lithographia Artistica Cartographica, Florence. The privately owned motorways are a great help if one is travelling long distances, for getting through some of the villages and small towns on the ordinary roads can be extremely frustrating. The motorways often stride over the countryside on concrete stilts and one has to admire the large scale thinking in their planning but, before starting ones journey, it is advisable check that the whole of the section one wishes to use has been completed as this is not clear from some maps

As in most Mediterranean countries, April is a good time to see the lowland flowers but one should come prepared to encounter heavy rain showers at this time of the year, especially in the northeast. To study the flora of Mount Etna it is necessary to go there later in the year for the snow can lie at relatively low levels until June, and there is a permanent cover at the summit.

13.2 The Terrain: The most obvious physical feature of the island is Mount Etna which rises to 3262m near the east coast, north of Catania. It is an active volcano that is nearly always smoking and erupts from time to time. The beautiful, snow-capped summit can be seen from many eastern parts of the island. It is not the only site of volcanic activity in the region, several of the Lipari islands to the north-west of Messina are active and a trip round them would make a good subject for a spectacular botanical survey. There are a number of small, non-volcanic islands off other parts of the coast which provide a home for some interesting plants not found on the main island.

A casual look at the map will suggest that, apart from Etna, the whole of the rest of Sicily is equally hilly, but there is a fairly well defined range of limestone mountains near, and parallel to, the north coast. From Messina in the east these are the Peloritani, Nebrodi and the Madonie which ends before Palermo. They merge into one another and have a number of peaks over 1800m that carry snow in the winter. Another high part is that of the Monti Iblei in the south east that is not so elevated but is composed of windswept and relatively treeless expanses of limestone hills.

Etna and the northern mountains attract much of the rainfall and the weather tends to be drier and sunnier as one goes towards the south west of the island.

13.3 Literature: There is practically no popular literature dealing specifically with the plants of Sicily. Until recently the main floras of the area were by Tornabene (1887) and (1889-1892); the first of these is on the plants of Sicily in general and the second comprises four volumes on the plants of Mount Etna. Both are in Latin, unillustrated and difficult to consult for they have been out of print for many years; though there is a reprint by the Dutch publishers Otto Koeltz (1973). Fortunately, a really excellent 3-volume flora of Italy by Pignatti (1982) is now available. It gives line drawings

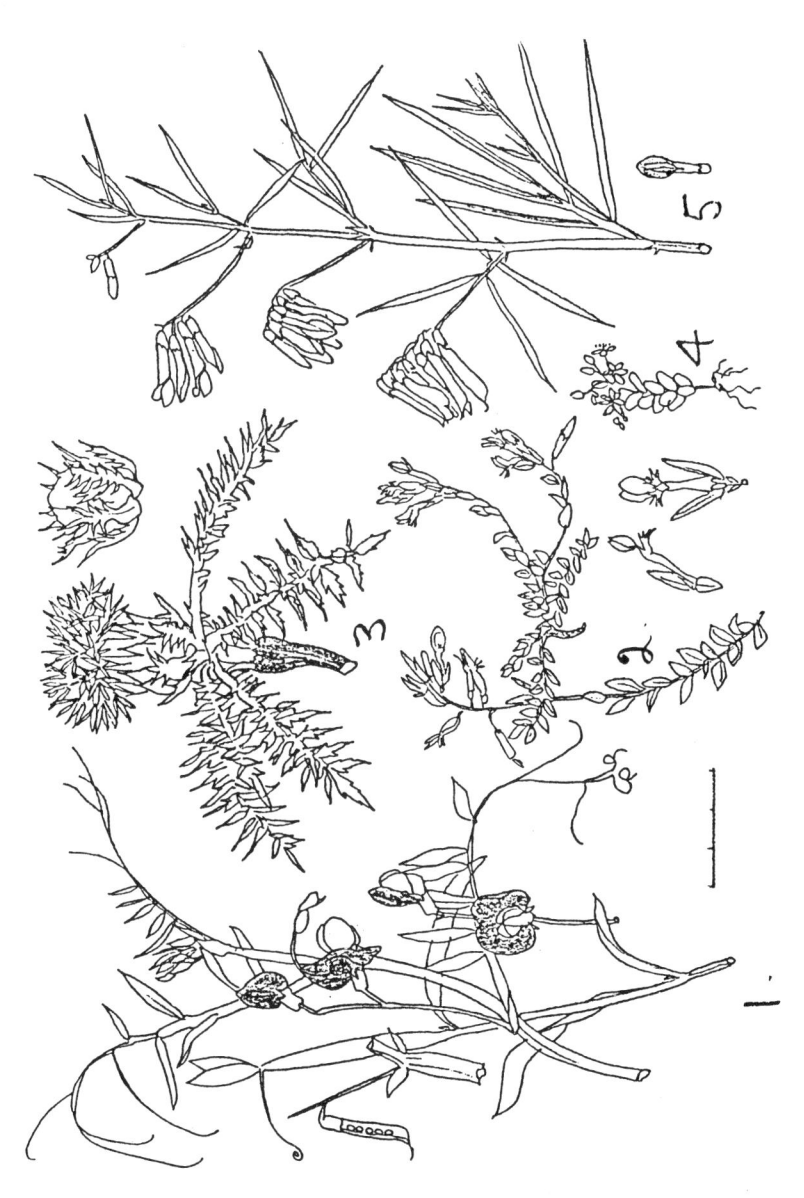

Sicily 1. Lathyrus articulatus 2. Polygala preslii 3. Carduncellus pinnatus
4. Sedum coeruleum   5. Vicia sicula

and distribution maps of all Italian species including, of course, those of Sicily. By systematically going through the maps in this work it is possible to build up a rough picture of the flora of the island.

13.4 Taormina Region: Most British visitors on package holidays to Scily will lodge at this picturesque small hilltop town half way between Messina and Catania or on the coast nearby at Giardini. Close to Taormina itself there is a small Greek amphitheatre from which one has a fine view of Mount Etna and it is a good place to start looking for plants typical of the area. They are likely to include the following species common to most parts of the western Mediterranean:

Anagallis foemina *Hedysarum coronarium*
Calicotome spinosa *Oxalis pes-caprae*
Cistus salvifolius *Teucrium fruticans*
Euphorbia rigida *Vicia hybrida*
Fedia cornucopiae

The hedysarum is frequently grown as a forage crop in Sicily and often occurs as an 'escape' from cultivation. Two less common weeds of arable land seen round here are *Achyranthes aspera* and *Coleostephus myconis*. The first of these is a curious member of the Love-lies-bleeding family Amaranthaceae with spikes of tiny insignificant flowers; it is an introduced annual originating in the tropics. The coleostephus is superficially very like a corn marigold with entire leaves and used to be included in the genus *Chrysanthemum*. The ray florets are usually yellow but white-flowered forms occur.

Two interesting and rare plants to be looked for here are *Centaurea tauromenitana* and *Antirrhinum siculum*. The centaurea, which is named after the town of Taormina, is only found there. In June and July it has pale yellow flower heads which are probably larger than those of any other Mediterranean member of the genus and Pignatti (1982) eulogizes about it as a beautiful plant in a beautiful setting. It seems to be close to *C. orientalis* which figures in Polunin (1980). The antirrhinum is a typical snapdragon with narrow leaves and is the only yellow-flowered species to be found wild on the island. Pignatti (1982) describes it as endemic to south west Italy, but it is also fairly common on Malta (14.4).

Down near the coast at Giardini the small annual *Linaria reflexa* is common on waste land. The flowers here are generally white with an orange blotch and violet striping on the upper petals but mauve-flowered forms are known. The ripening seed capsules bend downwards and inwards. It is not a common species but occurs also in Corsica and Sardinia. On the rocks by the main road to Messina one can see the annual blue cabbage *Moricandia arvensis* and snapdragons which may be forms of *Antirrhinum majus* that is reported as wild in Sicily but could be garden escapes. An interesting plant growing on the rocks here near the sea is *Scabiosa cretica*. It is a perennial with tufts of lanceolate leaves and leafless flowering stems of typical violet flowers. Though not a common species it is also be found on similar sites in the Balearic islands, Crete and Rhodes.

13.5 Around Etna: The highest slopes of Etna are inaccessible to the plant hunter in winter and spring for they are usually covered with thick snow until late April or May. The lower slopes are mostly cultivated with vines and other crops to take advantage of the fertile soils derived from the decaying lava but a drive round the volcano in spring is, nevertheless, worth while. The 120 road from Taormina goes westwards through Linguaglossa to Randazzo and from there the 284 leads to Adrano where one can take a side road through Belpasso and Trecastagni to Acierale on the coast. If, instead of turning off to Adrano one continues westwards along the 120 from Randazzo there is some wild countryside around Cesarò and Troina on the lower slopes of the Nebrodi Mountains. Here, even as late as April one may came across a pale-mauve flowered form of *Anemone coronaria* and late flowering specimens of *Narcissus tazetta* and *Iris planifolia*. The last of these is a juno iris with beautiful mauve-blue flowers. It normally flowers in January and February and is fairly plentiful in the hilly areas of Sicily though it is limited in distribution to Sicily, Sardinia and the south of Spain (8.5,8.6). From Cesarò the 209 road climbs northwards over the Nebrodi mountains to the pass of Femmina Morta (13.10).

13.6 Mount Etna: In spring one can not get to the higher reaches on Mount Etna because of the snow but there are some interesting plants flowering on the way up to the tourist pavilion such as *Dactylorhiza romana* and *Anchusa arvensis*. However, the plants of the volcano are very interesting and a new flora of the region being prepared by the University of Catania is anxiously awaited. A very brief summary of the species which grow here would include:

Acinos granatensis	*Genista arisata*
Adenocarpus complicatus	*Potentilla calabra*
Alyssum minutum	*Rosa sicula*
Astragalus siculus	*Saponaria sicula*
Brassica rupestris	*Viola aetnensis*
Euphorbia melapetala	*Viola parvula*
Genista aetnensis	

These are all described and illustrated in Pignatti (1982). The most well known of them is *Genista aetnensis*, the Mount Etna broom, a charming large shrub with spreading branches carrying yellow flowers. It is a popular and accommodating garden plant which blossoms in June and July when most other shrubs are past their best. The astragalus is a prickly species similar to *A. massiliensis*. Adenocarpus is another leguminous shrub, shorter-growing and more leafy than the genista and with sulphur-yellow coloured flowers. The saponaria is a tufted species with a head of 4-8 pink flowers. *Viola aetnensis* is a small perennial pansy, also found in the Gargano (12.4) and *V. parvula* a miniscule annual to be seen as far away as the Greek mountains (17.6) and Turkey.

13.7 Monti Iblei: This fairly extensive area of limestone hills, rising to some 900m, lies in the south-east of the island. To approach it from Taormina one must drive south through Catania and, owing to the traffic congestion there, this may well take 1½ hours. At the time of writing the proposed motorway along the east coast has not been completed but when this is done it will greatly facilitate the journey. South of Catania one crosses the Piano di Catania, which is fairly intensively cultivated with a number of crops, and one then begins to climb to the small town of Lentini. South of here there are high treeless grazing areas and some cereal crops where a number of interesting species grow along the roadsides and in uncultivated areas including:

Anemone hortensis   *Hermodactylus tuberosus*
Asphodeline lutea   *Iris planifolia*
Cerinthe major   *Muscari neglectum*
Gynandriris sisyrinchium   *Romulea bulbocodium*

The anemone, referred to sometimes as *A. stellata*, is a particularly fine form with flowers of dark pink, mauve or white. The asphodeline is widespread and seems to take the place of the usual white asphodelus species in Sicily.

A special plant to be looked for here is *Iris pseudopumila*, which Pingatti (1982) says is endemic to Sicily and the Gargano peninsula on mainland Italy, but Schöenfelder (1990) notes that it also grows in western Yugoslavia. It closely resembles the dwarf *I. lutescens*, which is widespread in France and Spain but does not grow in Sicily. It differs from it by having a longer perianth tube (3-5 times the length of the ovary). It often forms large groups with flowers that are either yellow or purple. Another plant which is a speciality in Sicily and occurs in quantity around here is *Bellevalia romana* - a muscari-like plant with very pale blue or dirty white flowers. It grows by roadsides and as a weed in cereal crops where the bulbs usually escape serious disturbance as they form below the depth of ploughing. It is sometimes accompanied by the rarer *B. dubia* which has fine blue flowers and is close to *B. hackeli* that grows in the Algarve around Cape St. Vincent (4.4).

Several orchids can be seen in the region, especially *Orchis*

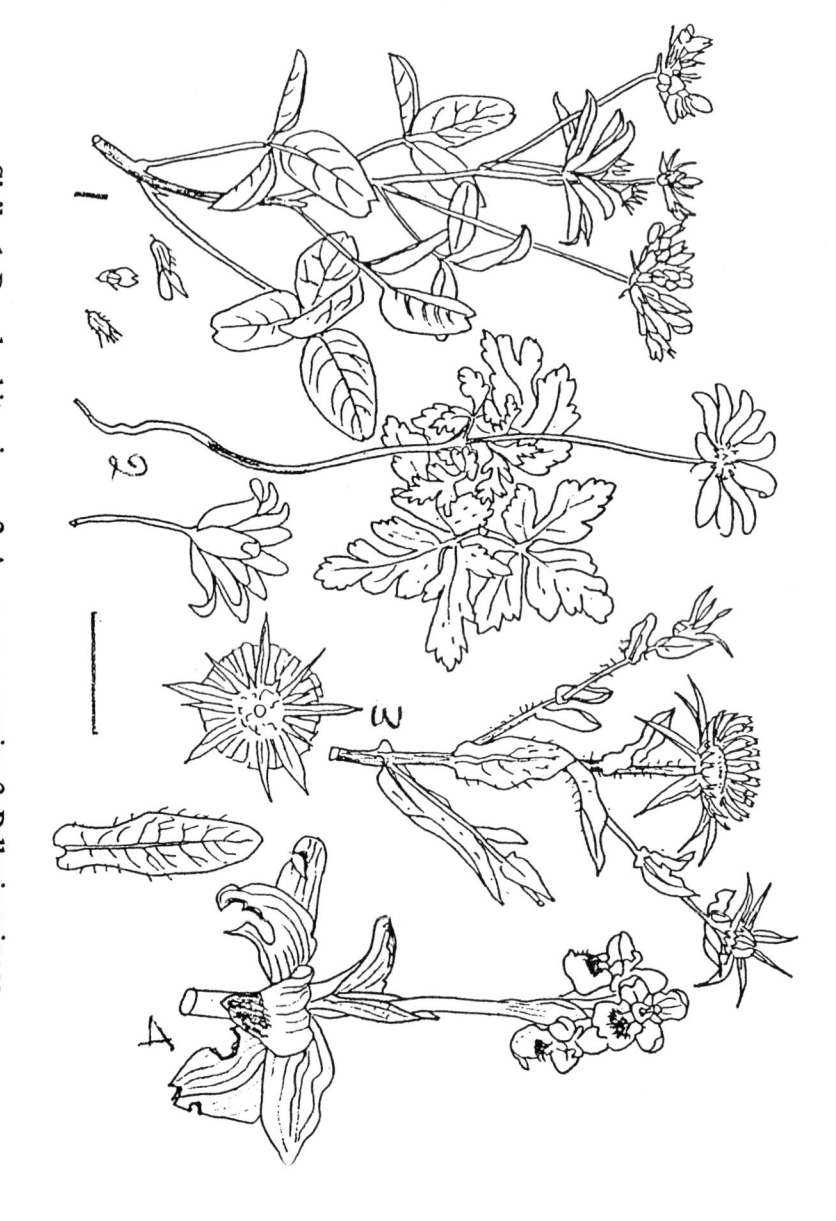

Sicily 1. Psoralea bituminosa 2. Anemone apennina 3. Pallenis spinosa 4. Ophrys tenthredinifera

lactea and *Ophrys tenthredinifera*, and along the roadsides there are occasional large groups of the giant orchid *Barlia longibracteata* flowering in March. Passing through the village of Sortino one may come across groups of *Ophrys sphegodes ssp panormitana*, which is sometimes classed as a sub-species of *O. spruneri*. Other special ophrys species worth searching for here include; *O. lunulata, O. pallida* and *O. oxyrrhynchos* (= *O. fuciflora ssp oxyrrhynchos*) but do not expect to see them in large quantities. Davis (1983) reports that the rare *Ophrys oxyrrhynchos* grows in the Iblei region near Syracuse. It has a relatively large upturned and backward bent appendage from the end of the lip.

Even the dedicated botanist will want to see some of the interesting classical sites whilst in the region, especially the Greek/Byzantine site of Palazzolo Acreide, the roman mosaics of Piazza Armerina and the city of Syracuse. Near the harbour in a fresh water spring at Syracuse grows the papyrus *Cyperus papyrus* imported and left by a homesick Egyptian princess in classical times.

13.8 **Around Cefalù:** This interesting old costal town lies roughly half way along the north coast between Palermo and Messina. It is an important tourist attraction and a good centre from which to study the flora of the Madone mountains. There is a fair amount of building to the south of the town but in isolated patches of scrub one may expect to see the following shrubby species:

Calicotome villosa *Erica arborea*
Cistus monspeliensis *Myrtus communis*
C. salvifolius *Rosmarinus officinalis*
Daphne gnidium *Teucrium fruticans*

Amongst them grow several herbaceous species, including:

Convolvulus althaeoides *Scopiurus muricatus*
Foeniculum vulgare *Tetragonolobus purpureus*
Galactites tomentosa *Urospermum dalechampii*
Psoralea bituminosa *Vicia incana*
Scabiosa atropurpurea

Most of these are widely distributed and common Mediterranean species. The tetraganolobus is the asparagus pea occasionally cultivated as an unusual connoisseurs' vegetable. It is a short-growing plant somewhat resembling bird's foot trefoil but with solitary or paired crimson flowers followed by 2-3cm four-winged pods. The scabious, frequently classed as a form of *S. maritima*, is the mournful widow or annual sweet scabious of our flower gardens. As a rule it has typical pale mauve flowers in the wild but in some specimens they are dark crimson as in selected garden forms. The vicia closely resembles the tufted vetch V. cracca - which is absent from Sicily - but has a rather denser inflorescence.

In some places one may see large drifts of the corn marigold-like plant *Coleostephus myconis* growing in olive orchards (13.4) and here they may be the uncommon form with pure white-flowered ray florets.

Two other interesting plants which grow around Cefalù are *Antirrhinum tortuosum* and *Lavatera olbia*. The first is a typical, tall-growing snapdragon (sometimes classed as a form of *A. majus*) with narrow hairless leaves and purple flowers that have a yellow or white marking. It is fairly widespread in Sicily and seems to prefer walls and rocky places. The lavatera is a robust shrubby plant to 2m with elongated inflorescences of handsome purple flowers. It is sometimes cultivated at home and gets into the British flora as having been naturalised in Epping Forest, Penzance and the Scilly Isles.

13.9 **The Madone:** Cefalù is a good centre from which to explore these mountains that are Sicily's highest after Etna. At one time they were largely covered with forests but these have mostly been felled leaving extensive areas of open rocky pastures on the higher slopes. The inhabitants of Palermo come skiing here in winter, for the tops of the peaks may be white with snow well into April.

At a short distance south of Cefalù, as the road commences to climb, one comes to the 'Santuario di Gibilmanna' and it is interesting to surmise that this name was derived from the Arabic which might be written on a map of North Africa as 'Djebel Manna' meaning the Hill of Manna. Between here and the village of Lascari there are plantations of trees of the manna ash *Fraxinus ornus* which are tapped to produce 'manna', a gum-like exudate used as a mild laxative. The Israelites of the Old Testament were not so unwise as to use this as 'food'. Although the manna ash does grow in Israel; the book of Exodos almost certainly refers to another plant, possibly a tamarix, the prickly leguminous shrub *Alhagi maurorum* or a species of lichen. Manna ash grows wild in many parts of the Mediterranean but is nowadays cultivated as a crop only in Sicily and parts of Calabria. In addition to the exudate the seeds have been exported to Egypt where they are prized for culinary and medical purposes.

Around here and the road to Collesano there are substantial thickets of *Rhus coriaria* - a suckering shrub similar to the North American species *R. typhina* frequently grown in our gardens for its brilliant autumn colouring. In places it is accompanied by *Robinia pseudoacacia* that has become naturalised in many parts of the Mediterranean. On the banks one can expect to see the stately spikes of *Acanthus mollis* and plants of *Gladiolus italicus*. Other plants to be seen by the roadside and sometimes in cultivated fields include:

Allium nigrum	*Cynara cardunculus*
Bellevalia romana	*Fedia cornucopiae*
Borago officinalis	*Galactites tomentosus*
Convolvulus tricolor	

The allium is a stocky species with leaves somewhat like a tulip and hemispherical heads of dusky mauve flowers. The bellevalia is especially common in Sicily and the convolvulus is the showy annual species often offered in seedsmans catalogues. The cynara is the very prickly wild cardoon.

Clumps of *Ampelodesmos mauritanicus*, a tall coarse grass also found on Majorca (8.8), can be seen in places and amongst them many plants of *Serapias vomeracea* accompanied by the pink *Centaurium erythraea* and occasionally *Antirrhinum tortuosum*, mentioned previously. Another plant to look out for is *Lathyrus articulatus* - an everlasting pea with medium-sized mauve flowers that have a pale coloured keel. The stems are winged and curiously 'articulated', as indicated by its Latin name.

The road leaving Collesano eastwards to the village of Munciarrati and on to Isnello and Gratteri is a profitable area for botanizing. There are patches of deciduous woodland and old cork oak plantations. Several orchid species can be seen here:

> *Ophrys fusca*   *Orchis italica*
> *Op. lutea minor*   *O. longicornu*
> *Op. incubacea*   *O. papilionacea*
> *Op. tenthredinifera*

The most outstanding of these is *Orchis longicornu* which grows in profusion round here and could be said to be the 'orchid of the region'. It is not a widespread species and is limited to the western end of the Mediterranean. It somewhat resembles a rather robust form of *O. morio* but is quite distinct when one sees live plants. The flower is usually dark mauve with a white stripe down the lip bearing a few dark spots and could be said to remind one of a gentleman in a dinner jacket.

Amongst other plants that are worth looking for here, along the roadsides and in clearings, is the Sicilian endemic *Symphytum gussonei* - a comfrey with pale yellow flowers. Two other species confined to southern Italy and found here are *Melittis albida*, a bastard balm similar to the common *M. melissophyllum* (not recorded from Sicily) and *Biscutella lyrata* - a rather uninteresting buckler mustard. The uncommon, and attractive, *Centaurea napifolia* with partly winged stems and reddish flower heads grows together with another composite that has branched stems of striking pink capitulae and may be a form of *Crepis praemorsa* or *C. froelichiana*.

The old cork oak plantations and deciduous woodland around here hold several botanical treasures:

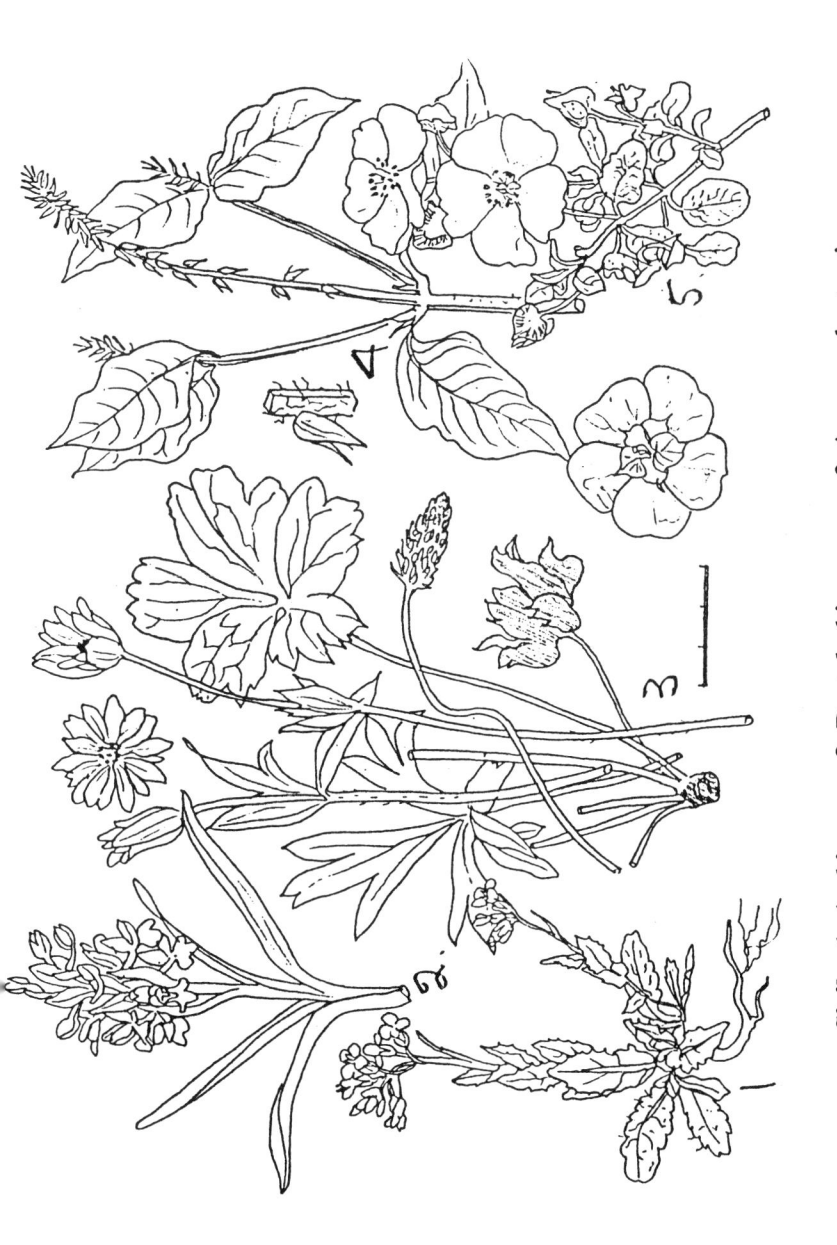

Sicily 1. Arabis rosea 2. Dactylorhiza romana 3. Anemone hortensis 4. Achyranthes aspera 5. Cistus salvifolius

Allium triquetrum
Anemone apennina
A. hortensis
Cyclamen repandum

Doronicum orientale
Lamium flexuosum
Orchis provincialis
Paeonia mascula

The lamium is an uncommon species of dead nettle with white flowers that have a dark-pink and hairy 'hood'. The paeony may well be the rather rare sub-species *russii* that is more hairy than the type and here it sometimes has white flowers.

A good road to take for crossing the higher mountains is that between Munciarrati (near Collesano) and Polizzi Generosa. This leads one between Monti dei Cervi 1795m, M. Carbonara 1979m and M. San. Salvatore 1910m and the views can be impressive. Approaching from the north one can get refreshment at Piano Zucchi and the landscape in March sometimes has a Swiss Alp-like appearance, with snow capped mountains and grazing cattle carrying large bells. Higher up one comes to the Piano di Bataglia at 1600m with near alpine conditions and a number of special plants grow in the stony landscape:

Anthemis montana
Arabis alpina
Astragalus nebrodensis
Bellis margaritaefolia
Cerastium tomentosum
Colchicum bivonae

Euphorbia rigida
Erysimum bonnanianum
Iris pseudopumila
Linum punctatum
Onosma echioides
Valeriana tuberosa

The astragalus is a typical, very-spiny plant with white flowers that is endemic to the Madone and Nebrodi but difficult to distinguish from some other similar species. The bellis resembles a common daisy with rather large flower heads and the underside of the ray florets tinted purple. It is limited to Sicily and the southern tip of Italy, where it is called 'pratolina calabrese', the Calabrian daisy. The erysimum is also an endemic - a neat short wallflower with yellow flowers. The iris (13.7) grows in large patches here and produces a beautiful effect. Most of the flowers are either mauve or yellow but a few interesting bicolour forms occur. The colchicum flowers in autumn and closely resembles *C. autumnale* but has narrower leaves. The linum is yet another endemic and a very beautiful sub-species. In many respects it resembles a typical blue flax with small leaves, somewhat like *L. alpinum* but the stems lie flat on the ground and in some specimens radiate from the rootstock like spokes of a bicycle wheel. The 25mm flowers, are of azure blue. It is found isolated in Sicily and far away on Mount Chelmos and the Pindhos mountains of northern Greece (16.9). *Onosma echioides* is typical of the genus and has very pale yellow flowers.

It is possible to walk to some of the summits from here and botanizing could be very interesting. Another endemic to look for on the higher slopes is *Allium nebrodense* - a small garlic with yellow flowers in July and August that resembles *A. flavum*.

Amongst scrub lower down one can find *Hermodactylis tuberosus* and the white form of the common primrose. Eventually one arrives at the small town of Polizzi Generosa where there are hazel orchards. Not far from here is the Vallone della Madonna degli Angeli where the Sicilian endemic *Abies nebrodensis* has its last foothold. It is somewhat similar to *A. alba* (= *A. pectinata*) that grows predominantly in the Alps but it is a smaller tree with shorter needles and well adapted to drought conditions. Unfortunately the original specimens have been decimated by fire and at present there are only some 23 specimens left in the wild though the species presumably covered large areas around here in the past.

One can return to Collesano by the lower road through Sicillato. Along the wayside here are fine bushes of *Euphorbia dendroides* and clumps of *E. characias* amongst which grow patches of the showy annual *Convolvulus tricolor* and the rather less striking *Convolvulus cantabrica* that has pale pink flowers and, to add to the genus, one may also find *C. althaeoides*. On one occasion I found plants of *Tanacetum cinerariifolium* like an ox-eye daisy with finely divided leaves. This is a relict of cultivation when it was grown as a crop to produce the insecticide pyrethrum. Near here there are extensive fields of *Hedysarum coronarium* grown for fodder in much the same way as alfalfa; indeed, is sometimes referred to as Italian sainfoin. It has brilliant carmine flowers and as the field boundaries often carry many plants of the yellow-flowered *Asphodeline lutea*, the combination gives a very colourful effect.

13.10 The Nebrodi Mountains: These stretch eastwards from the Madone and in some parts retain their ancient covering of dense deciduous woods of oak and beech. They must be some of the most southerly of all beech forests. A good way to get an idea of the typical vegetation is to cross from Santa Agata di Militello on the north coast along the 117 road to Cesarò. Climbing out of S. Agata the oak woods by the roadside are carpeted in spring with *Cyclamen repandum*, the lesser celandine *Ranunculus ficaria* and *Cyclamen hederifolium* in the autumn. As the road rises the woods are almost entirely of beech with a few trees of holly and this combination continues to the top of the pass of Femmina Morta 1524m. Under the trees and by the wayside grow:

Anemone apennina	*Orchis lactea*
Cyclamen repandum	*Primula vulgaris*
Dactylorhiza romana	*Ranunculus ficaria*
Daphne laureola	*R. millefoliatus*
Doronicum orientale	

A small road leads eastwards from the pass towards Monte Soro 1847m. Here are scattered, stunted beech trees in places and some charming alpine meadows where, in spring, carpets of *Romulea bulbocodium* asociate with the yellow flowers of *Ranunculus millefoliatus, R. ficaria,* mauve violets and thousands of the tiny blue stars of *Scilla bifolia*. In some places substantial groups of the white form of the common primrose mingle in the grass with *Polygala preslii* - a very local endemic milkwort with rather large lilac coloured flowers. If the weather is fine one may be rewarded with impressive views of Mount Etna from here.

A crocus growing in this area is probably the uncommon autumn-flowering species *C. longiflorus*. One should also look for the endemic *Fritillaria messanensis* and snowdrops which are

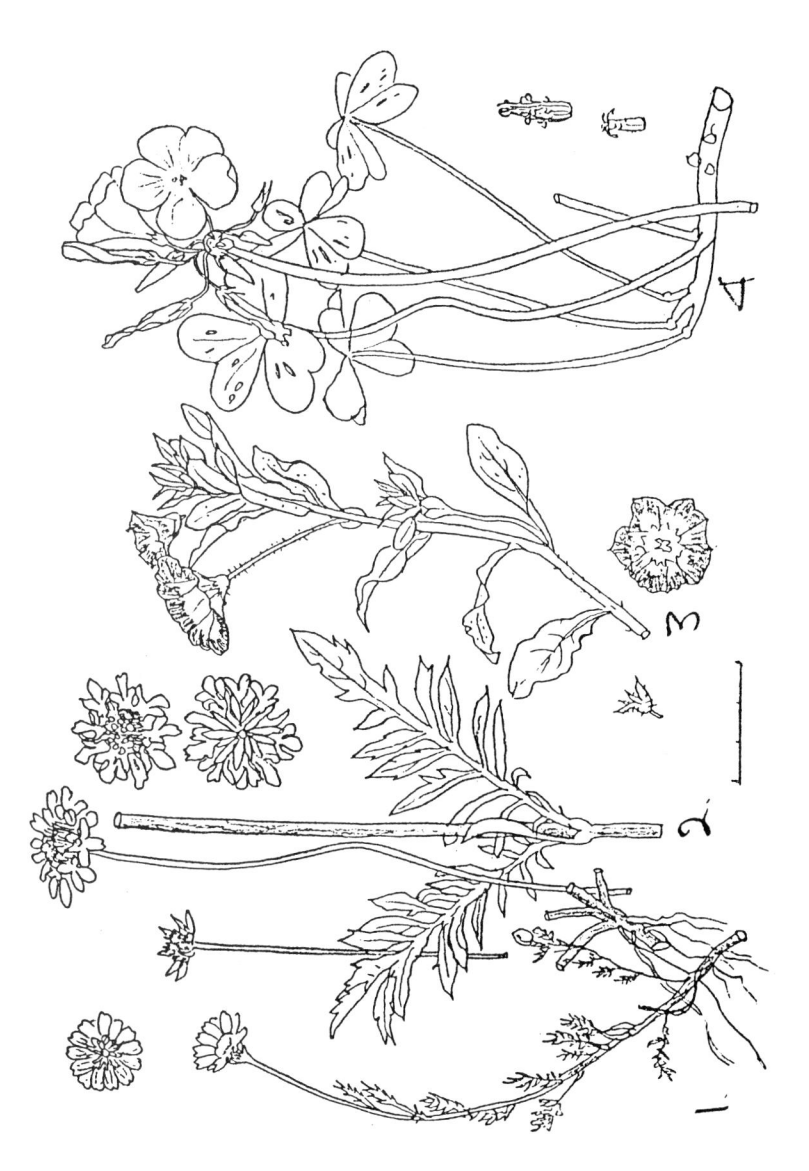

Sicily 1. Anthemis montana montana 2. Scabiosa maritima
3. Convolvulus tricolor 4. Oxalis pes-caprae

variable here - a form resembling *Galanthus reginae-olgae* has been recorded from the district.

13.11 Cefalù Southwards: To get to the classic sites in the south from Cefalù one can either travel by the 121 road from Palermo or take the A19 motorway to Enna. The more one uses the motorway, the less frustrating is the driving and the shorter the time taken to cover a distance. Much of the centre of the island is cultivated and of lesser interest for botanizing but an occasional stop on the way to Agrigento or Piazza Armerina may be rewarding. In some places one can find *Lavatera trimestris* growing by the roadside. This beautiful annual mallow, with large shell-pink flowers, has become popular in our gardens and even as a cut flower in florists. In the wild it is nearly as good as the selected varieties such as 'Silver Cup' offered in seedsmans' catalogues.

In a wood near to Piazza Armerina, with its impressive Roman mosaics, grows *Tulipa sylvestris*. Under the shade of trees the plants are quite tall with pointed yellow tepals bent flat or backwards like those of a cyclamen. Their fragrance is superb. The ground is carpeted underneath with *Anemone hortensis* and *Geranium lucidum*. The tulip is classed as a British native, found sparingly in the south of England. It can easily be grown in the home gardens but tends to be shy-flowering.

Approaching the town of Agrigento, groups of the Spanish thistle *Scolymus hispanicus* grow by the roadside and amongst the temples a few plants of a much finer species of the genus, *Scolymus grandiflorus*, with, as the name suggests, larger flower heads and leaves with 'milky' veination. Other interesting plants amongst the ruins include *Antirrhinum tortuosum* and the European palm *Chamaerops humilis*.

13.12 Piana de Albanese: The countryside around this village, some 10km to the south of Palermo and close to a small lake, is of considerable botanical interest. One may approach it from the village of Marineo which lies some 15km to the east and not far from the main 121 road that leads from Palermo to Agrigento on the south coast. In places the roadsides are thickly covered with *Lathyrus odoratus*. This is the wild sweet pea from which all our cultivated varieties have been derived. Although there is some uncertainty about its origin, it is thought to be endemic to Sicily and the extreme south of Italy but has become naturalised in other parts, especially around Lake Como. The wild form has flowers that are nearly as large as garden varieties but the colour is less variable and generally mauve with a darker, and redder, standard. Wild plants always have a very strong perfume. Other plants that may be seen in these limestone hills include the very variable *Dianthus sylvestris* and the curious little legume *Coronilla scorpioides*.

The district is rich in orchids with many serapias, some fine specimens of *Ophrys bertolonii* and the common bee orchid *Ophrys apifera*. Other species found here include *Ophrys spruneri ssp panormitana* (13.7) and the surprisingly similar *Ophrys sphegodes ssp sicula*. *Orchis longicornu* also grows here but the main specialities to be searched for are: *Ophrys lunulata, Op. pallida* (which resembles a small flowered *Op. fusca* with a recurved lip) and *Op. fuciflora ssp. oxyrrhynchos*. There is a variety of the last of these that has a yellow lip and is called *lacaitae* but like the rest of these it is not common and has to be diligently sought after; I have never seen a distinct form of it but only specimens with an entirely brown lip that turns a dull straw-yellow with age. Clearly there is a great deal of variation and many treasures to be sought by the avid photographer.

Past the small lake of Scanzano and heading towards Piana de Albanese one passes over hilly country where the conical plants of *Echium italicum* are a prominent feature. A species to be noted here is the uncommon *Vicia sicula*, resembling an orobus rather than a vicia. It has an upright growth habit and no tendrils. The leaves consist of 2-6 narrow, pointed leaflets and the dull, brown-purple elongated flowers are carried in hanging clusters up to about twenty in a group. Other plants that may be seen here include:

Calendula suffruticosa   *Pallenis spinosa*
Carduncellus pinnatus   *Sedum coeruleum*

The carduncellus is a rather rare, neat, dwarf thistle also found in North Africa. It has a rosette of prickly basal leaves with large, stemless, purple flower heads and could be confused with *Atractylis gummifera*. The sedum is a tiny annual with reddish foliage and pale blue or mauve flowers; it is sometimes offered in seed catalogues and makes an attractive small pot plant.

In rocky areas with scattered pine trees lower down one may find the uncommon *Orchis quadripunctata var branchifortii* that is limited to Sicily and Sardinia. Anybody who has seen the type species will recognise it without difficulty, for it has a very much reduced lip. It is afforded specific status as *Orchis branchifortii* by Buttler (1991). *Iris planifolia* (13.5) grows in abundance here together with *Sedum coeruleum* on the rocks and one should look for *Cymbalaria pubescens* - like a weak-growing, hairy form of our ivy leaved toadflax but an endemic limited to this small area of Sicily.

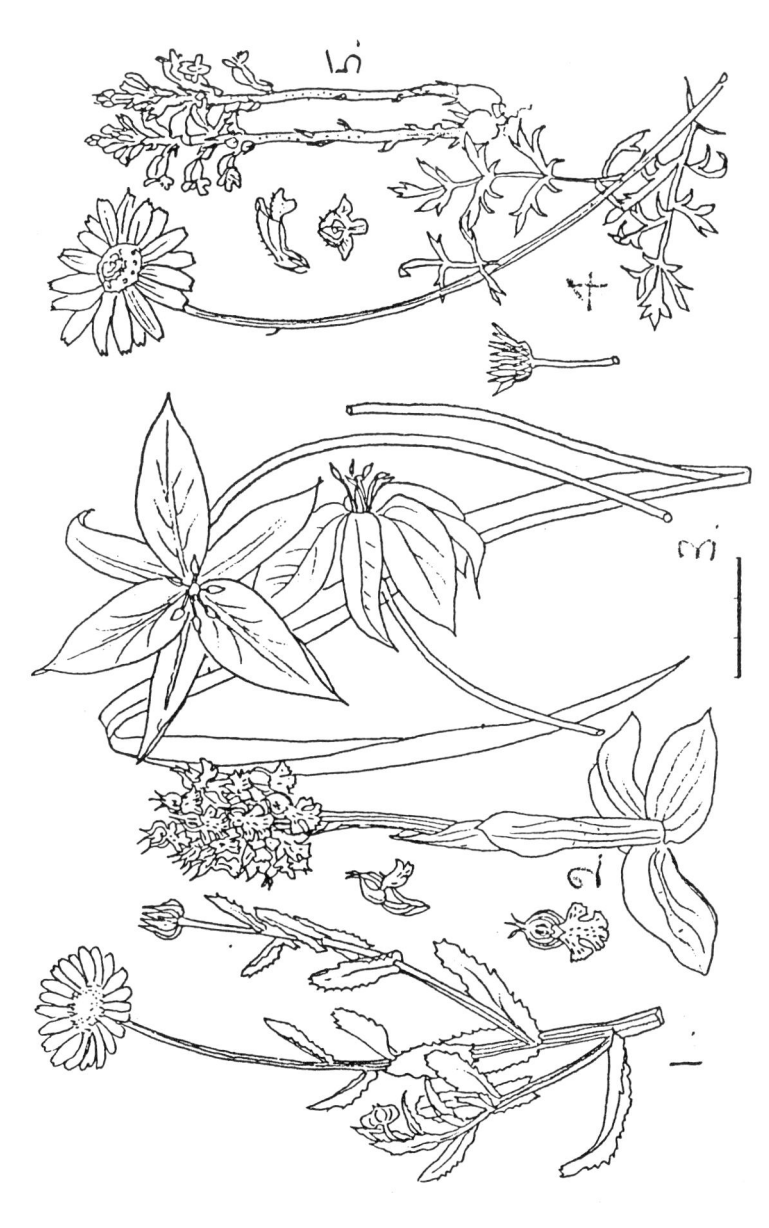

Sicily 1. Coleostephus myconis 2. Orchis lactea 3. Tulipa sylvestris 4. Tanacetum cinerarifolium 5. Orobanche ramosa

Sicily 1. Convolvulus cantabrica 2. Tragopogon porrifolius 3. Linaria reflexa var. castelli 4. Antirrhinum tortuosum

14. THE MALTESE ISLANDS

14.1 This group of small islands lies about 90km south of Sicily and 300km east of Tunisia. Malta, some 27km long by 14km wide, is the largest and most populous; Gozo is about one third of its area. The other two islands, Comino and Cominetta are much smaller and virtually uninhabited.

Malta is a popular holiday resort for British visitors. It is steeped in history and has a wealth of fine buildings so most visitors go there for the sunshine and the history. On first appraisal, the native flora does not seem to offer much to the plant enthusiast but there are a few endemic plants, notably the Maltese rock-centuary, previously called the Maltese knapweed *Palaeocyanus crassifolius* (O = *Centaurea crassifolia*) and *Cremnophyton lanfrancoi*, which is a modest halophyte related to atriplex. Both of these are monotypic genera - ie. genera with only one species. Another special plant of the region is the rare conifer *Tetraclinis articulata*. In addition, the islands are rocky, windy and almost treeless, and the very severity of the terrain for plant life makes the area interesting to the enthusiast. Most species are to be seen in flower during early spring and others in the autumn. During the summer much of the wild vegetation is dried up due to the lack of rainfall from May to August and to the shallow depth of soil.

The Maltese have a unique language which is structurally Arabic but with about fifty percent Latin language words, especially Italian, and it is written in Roman lettering. However, nearly all the inhabitants also speak fluent English and notices and shop signs are almost invariably in English. The visitor from Britain will immediately feel at home. Getting around the island also presents no difficulty. Distances are small and car hire is relatively cheap. Traffic drives on the left-hand side of the road but much of the area is 'built up' and signposting can be a confusing or non-existent hazard for the visitor. It is not difficult to cover the island cheaply by using the buses, with a little foot-slogging and a large-scale (1:38,000) map.

14.2 The terrain: It is impossible to describe the terrain of the Maltese islands without reference to the geology. The basic structure is a more or less horizontal stratum of hard coralline limestone on top of which is a layer of much softer limestone called globigerina with a few regions of another hard limestone on top of this and some sticky blue clay in the sandwich. The globigerina is a uniform and rather soft material that is easily worked but hardens somewhat on exposure. It is abundant, easily quarried and is used to produce most of the magnificent buildings of the island. The much harder coralline limestone is rarely employed for building and where it is exposed it forms rather barren pitted surfaces that defy cultivation.

Malta, east of a line roughly from Bugibba southwards through Rabat to the coast, has a surface mainly of globigerina whilst west of this line is largely coralline limestone. Gozo is mainly of the soft limestome but there is an area of the harder rock in the north east. The two remaining islands are composed solely of hard limestone. Much of the area of soft limestone is intensively cultivated and terraced, though the soil is nowhere very deep and the main plants of botanical interest here are found around the coasts or inland as arable weeds. The hard, pitted surfaces of the coralline limestone carry an interesting flora of species able to withstand drought and occasional periods of waterlogging. They also have to be able to thrive under windy conditions and to be tolerant to salt spray, for no part of the island is far from the sea and strong winds are frequent.

In the south west of Malta there are the Dingli sea cliffs and here is the higest point of the island at 251m - the land slopes away to near coastal level in the north east. The western section of Malta, composed of coralline limestone, has a number of geological faults which have given rise to a series of parallel ridges running roughly north-east/south-west. From east to west they are; Victoria Lines, Wadija Ridge, Bajda Ridge, Mellieha Ridge and the Marfa Ridge. These are of special interest botanically; the valleys between them are mostly cultivated.

14.3 Literature: By far the most useful book is 'A Flora of the Maltese Islands' by Haslam and others (1977). It is probably more academic than the casual visitor would expect to buy, but it is not expensive and at present is readily available in the larger towns such as Valetta and Sliema and from a few specialist bookshops in Britain. It has some line illustrations and the bonus of a rather full glossary of botanical terms. However, it is not fully updated and may soon be out of print. Plans are developing for a new and improved flora of the island.

The flora of Malta is also, surprisingly, included in Pignatti's excellent three volume 'Flora d'Italia' (1982) which is, of course, in Italian. The 'Red Data Book for the Maltese Islands' (Schembri P. & Sultana J. 1989) gives up-to-date information on the flora and wildlife of the islands and has some useful illustrations of the rare and endemic plants.

14.4 Around Valetta: All visitors will want to see the splendors of this city and its surroundings and there are a few plants to be discovered in the vicinity. The rocky coastline is home for the late summer flowering *Inula crithmoides* and some magnificent specimens of *Lavatera arborea* which can grow to 3m tall, with impressive trunks and producing their lilac-purple flowers in March to June. Walls of old buildings are often adorned with plants of the caper *Capparis spinosa* from which the flower buds are collected to produce 'pickled capers'. The beautiful white or pink flowers open during the heat of summer. Plants of this species in Malta are the variety *inermis*, that has no thorns, suggesting that they were originally introduced by man - though the species commonly grows wild in most of the Mediterranean region. Another shrub that inhabits walls is *Nicotiana glauca* with tubular yellow flowers. It is a species introduced from South America that is widespread throughout the Mediterranean. The old walls also carry antirrhinum plants which may be either *A. tortuosum* or *A. siculum*. The former has mauve-red flowers whereas the other, with smaller pale-yellow flowers, is much less common but hybrids that have features of both species can be found. Both species also grow in Sicily and the *A. tortuosum* there is usually much taller and more robust (13.4).

Between Sliema and Valetta lies Manoel Island, which looks very green from a distance. The shrubs which give this verdant appearance are *Acacia cyanophylla*, which have been planted and the ground below them is carpeted with *Oxalis pes-caprae*. The small monk's cowl *Arisarum vulgare* is one of the few native species which competes succesfully here with this ubiquitous weed from South Africa. Haslam et al.(1977) quote that the winter-flowering *Allium chamaemoly* can be found on this site but it has now probably succumbed to the oxalis. Waste places within the city carry a number of wild plants such as:

Beta maritima	*Mercurialis annua*
Calendula arvensis	*Oxalis pes-caprae*
Conyza bonariensis	*Pancratium maritimum*
Ecballium elaterium	*Reichardia picroides*
Erodium cicutarium	*Sonchus oleraceus*
Hyoscyamus albus	

The conyza is an insignificant member of the compositae (Asteraceae)- a widespread weed from tropical America. The pancratium is not common on Malta and may be an escape from cultivation.

14.5 The Dingli Cliffs: A convenient route from Sliema or Valetta takes one the 15km to the Dingli Cliffs past the fine church at Mosta and the interesting old capital Mdina near Rabat. A road runs paralell to the coast along the top of the cliffs where there is typical coralline plateau flora. Shrubs here are much more scattered than one expects to see in the average garigue, they include:

Erica multiflora	*Rhamnus oleoides*
Euphorbia dendroides	*Teucrium fruticans*
Phlomis fruticosus	*Thymus capitatus*

The erica is an interesting and showy winter-flowering species with terminal clusters of mauve-pink flowers; unfortunately it

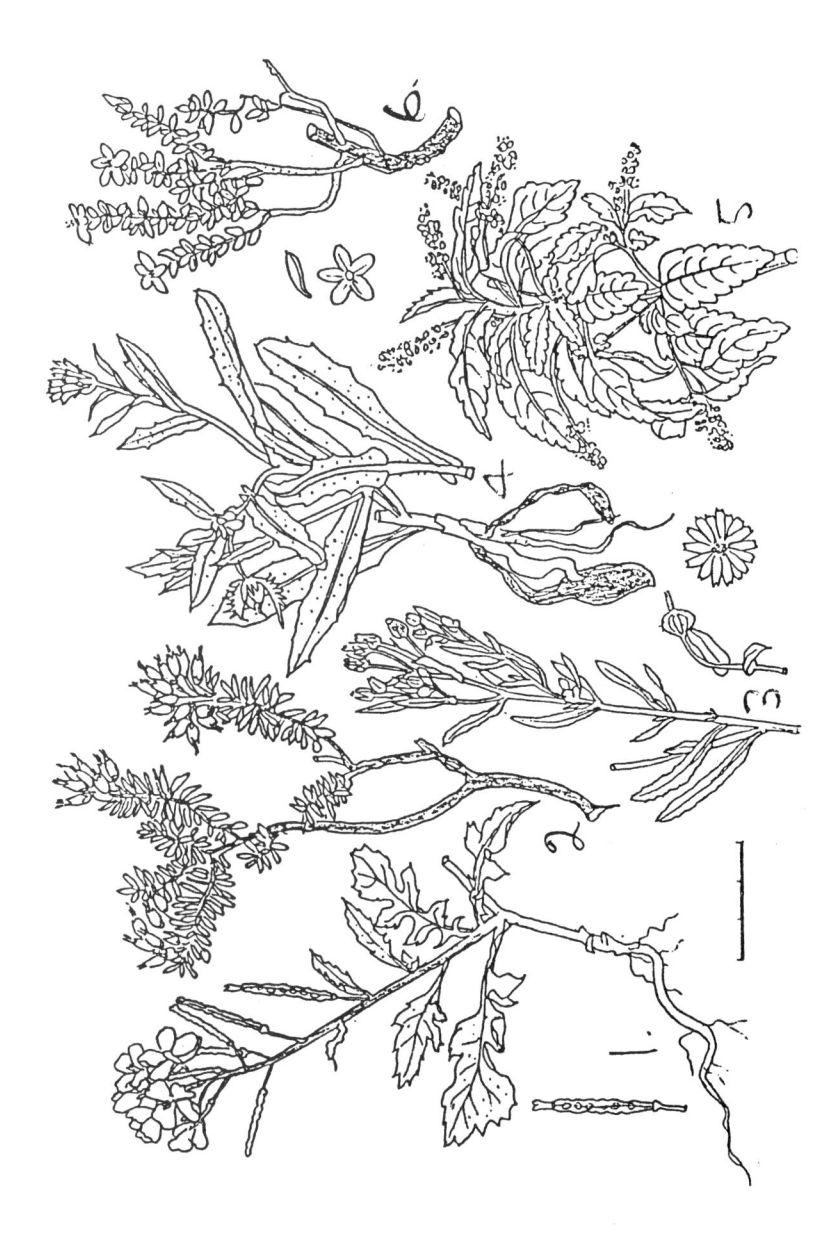

Malta 1. Diplotaxis erucoides 2. Erica multiflora 3. Conyza bonariensis 4. Calendula arvensis 5. Mercurialis annua 6. Hypericum aegypticum

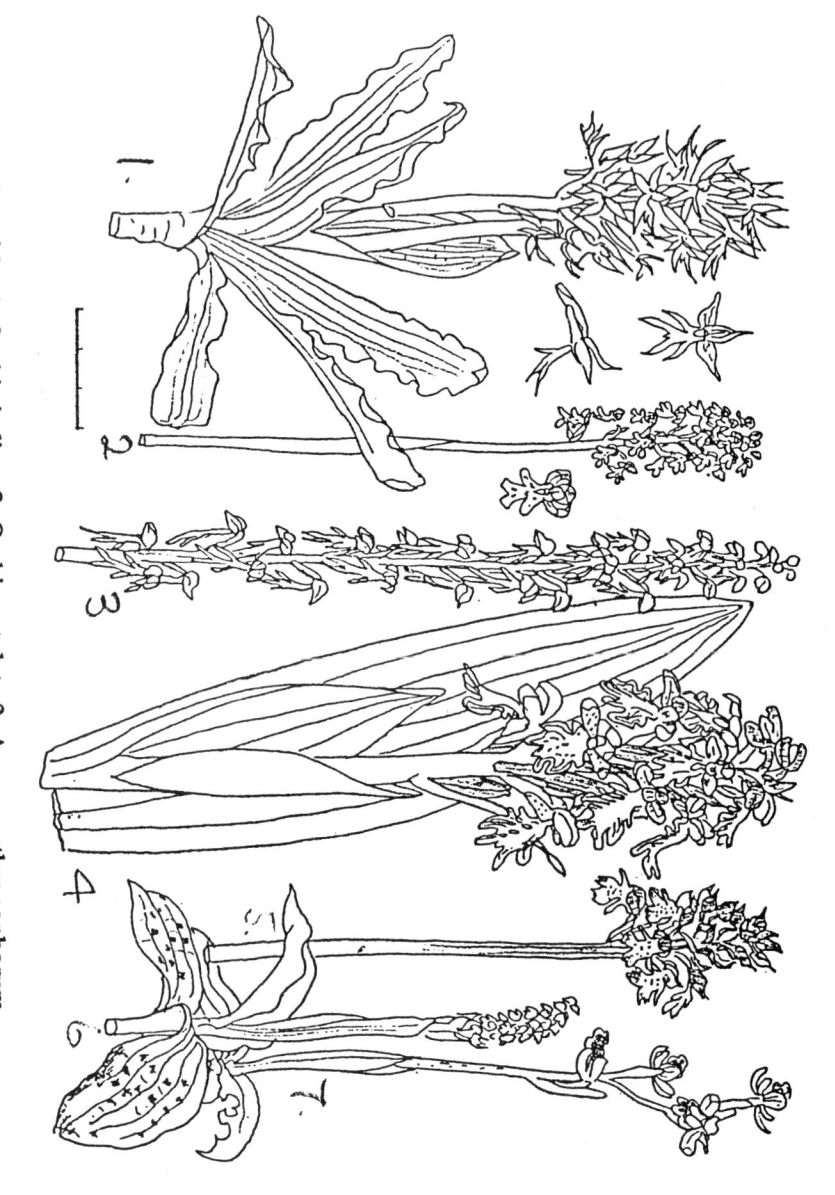

Orchids 1. Orchis italica 2. Orchis ustulata 3. Aceras anthropophorum 4. Barlia robertiana 5. Orchis tridentata 6. Neotinea maculata 7. Ophrys fusca

is not sufficiently hardy for cultivation in most parts of Britain. Amongst the shrubs grow a number of herbaceous species including:

Anthyllis vulneraria   *Plantago coronopus*
Asparagus aphyllus   *Psoralea bituminosa*
Cynara cardunculus   *Ranunculus bullatus*
Euphorbia melitensis   *Ruta chalepensis*
Ferula communis   *Salvia verbenaca*
Fumana thymifolia   *Sedum sediforme*

The euphorbia, a small-growing shrub with blue-green leaves and the older branches persisting as spines, is included under *E. spinosa* by Pignatti (1982). It is rather similar to *E. acanthothamnos* which is common in the eastern Mediterranean but not found in the west. *Sedum sediforme* has pointed fleshy leaves 1-2cm long and heads of greenish-white flowers, it is widespread throughout the Mediterranean. The rest of the above are also widespread.

Several bulbous and tuberous rooted species grow here, notably *Arisarum vulgare, Asparagus aphyllus, Asphodelus aestivus, Narcissus serotinus* and *Urginea maritima*. The last two of these flower in the autumn and the urginea here is sometimes segregated as *U. pancration*. One may also be lucky, in October to December, to see the rare *Crocus longiflorus* that has mauve, yellow-throated flowers with a scent said to resemble that of primroses. Its only other sites are in Sicily (13.10) and the extreme south of Italy. Another exciting plant found here is the endemic Maltese rock-centuary *Palaeocyanus crassifolius* which has somewhat fleshy leaves that are nearly all basal. The more or less unbranched 50cm flowering stems carry heads of all tubular flowers that are mauve or very occasionally white. Taxonomically this species is extremely interesting for it is close to serratula and centaurea and may thus represent a primitive form from which these two genera have evolved. Unfortunately it usually grows high up and out of reach on the cliffs but can occasionally be seen cultivated in Maltese gardens. In 1971 it was declared Malta's national plant.

Several orchids also occur here, especially *Anacamptis pyramidalis, Ophrys bombyliflora, Orchis saccata, Spiranthes spiralis.* The anacamptis should be examined carefully, for a similar but early-flowering species with pale or white flowers is found in Malta and sometimes referred to as *A. urvilleana* though Buttler (1991) seems to include it as a synonym for *A. pyramidalis.* It is, however, distinct and a diploid whereas *A. pyramidalis* is tetraploid. On the clifs and in the so-called 'rdum' area between the layers of different limestone one may also see the heath-like small shrub *Hypericum aegypticum* and *Crucianella rupestris.*

14.6 The Western Coralline Plateaux: The coast road westwards from Sliema and Valetta takes one past Salina Bay to St. Paul's Bay from where one has easy access to the main coralline ridges. The plant associations vary slightly from place to place but one may reasonably expect to see the following species of small shrubs:

Cichorium spinosum	*Phagnalon rupestre*
Erica multiflora	*Senecio bicolor*
Lavatera arborea	*Teucrium flavum*
Lonicera implexa	*T. fruticans*
Micromeria microphylla	*Thymus capitatus*

The senecio (= *S. cineraria*) is found also in parts of southern Italy and the south of France (10.14). Its leaves are intensely greyish-white, especially on the undersurface, and cultivated forms are much in demand for plant bedding schemes. They are sometimes referred to in seed catalogues as 'cineraria' foliage varieties or occasionally as 'dusty miller' and have become naturalised in some of the warmer parts of Britain. *Teucrium flavum* is not a very well known species; it has yellow flowers flushed with mauve but occurs in a range of different forms. Some other small shrubs which are rare in Malta but common in other places throughout the Mediterranean include: *Cistus incanus, C. monspeliensis, Myrtus communis, Rosmarinus officinalis.* A large number of herbaceous species and annuals may be found growing between the shrubs, including:

Asperula aristata *Lobularia maritima*
Anthyllis tetraphylla *Notobasis syriaca*
Atractylis gummifer *Reichardia picroides*
Calendula arvensis *Ruta chalepensis*
Carlina lanata *Sanguisorba minor*
Carthamus lanatus *Scabiosa atropurpurea*
Chiliadenus bocconei *Silene colorata*
Foeniculum vulgare

The chiliadenus (O = *Jasonia glutinosa*) is a Maltese endemic. It is a glandular small, well-branched, shrubby composite without ray florets and small yellow disc florets - a rather uncommon but insignificant-looking plant. Amongst these one may expect to see several bulbous species:

Allium roseum *Leopoldia comosum*
A. subhirsutum *Romulea columnae*
Colchicum cupani *Ornithogalum narbonense*
Gynandriris sisyrinchium *Scilla autumnalis*

The romulea flowers in February to March and is variable and may cover some four sub-species. The scilla and colchicum bloom from September to December. Some 23 species of orchid are recorded from Malta some of them are rare but one may reasonably expect to see:

Anacamptis pyramidalis *Orchis lactea*
Ophrys bombyliflora *O. saccata*
O. fusca *Serapias parviflora*
O. coriophora

The anacamptis may be the species *A. urvilleana* (14.5). *Orchis lactea* quoted here could well be the recently recognised

O. conica. *Ophrys pallida*, which resembles a small form of *O. fusca* in which the end of the lip is bent inwards towards the stem has been recorded from Malta but not seen recently and the record may well have been a mistaken identity though it is to be found not far away in North Africa and Sicily. The much more impressive *Ophrys lunulata* is also recorded from Malta, though it also is said to be rare here and essentially a species from Sicily and Sardinia. Again, its reporting may be a case of mistaken identity for it resembles some forms of *Ophrys sphegodes*.

Some of the hollows in the coralline limestone become filled with water to form pools in winter and dry out entirely during the summer. They frequently provide a home for *Crassula vaillantii, Damasonium bourgaei, Elatine hydropiper v. gussonei*. These are small or minute flowering plants. The first is very small and with tiny white flowers and, although it is a crassula, the leaves are barely succulent. The damasonium is monocotyledenous and has white, three-petalled flowers which are followed by star-shaped fruits. The elatine is the 'waterwort', a tiny plant with pink or white flowers and opposite leaves; the type form grows in Britain and as far north as Scandinavia. The damasonium and elatine described here are more or less confined to the smaller islands of the western Mediterranean.

14.7 **Sandy coastal areas:** These are not common in the Maltese Islands but there is some sand with dunes near Mellieha Bay situated at the extreme north west of Malta and at Ramla Bay in the north of Gozo. At Ghadira there is a nature reserve surrounded by a high wire fence and acacia bushes. It has a small lake, interesting for its water fowl and access is possible on weekends by arrangement. In these sandy areas grow several typical shore plants such as:

Cakile maritima	*Medicago marina*
Eryngium maritimum	*Ononis natrix*
Euphorbia parialis	*Pancratium maritimum*
E. terracina	*Pseudorlaya pumila*
Glaucium flavum	*Scolymus hispanicus*

Most of these are widespread though *Euphorbia terracina* is not a common species. It closely resembles the well-known *E. parialis* but the leaves are slightly narrower and their margins have minute serrations that can be seen with a hand lens. The pseudorlaya is a very small plant allied to the carrots (*Daucus sp.*); it has interesting spiny fruits. The scolymus is an attractive prickly member of the genus with capitulae of yellow flowers.

Other species commonly seen near the shore include *Beta maritima, Cichorium spinosum, Limonium zeraphae, L. melitensis* and *Lygeum spartum*. The last of these is the esparto grass or albardine, a rush-like grass that has a boat-shaped sheath enclosing the flowers. It is a common feature of Andalucia (7.4) where it is used for thatching and other purposes and at one time it was imported to Scotland for the production of a high quality paper.

14.8 Cultivated areas: About two thirds of the eastern part of Malta are cultivated with extensive terracing, some of which probably dates back to prehistoric times. Practically the whole of Gozo is used for crop production and it has a reputation for looking greener than Malta. In the spring the annual, arable weeds are very colourful, especially *Brassica rapa, Chrysanthemum coronarium, Diplotaxis erucoides, Oxalis pes-caprae, Silene colorata*. The diplotaxis is a variable species, sometimes grown as a vegetable, and the form here has rather large pale mauve or white flowers that remind one of 'Lady's smock' (*Cardamine pratensis*). The same species in Tunisia (25.4) has much smaller flowers. Many other annual weeds may be found in smaller quantities and some less common herbaceous perennials such as *Delphinium halteratum* and *Amaranthus retroflexus* - a member of the love-lies-bleeding family which hails from North America. These perennials usually flower later in the summer. Bulbous and tuberous-rooted species which occur in fields and headlands include:

Allium nigrum	*Narcissus tazetta*
Asphodelus aestivus	*Ornithogalum arabicum*
Leopoldia comosa	*O. narbonense*

The narcissus is frequently planted in gardens and at the neolithic sites throughout the islands but it seems to be the diploid form and may well be a genuine wild element of the flora. *Ornithogalum arabicum* is a showy and uncommon species with large, scented, white flowers that have a conspicuous shiny black ovary; it is sometimes cultivated for ornament in Britain but is not fully hardy in colder areas.

14.9 **Valleys:** These are generally marked on the map as 'wied' and not all of them are accessible. They usually afford shelter for the few naturally-growing trees on the islands such as *Ceratonia siliqua* and *Quercus ilex*, sometimes with the following lower-growing shrubby species:

Anagyris foetida   *Rhamnus oleoides*
Crataegus azarolus   *Vitex agnus-castus*
Pistacia lentiscus

A rare plant to be found in this environment is *Tetraclinis articulata*, a strange conifer allied to *Thuja* and the only species of the genus. Under favourable conditions it can grow to a tree but is usually seen as a bush with weeping, tamarix-like branches. At one time it was common throughout the island and its Maltese name 'ghargar' occurs in various forms in place names throughout Malta. Unfortunately it is extremely rare nowadays but a remarkable find of a wild grove is recently recorded, showing what interesting plants may still be discovered in one of the most populated areas of the Mediterranean. The species is sometimes cultivated for ornament in the islands and in North Africa for its wood and resin, which is used to produce a kind of varnish; it is not hardy in Britain but fairly easily grown as a pot plant. Amongst the vegetation growing under the shrubs one may occasionally encounter *Scilla peruviana*. This species is mainly a native of southern Spain, Portugal and North Africa and the form here is sometimes described as *Scilla sicula*. It does not hail from Peru, this curious nomenclature seems to have been one of the great Linnaeus's few mistakes.

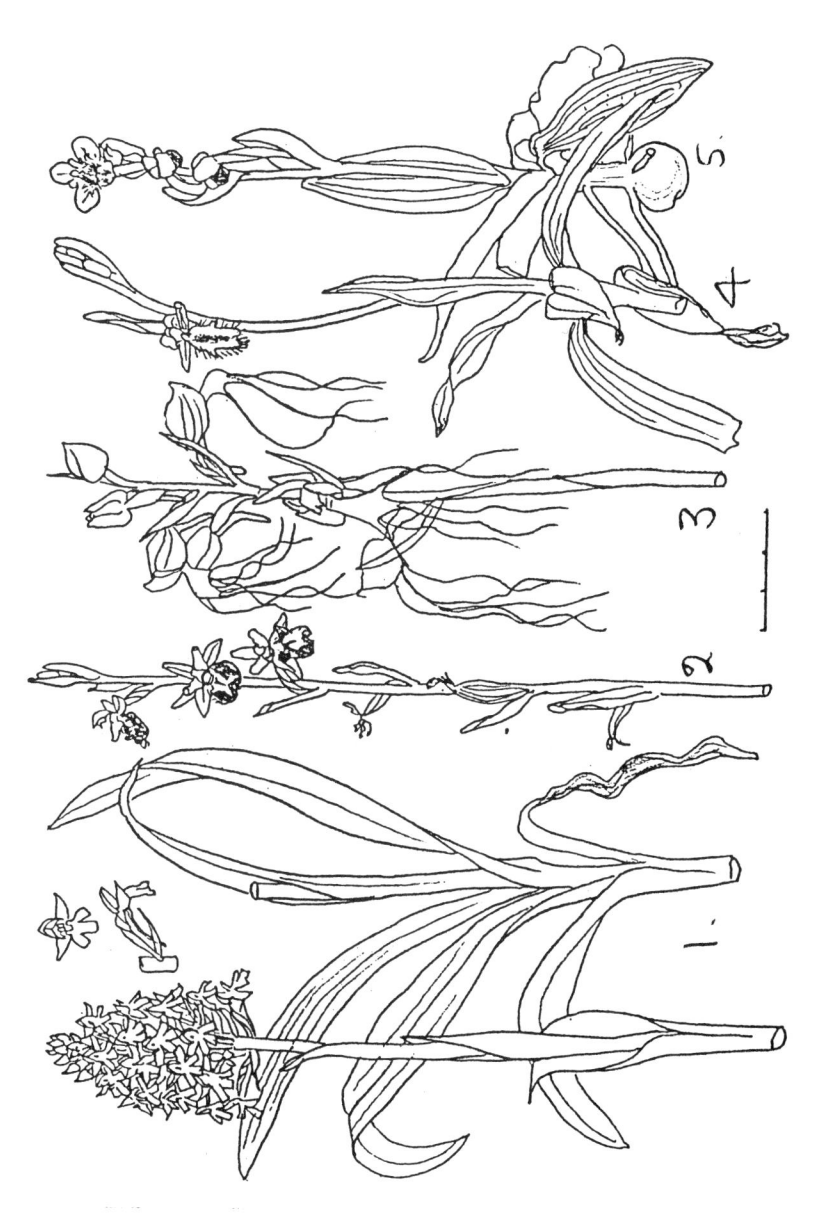

Orchids 1. Anacamptys pyramidalis 2. Ophrys sphegodes atrata 3. Comperia comperiana 4. Ophrys vernixia 5. Ophrys bombyliflora

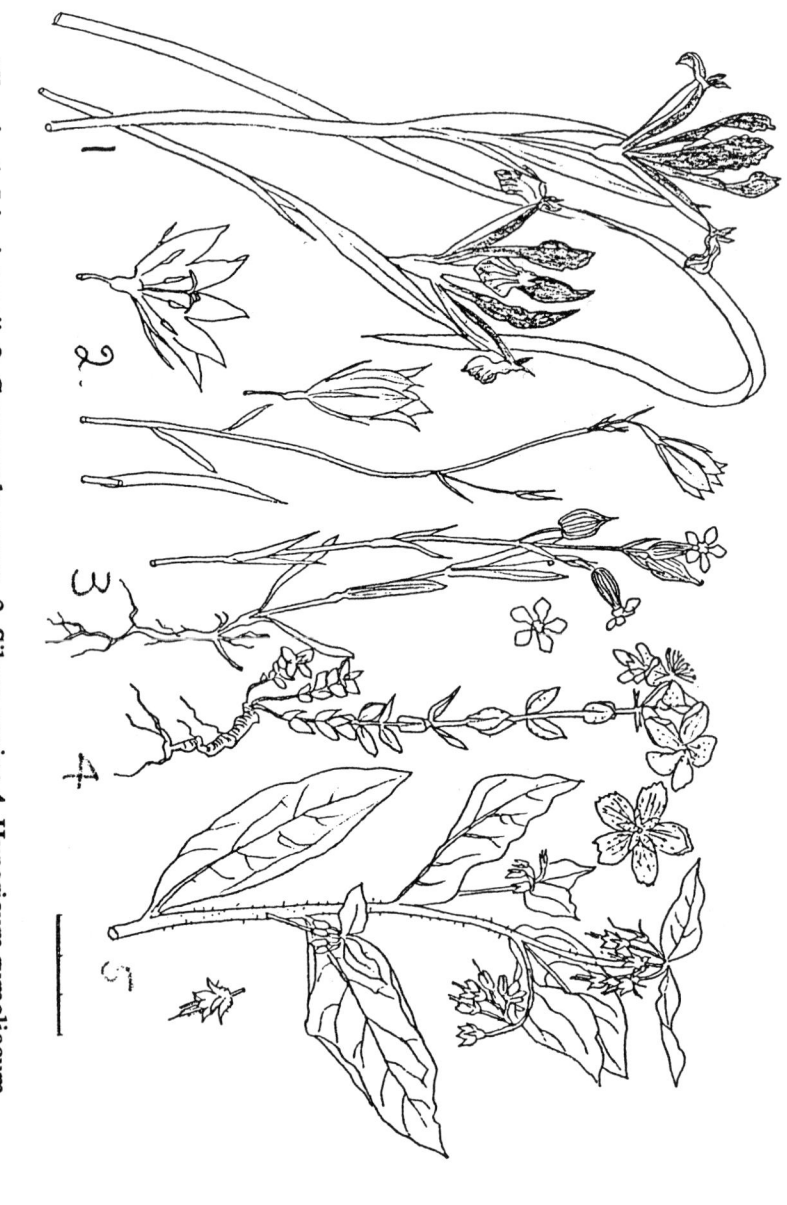

Albania 1. Iris sintensii 2. Campanula sparsa 3. Silene conica 4. Hypericum rumeliacum 5. Symphyum ottomanum

15. ALBANIA

15.1 Albania, called Shqipëria by its inhabitants, became a hard-line Communist state in 1944 and for many years it was closed to visitors. Since 1982 tourists were admitted in limited parties arranged through Albturist. This chapter is based on an organised botanical tour in May 1989. Since then the Communist regime has collapsed and until the situation there has stabilised politically and economically, it will be difficult to study the flora in more detail.

It is a small country, about half the size of Scotland, and sandwiched between Greece and former Yugoslavia. The coach tour, with stops for plant hunting, left from Lake Ohrid (Liqenii Ohrit, in Albanian) on the eastern border, travelled through to the south of the country and then northwards by the Mediterranean coast to Durres and back via the capital Tirana to Lake Ohrid. There was no opportunity to see the northern part of the country which is reputed to be of special botanical interest and the mountains that border Greece in the south were politically out-of-bounds.

15.2 **The Terrain:** Practically the entire country is mountainous with peaks rising to 2,693m at Mt. Jezercë in the north and others over 2,000m in the east and south. The Grammos mountains on the Greek border in the south have extensive outcrops of serpentine rock which is toxic to many plants, and carry an interesting flora of tolerant species. Isolated outcrops of serpentine occur in other parts of the country asociated with ores of less common metals such as chromium and nickel which are mined as important sources of export.

Winters are often quite severe in the central and eastern areas with snow lying on the hills well into summer, so this region can best be described climatically as Balkan. The western coastal areas enjoy a Mediterranean climate but the rainfall is rather high. Indeed, northern Albania and the adjacent parts of Montenegro and Hercegovina have some of the highest annual precipitation in the Mediterranean.

15.3 **Literature:** As one might expect, there is practically no up-to-date popular literature in English on the flora of the region. The most useful volume is that of Polunin (1980) which describes many species to be found in the area but does not deal with them specifically in Albania. Flora Europaea (Tutin et al. 1964-72) includes Albania. For the enthusiast there is a relatively new and substantial flora of the country by Demiri (1981) which may be obtained at a very modest price in the country. It is, of course in Albanian which is said to be based largely on ancient Illyrian and will be quite incomprehensible to most readers. Nevertheless, the botanical names are recognisable and there are 2484 line drawings, many of which show a remarkable similarity to those in the Italian flora by Pignatti (1982)!. Another very useful publication for the visitor is a booklet on the orchid species of Albania by the Swiss authors Gölz and Reinhard (1984). It is one of a series dealing with the mapping of the distribution of Mediterranean orchids under the auspices of OPTIMA but is not readily available to most readers.

15.4 **Podradec to Ersekë:** Stopping near the shores of Lake Ohrid within Albania we saw a number of familiar plants growing by the roadside including: *Convolvulus althaeoides, Dictamnus albus, Helleborus cyclophyllus, Iris sintenesii* and *Linaria peloponnesiaca*. The dictamnus seems to be fairly widespread in Albania - it is the burning bush that is quite popular in our home gardens. The iris is a non-bearded species with narrow leaves and purple flowers, rather similar to the more familiar *I. graminea* but with stems that are less flattened and not winged. The helleborus is the Greek hellebore with green flowers and not unlike *H. odorus* which also grows in Albania. The linaria is a tall species with dense spikes of pale yellow flowers. The box *Buxus sempervirens*, scorpion senna *Coronilla emerus* and danewort *Sambucus ebulus* were also common in the vicinity.

Near Korçë, some 40km south of Lake Ohrid, we visited a small ski slope. There was no snow at the time and around the cleared areas grew thickets with the cornelian cherry *Cornus mas* carrying its tiny yellow star-shaped flowers and *Pyrus elaeagrifolia*

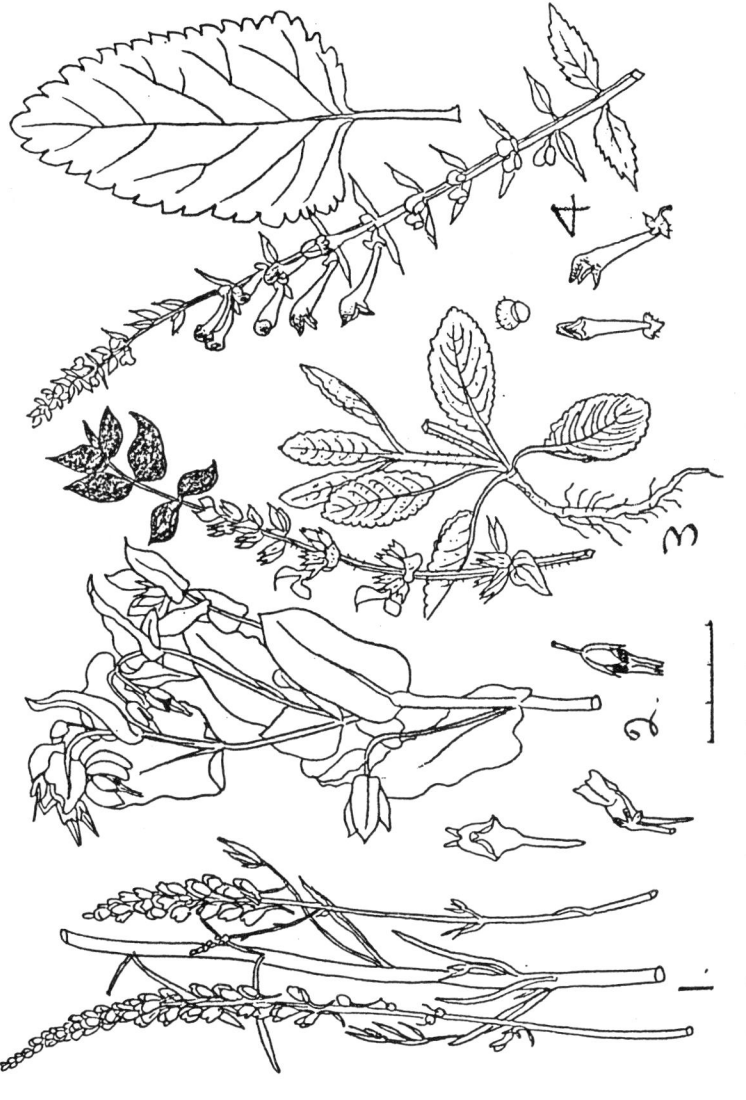

Albania 1. Linaria angustissima 2. Cerinthe minor 3. Salvia viridis 4. Scutellaria altissima

- a small spiny tree, somewhat resembling the willow-leaved pear *Pyrus salicifolia* that is frequently seen in our gardens. In the grass and under the trees grew:

Acanthus spinosus	*Convolvulus lineatus*
Ajuga genevensis	*Eryngium campestre*
Anchusa undulata	*Helleborus cyclophyllus*
Arum italicum	*Scrophularia peregrina*
Campanula sparsa	*Silene conica*

The arum is particularly interesting since it seems to be the sub-specie *Arum italicum byzantinum* with spathes tinged red round the margin and smaller than the type. Unlike the usual form, the leaves tend to wither at flowering time. Demiri (1981) does not record it for Albania.

Leaving the town southwards the road passes through the plain of Korçë which lies at about 800m. It is extensively cultivated with cereals and other crops and much of the work is done by hand but on a large scale. The few tractors and other pieces of agricultural equipment we saw were well-used Chinese models. Climbing out of the plain one passes through pine woods and large areas of typical Balkan natural woodland; rather small trees of various oaks, hop-hornbeam *Ostrya carpinifolia* and the shrubby oriental hornbeam *Carpinus orientalis*. The oaks include the Valonia oak *Quercus aegilops* (= *Q. macrolepis*) that has 'cups' up to 5cm diameter covered with large scales. This species is occasionally grown as a crop in Albania for it is an excellent source of tannin used in the preparation of leather. The common juniper is spread around the edges of the woods and as an undergrowth in places.

15.5 Ersekë to Gjirokaster: Stopping outside Ersekë there is a good view of the Grammos mountains which rise to 2,523m on the Greek border. In the short turf here grew the attractive *Hypericum rumeliacum* that has yellow flowers with the margins of the petals and sepals spotted with blackish dots. There were species of astragalus, one of them probably *A. spruneri*, and the uncommon *Thymus cherlerioides* which is a typical Balkan species.

Passing rocky hillsides with box *Buxus sempervirens* and *Acer monspessulanum* one can see primroses with pale yellow or nearly white flowers, *Ajuga genevensis* and hellebores. Near the town of Leskovik there are good views of the snow-capped Mt. Papingut 2,485m. In the short turf grew *Linum flavum*, along with the hypericum already mentioned and which it superficially resembles. There were gladioli here, probably and appropriately *G. illyricus* here in the old country of Illyricum, but the species are not easy to identify with any degree of certainty. Other plants included *Vicia dalmatica* like a large-flowered *V. cracca, Convolvulus arvensis* and *Salvia viridis*. The last of these is often grown as an 'annual' in our gardens where it is called *Salvia horminum*. There are pine woods in places here and along the edges one could see the following woody species:

Cercis siliquastrum	*Platanus orientalis*
Cistus creticus	*Rhus cotinus*
Erica arborea	*Spartium junceum*
Myrtus communis	

The rhus is the wig tree, smoke tree or Venetian sumach, sometimes called *Cotinus coggygria*. It is a well appreciated garden shrub in Britain, especially the form with purple leaves. The fragrant orchid *Gymnadenia conopsea* grows here and other orchis species which have been recorded for the area include *Orchis militaris, O. morio, O. purpurea, O. tridentata*.

The road turns north westwards to follow along the valley of the River Viose that flows down from the Pindos mountains of northern Greece. One begins to feel the effect of the Mediterranean here near the town of Përmet where the crops include not only cereals but also tobacco and figs. At Këlcyrë one turns westwards through the Gryka pass. At the entrance there were large groups of the shrubby Christ's thorn *Paliurus spina-christi*, occasionally cultivated and said to be hardy in the London area. Its spiny flattened, horizontal branches are covered with tiny yellow flowers in spring and followed by strangely-shaped fruits sometimes said to look like minature wide-brimmed hats. There were also bushes of *Phlomis fruticosa, Punica granatum* and *Spartium junceum* - all typical Mediterranean species. In the Gyrka pass itself the rocky terrain was almost completely covered in places with *Salvia officinalis*, the wild form of common sage grown in our herb gardens at home. At the end of the pass the road turns southwards to Gjirokaster past the town of Tëpelene which Byron visited in 1809 when it was a stronghold of Ali Pasha.

15.6 **Gjirokaster to Serandë:** Gjirokaster is an interesting town with old stone buildings of a special local style well illustrated by the artist Edward Lear when he made a journey there in 1848. A typical example is the alleged birthplace of Enver Hoxa, designer of the People's Republic of Albania. Between the stones around this house grows *Capsella grandiflora* a sub-species of the common shepherd's purse with relatively large scented flowers. It is primarily an Epirotic form but does occur rarely in northern Italy. The town boasts a castle, part of which is given over to a rather sinister museum of war equipment. However, there are a number of interesting plants growing within its walls and amongst the surrounding rocks:

Clematis flammula   *Leopoldia comosa*
C. vitalba   *Lysimachia atropurpurea*
Digitalis lanata   *Trifolium ochroleucon*
Dorycnium hirsutum   *Umbelicus horizontalis*
Erysimum sylvestre   *Vicia lutea*

The digitalis is a foxglove with tall narrow spikes of closely spaced, rather small, greyish-yellow flowers. It is a perennial that grows to 2m and is sometimes cultivated in 'wild' gardens at home. The erysimum is a small wallflower-like plant with light yellow flowers. The recently published Mediterranean-Checklist classes it as having been reported from Albania in error though it does seem to

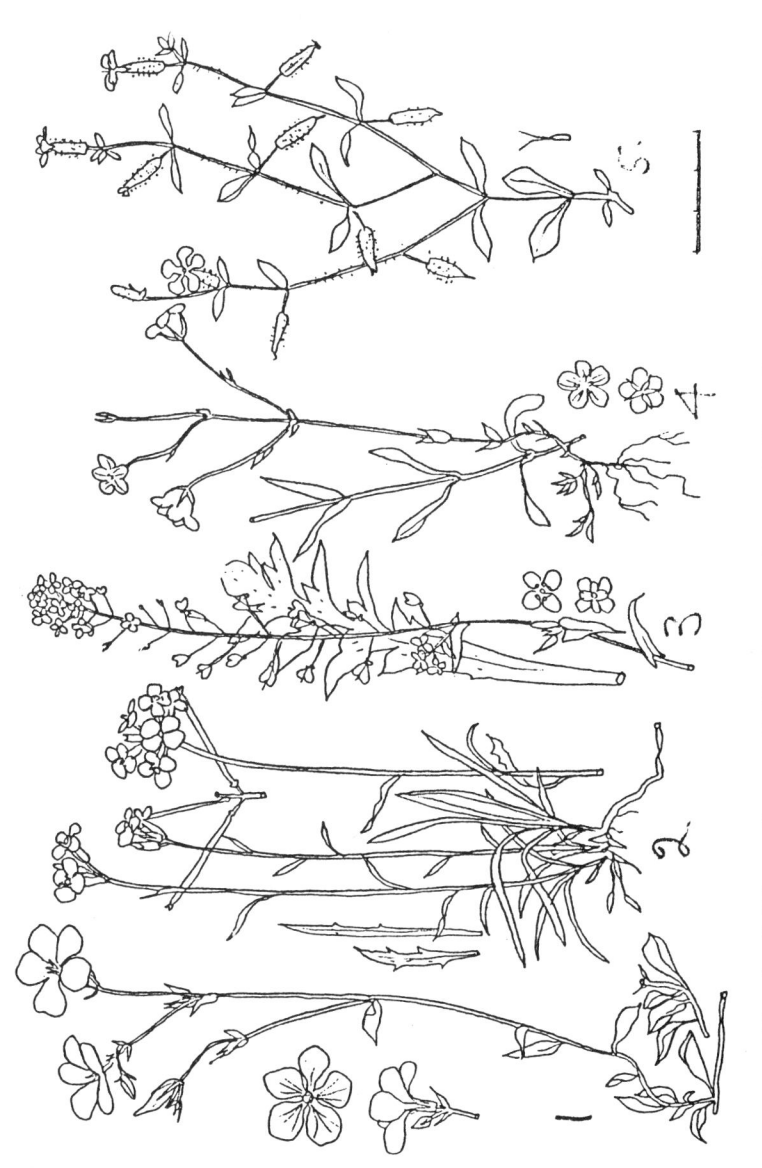

Albania 1. Linum flavum 2. Erysimum sylvestre 3. Capsella grandiflora 4. Moenchia mantica 5. Saponaria calabrica

occur here and is included in Demiri (1981). However, the genus is difficult and this may be a case of mistaken identity! The lysimachia is unlike others of the genus with spikes of small dark-red flowers which seem to remain unopened and leaves with very wavy margins. It is widespread throughout the Mediterranean but often overlooked. Other plants found in grassy areas around Gjirokaster were:

Alcea pallida   *Echium plantagineum*
Anchusa cretica   *Euphorbia characias*
Arum italicum   *Micromeria graeca*
Asphodeline liburnica   *Rumex pulcher*
Bunias erucago   *Silene cretica*
Campanula drabifolia   *Stachys germanica*
Crepis rubra   *Tordylium apulum*

The identification of the campanula is not certain for neither Polunin (1980) nor Demiri (1981) record it from Albania. It is an annual, widespread in southern Greece but could be confused with *Campanula erinus*. The crepis is also an annual with large pink 'dandelions' and occurs throughout the north-eastern Mediterranean and is especially common in the Gargano peninsula of Italy (12.7). It should not be confused with the somewhat similar *C. incana* or 'Greek dandelion' which is perennial and has smaller capitulae.

The road south out of Gjirokaster to Sarandë turns westwards over the Muzine pass. Here are large areas of *Phlomis fruticosa* and *Salvia officinalis* with bracken in places. In some grassy areas one may see what seems to be *Verbascum arcturus* (= *Celsia arcturus*), a short-growing species with yellow or copper-coloured flowers and violet hairs on the filaments of the stamens. It is said to be a Cretan endemic (Sfikas 1987) but a very similar plant is common on Corfu which, as the crow flies, is not far from the Muzine pass in Albania. However, this may be just another case of mistaken identity and needs verification. With it grows the rather uncommon *Moenchia mantica*, rather like a greater stitchwort with bowl-shaped white flowers. From the pass the road leads down to the coastal town of Sarandë.

15.7 The West coast: Serandë is a seaside resort in the south of the country and overlooks Corfu. It seemed strangely deserted in the middle of May 1989 but it has potential and can now be visited on a day trip from Corfu. In the surrounding countryside one can see a rather fine salvia with tall inflorescences of rather large lilac-coloured flowers, probably a form of *Salvia virgata* or posible *S. amplexicaulis*. Another plant here is the goat's rue *Galega officinalis* which is not uncommon in British gardens. It is occasionally grown as a fodder crop and some plants in Albania may be escapees from cultivation. It derives its name from the belief that it increases milk flow in goats.

Near the coast, some 20km south of Serandë lies the small town of Butrint (or Butrini. Buthrotum in Latin) where there is an important classical site with extensive Illyrian, Greek, Roman and Byzantine remains. It lies in a pleasant wooded area near a large brackish lake which has a narrow passage to the Ionian sea and dominated by a fortress built by Ali Pasha - the Albanian 'brigand' who ruled here and much of what is now northern Greece in the time of the Ottoman empire. Apart from some of the plants already mentioned, several other interesting species may be seen:

Allium subhirsutum   *Pallenis spinosa*
Anacamptis pyramidalis   *Prunella laciniata*
Anagyris foetida   *Ruscus aculeatus*
Berteroa obliqua   *Scutellaria altissima*
Blackstonia perfoliata   *Securigera securidaca*
Centranthus longiflorus *Trigonella foenum-graecum*
Fumaria capreolata   *Vicia lutea*
Gynandriris sisyrinchium *V. melanops*

The berteroa is an insignificant small annual with white flowers. It resembles an alyssum and is sometimes included in that genus. The scutellaria, by contrast, is a rather tall growing (50cm) and impressive member of the genus often called scullcaps. They derive their name from the curious shape of the calyx. *Securigera* has been included in *Coronilla* and closely resembles that genus. It has yellow flowers and the seed 'pods' stand in a group like the spokes of an inside-out umbrella and their ends are curved like a shepherd's crook. The trigonella, a small legume with trifoliate leaves and yellow flowers is fenugreek, one of the main constituents of Indian curry. Its presence can be recognised at a distance on a warm day on account of its familiar aroma. It probably originated in Asia and has escaped from cultivation.

In addition to the the anacamptis another orchid seen here was *Epipactis microphylla* which is recorded for Albania by Göltz and Reinhardt (1984) but not for this region. It is a small, unimpressive member of the genus. This southern part of the country is home to many orchid species including:

Barlia robertiana	*Ophrys sphecodes*
Cephalanthera rubra	*O. tenthredinifera*
Himantoglossum caprinum	*Orchis italica*
Ophrys bombyliflora	*O. lactea*
O. ferrum-equinum	*O. morio*
O. helenae	*O. pauciflora*
O. lutea	*O. provincialis*
O. mammosa	*O. quadripunctata*
O. oestrifera	*O. simia*
O. reinholdii	*O. ustulata*

The himantoglossum is the sub-species of the lizard orchid most commonly found in the Balkans. It is a slender plant with flowers that have a pinkish lip that is somewhat longer than in the nominate form. *Ophrys oestrifera* is the name sometimes used for *Ophrys scolopax ssp. cornuta* and often called the horned orchid. In order to see these and many other interesting plants of the region it will be necessary to have freedom to wander from public roads.

In a wooded area near here several serapias plants were seen and on superficial examination were identified as *S. cordigera* but later seemed to be *S. vomeracea ssp. orientalis* though this is not recorded from the Balkans. A plant that grew with them was *Symphytum ottomanum* - a comfrey with small white flowers.

The road northwards from Butrint and through Sarandë more or less follows the coast and nearing Vlorë it rises to give a good view of the sea. In the rocky turf there are many interesting plants to be seen, including:

Haplophyllum coronatum   *Micromeria juliana*
Linaria genistifolia   *Silene behen*
Malcomia maritima   *Stachys oxymastrum*

The haplophyllum is a rue-like plant with a flat inflorescence of yellow flowers. The linaria is an attractive plant with relatively large yellow flowers. It is a variable species that can grow to 1m tall, but here it was no more than 30cm.

There are good views from here on a sunny day and the road rises to over 1000m at the Llogora pass called Qaf e Llogaresë in Albanian. This is a charming place with a distinct 'alpine' feel and easy access to a good plant hunting area. There are groups of Pinus nigra, or it may be the Bosnian pine *P. leucodermis*. Another conifer that grows with it is *Abies borisii-regis*, a somewhat variable robust tree intermediate between *A. cephalonica* and *A. alba* and thought by some taxonomists to be a hybrid between the two. Several interesting smaller species may be seen amongst the short turf:

Anthemis tenuiloba   *Parentucellia viscosa*
Cerastium alpinum   *Silene italica*
Hermodactylus tuberosus   *Tordylium apulum*
Onosma sp.   *Viola speciosa*

The anthemis is a variable species, usually found in the mountains. The form here is a particularly attractive one with broad white ray florets - some forms have no ray florets at all. The viola is a mountain pansy with a fairly long spur and finely divided stipules and occurs in yellow, mauve and bicolor forms. Its identiy is not certain for it is a difficult genus. Speciosa seems to fit in best with

Demiri (1981) but it is very similar to the *V. gracilis* (= *gracilis*) group and quite like V. aethnensis found in the Gargano and on Mount Etna (12.4,13.6). A possible connection with the latter would not be surprising since there is good evidence to suggest that Albania was joined to southern Italy during the tertiary period. Just to confuse the issue - could it be *V. magellensis* (16.7). Onosma is an even more difficult genus than viola and it is not easy to commit oneself to a name for the species found here. It is yellow flowered and may be *Onosma arenaria*.

Several orchids were seen, especially *Orchis pauciflora, O. quadripunctata, Ophrys ferrum-equinum, O. lutea* and also *Orchis lactea, Ophrys bombyliflora, O. sphegodes mammosa*. A real prize found here was, *Orchis albanica*, first described in 1983. It is similar to *Orchis morio* but is weaker growing and has smaller flowers with a somewhat wedge-shaped lip. The flowers are usually pale mauve or nearly white. It has been found mainly along the western and northern edges of Albania and there is one record of it from a site near the coast in Montenegro so it just misses being an Albanian endemic. Göltz and Reinhard who named it also found a hybrid between it and *O coriophora* near Vlorë which they named *O. x paparisti*.

The road from the Llogora pass goes down steeply northwards to near the coast at Vlorë and then proceeds at a low level to Durres. Much of the region here is cultivated though one may see the marsh orchids *Orchis laxiflora* and *O. palustris* growing in ditches. One may also have the good fortune to find *Orchis albanica* which is reported from here and starts to flower a few weeks earlier than the marsh orchids. The Dutchman's pipe *Aristolochia rotunda* abounds here in moist shady places.

Our tour took us eastwards from Durres, through Tirana and Elbasan back to Lake Ohrid and Yugoslavia. The capital Tirana (Tiranë in Albanian) has many Russian-style buildings and in 1989 the plantings of street trees, including cedars and other conifers, was very impressive. Reports suggest that since the collapse of an effective economy many of them have been felled for firewood. By the roadside past Elbasan one could see the creeping gromwell

Buglossoides purpurocaerulea (O = *Lithospermum purpurocaerulea*), which is rare in southern Britain. Another plant here is *Salvia ringens*, an especially imposing member of the genus which grows to 1m tall and has branched stems with extra large flowers that are dark blue with a white mark on the lip. It is hardy in Britain but, strangely, one rarely if ever sees it in cultivation. Like many of the other species mentioned in this chapter, it is also found in northern Greece and grows near Mount Olympus (16.4).

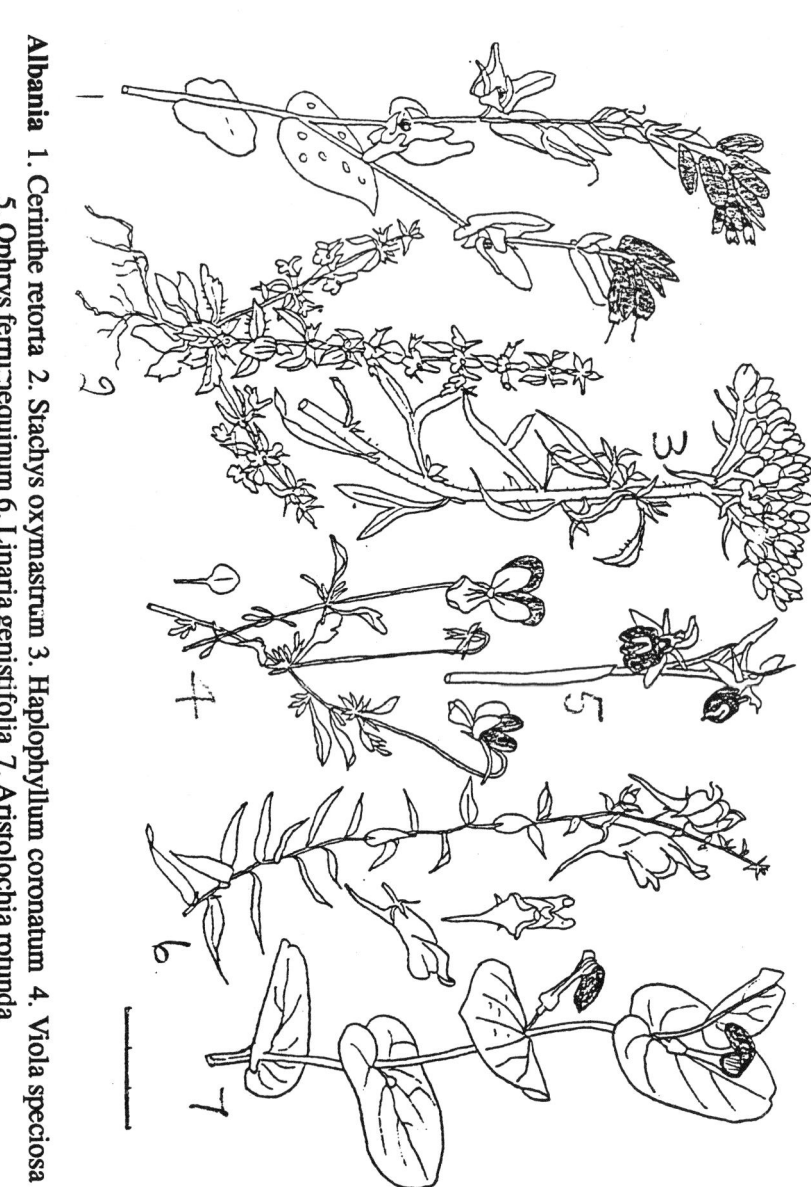

Albania 1. Cerinthe retorta 2. Stachys oxymastrum 3. Haplophyllum coronatum 4. Viola speciosa 5. Ophrys ferrumequinum 6. Linaria genistifolia 7. Aristolochia rotunda

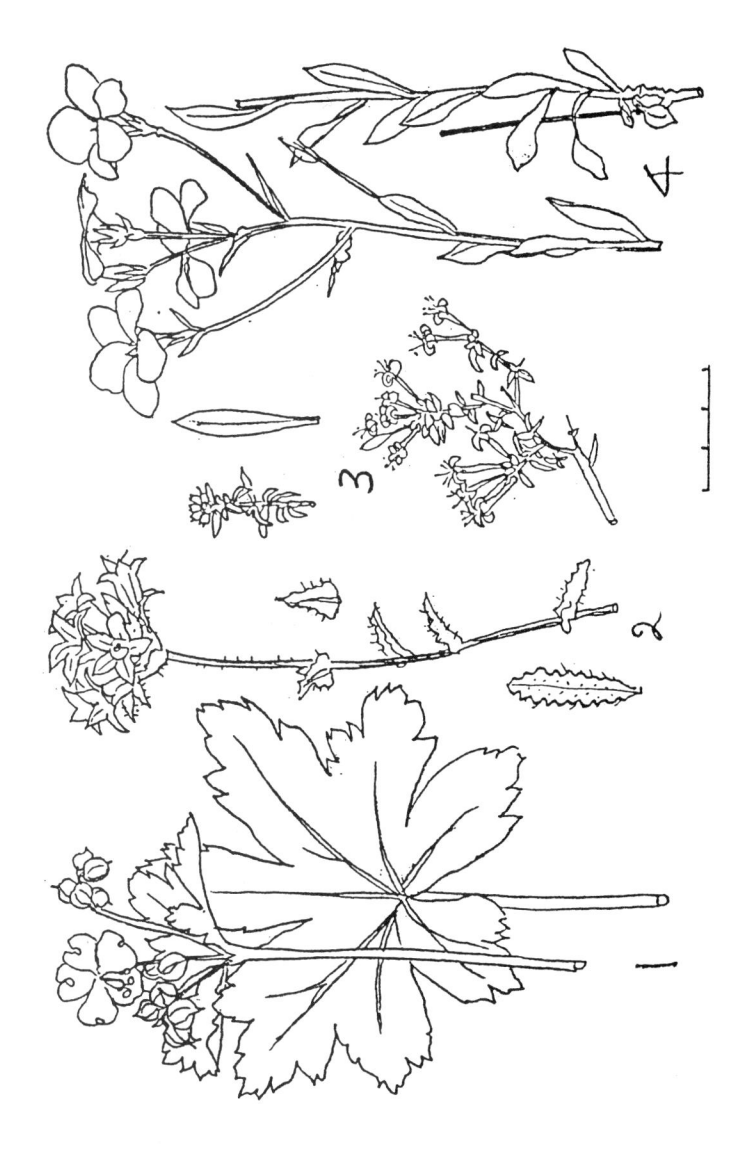

N. Greece 1. Geranium macrorrhizum 2. Campanula lingulata 3. Putoria calabrica 4. Linum thracicum

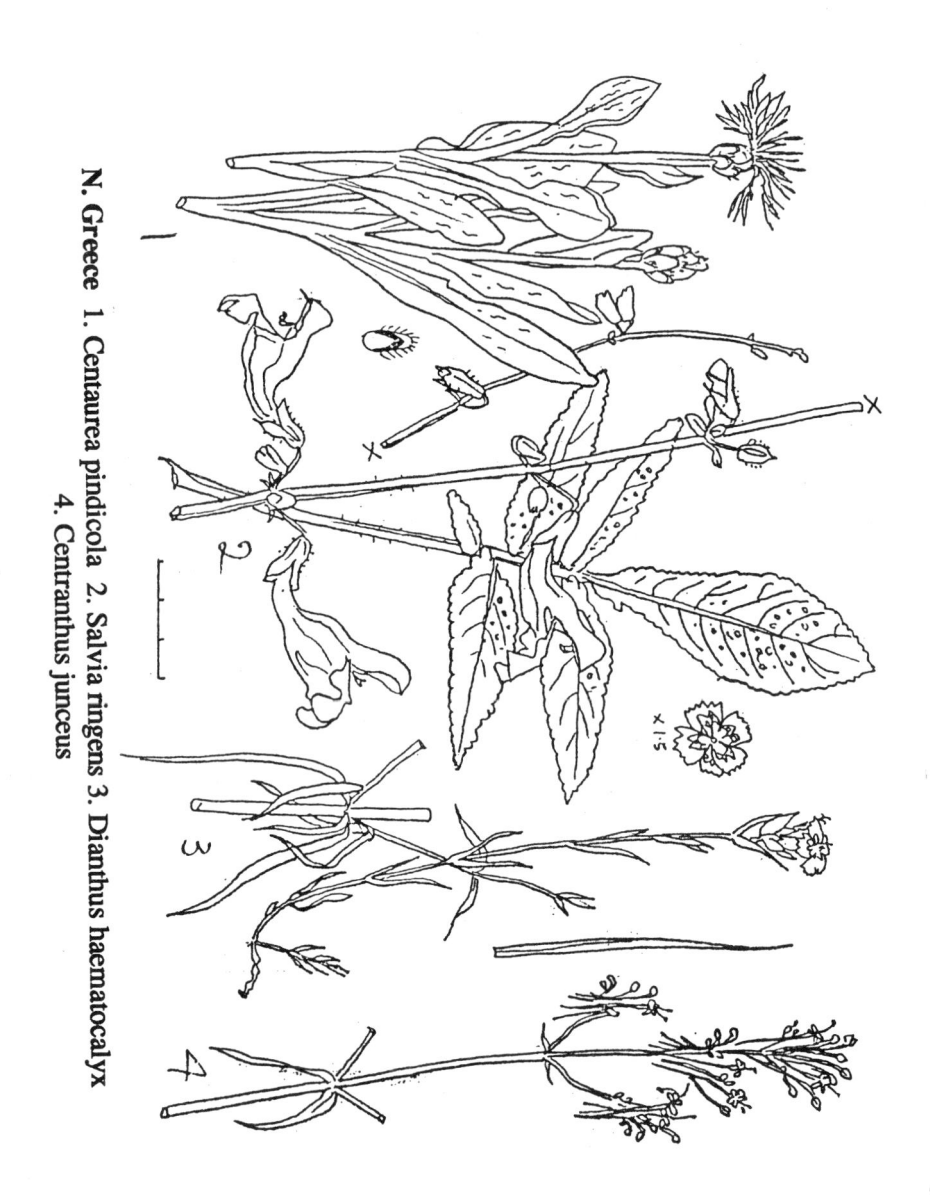

N. Greece 1. Centaurea pindicola 2. Salvia ringens 3. Dianthus haematocalyx 4. Centranthus junceus

16. Northern Greece

16.1 Much of northern Greece is mountainous and differs considerably in character from the Peloponnese. It has many peaks over 2,000m. which are treeless and with snow sometimes lasting into July. Below 1,800m the mountain slopes are covered with vast forests where bears, wolves, lynx and jackals still thrive. Still lower down there are dense stands of maquis and it is only in the coastal regions that familiar Mediterranean plant species will generally be encountered.

The mountain summits are of special interest and some have still not been adequately explored botanically, so they present a challenge to the dedicated enthusiast. Many are difficult and arduous to climb on account of the lack of roads and density of the forests and getting to the top may well be a matter for an expedition rather than an average plant hunting sortie. Fortunately, it is relatively easy to scale the two best known mountains Olympus and Parnassos. The enthusiast who does not have the time or physical ability to climb to the summits need not despair, for much can be seen by making sorties from the higher roads and villages. Indeed, a far greater number of different species will be encountered in this way than at the summits. Many of the plants to be seen can best be described as Balkan or Northern European and the first-time visitor who is familiar with the general Mediterranean flora may feel at a loss to identify species that grow there.

Since it is the plants of mountain slopes and summits which are the most interesting in northern Greece, the best time to go there is from the last week in May until the beginning of July, for spring comes later here than on the coasts. However some bulbous species flower in the maquis and near the coasts during March and April and a few rare species may not be in flower on the summits until the snow melts there in mid July. There are several ways of getting to northern Greece from Britain but the most convenient is to fly to Thessaloniki or Athens. It may also be worth considering flying to Corfu and then taking the ferry or hydrofoil to Igoumenitsa. It is quite possible to cover the region travelling by bus but this is time-consuming and the most convenient method is undoubtedly by hired car.

At the time of writing, a useful map of the region is that produced by Michelin, but there are others. One needs to be flexible about the spellings of the place names for there is no generally accepted way of transcribing the Greek lettering into Roman script and, for example, sometimes an 'h' follows a 'd' as in Pindhos or Pindos (sometimes Pindus). Also the colloquial or 'demotic' Greek is used frequently instead of the more correct written form so, for example, we get Metsovo (demotic) for Metsovon.

16.2 **The terrain:** The mountains of northern Greece run approximately in a north-west to south-east direction and are part of the Dinaric Alps which stretch from 'Yugoslavia' to the Peloponnese. They are largely composed of limestone and were formed by immense pressures on the region some 50 to 100 million years ago. In places there are intrusions of serpentine - a silicate of magnesium which is highly alkaline. This material is toxic to many species of plants and often carries a specialised flora. The most extensive region of this kind is Mount Smolikas 2637m, the second highest mountain of Greece, lying in the north-west near the border of Albania; but there are smaller pockets elsewhere.

The Pindhos or Pindus mountains form a more or less continuous chain down the western part of northern Greece and are sometimes referred to as 'the backbone of Greece'. They are mostly heavily wooded for they attract the greater part of the precipitation coming predominately from the west. Towards the eastern part of mainland Greece the chain of mountains is more broken. In the north, Mount Vermion 2052m is separated from the massif of Mount Olympos 2917m and further south the smaller peaks of Ossa 1978m and Pilion 1651m until one comes to Parnassos 2457m lying near to the Gulf of Corinth. In general these eastern mountains have about half the annual precipitation of their western counterparts. The rainfall is also lower in the eastern part of Macedonia which includes the three-pronged region of Halkidiki, famed for its tourist beaches, and the interesting mountains of Falakro 2111m and Boz Dag 2232m -

an extension of the Rhodope mountains of Bulgaria. In between these mountain masses there are several more-level regions such as the Plain of Thessaly, which is important for cereal production, and the wetland area of the Evros Delta, west of Thessaloniki, where rice and cotton are grown and there is a wildlife reserve noted for its bird life.

16.3 **Literature:** The most useful single volume to take on a visit to northern Greece is 'Flowers of Greece and the Balkans' by Polunin (1980). General books on the flora of the Mediterranean, including the recent valuable work by Blamey and Grey-Wilson (1993) do not deal in the same detail with the complex and extensive Balkan flora.

Anyone intending to climb Mount Olympus to see the plants there should at least consult 'Wild Flowers of Mount Olympus' (Strid 1980) before leaving. It is a very comprehensive and well illustrated work on the topic, but is rather large and too valuable to carry up the mountain. Flora Europaea covers Greece and the recent two volume 'Mountain Flora of Greece' by Strid and Kit Tan (1986,1991) is the authoritative work on the subject.

16.4 **Mount Olympus:** Greek-speaking regions have several mountains called Olympus (Olymbos) but the one referred to here is Thessalian Olympus, the highest in Greece. It towers out of the plains to 2917m and is the proverbial 'Home of the Gods'. It is here that Zeus hurls his thunderbolts and the weather is changeable and unpredictable so the climber may be glad of the shelters provided by the Greek Climbing Club (Information from Litochoron Central Square, Litochoro).

The traveller will probably approach Olympus from Thessaloniki, taking the E75 road towards Athens. It passes through the wetlands leading to the Evros Delta but is not particularly interesting botanically. However stopping in one of the parking places one may encounter *Solanum elaeagnifolium* which grows to about 30cm and, in June and July, has violet-blue flowers like those of a potato plant. It is an introduced 'weed' from Peru where its fruits were sometimes used to produce a kind of soap for washing clothes. In parts of north-east Greece it may cover extensive areas of waste ground and it even ventures into the sand on upper reaches of the beach acompanied at times by *Portulaca oleracea*, a small creeping succulent plant with yellow flowers. Another plant sometimes found growing in these situations is *Centaurea melitensis* which closely resembles the more widespread *C. solstitialis* but is usually taller and more robust. It is sometimes referred to as the Maltese star-thistle though it is not particularly common on Malta.

The best way to approach Olympus is from the village of Litochoro (Litochoron) which lies a few kilometers to the west of the main road. Opposite to this turning, there is a road to the east which leads to Litochoro station and the beach. Near the sea in a marshy area by the Enippeas river grows the summer snowflake *Leucoium aestivum* flowering in spring. Other plants with it include:

Nuphar lutea	*Periploca graeca*
Nymphea alba	*Ranunculus velutinus*
Orchis palustris	

Much of the lower slopes of Olympus, especially on the northern and eastern sides, are covered by extensive tracts of dense maquis which is accessible in places due to grazing. The main shrubby species here are:

Arbutus andrachne	*Juniperus oxycedrus*
A. unedo	*Paliurus spina-christi*
Cercis siliquastrum	*Pistacio terebinthus*
Cotinus coggyria	*Pyracantha coccinea*
Erica arborea	*Quercus ilex*

The cotinus (O = *Rhus cotinus*), the 'smoke tree' of our gardens, is particularly common and attractive here. The paliurus occurs mainly around the lower slopes and is also found by roadsides down to the coast. Herbaceous species growing amongst the shrubs include:

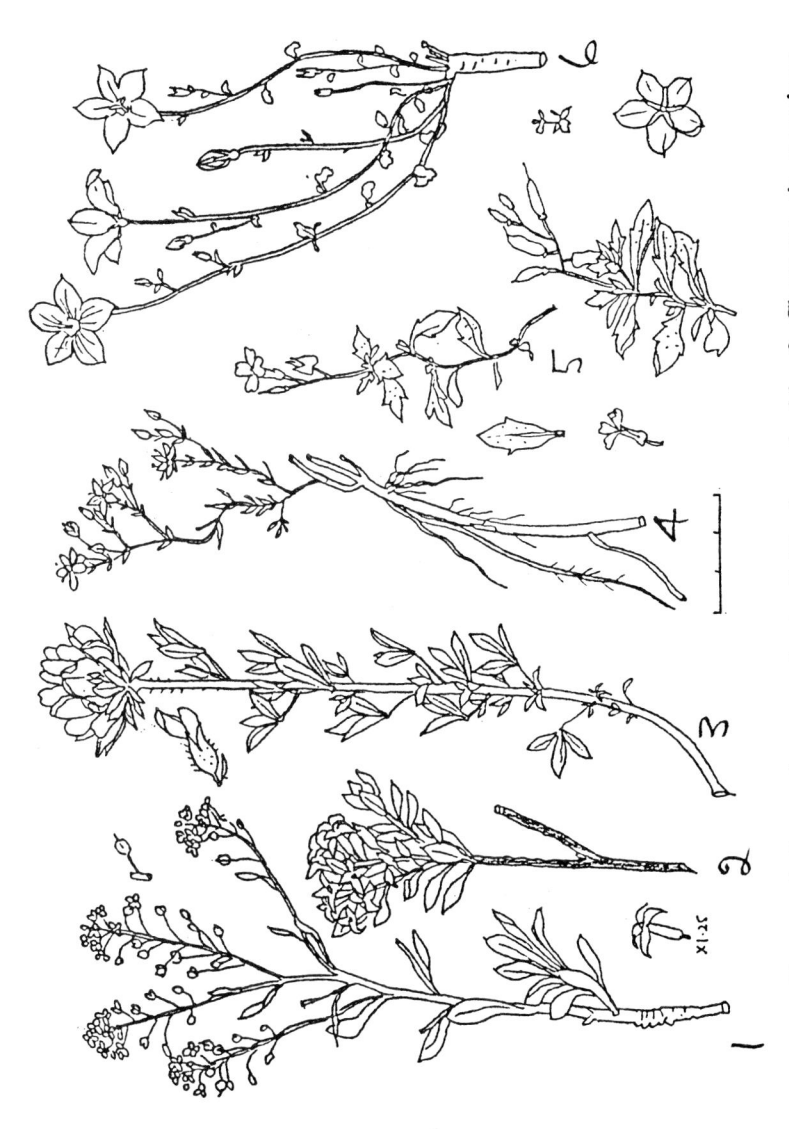

N. Greece 1. Leptoplax emarginata 2. Daphne oleoides 3. Chamaecytisus supinus
4. Minuartia baldacci 5. Aubretia thracica 6. Campanula hawkinsiana

Acanthus spinosus　　　　*Crupina crupinastrum*
Anthemis tinctoria　　　　*Dorycnium pentaphyllum*
Carduus thoermeri　　　　*Echium italicum*
Cistus incanus　　　　　　*Onopordon illyricum*
Coronilla varia

The carduus is an attractive and uncommon species allied to the musk thistle *C. nutans*, with large nodding purple flower heads. Some of the interesting monocots and tuberous and bulbous-rooted species growing with the above include:

Allium subhirsutum　　　　*Fritillaria messanensis*
Anemone blanda　　　　　　*Iris reichenbachii*
A. pavonina　　　　　　　　*Ranunculus rumelicus*
Crocus cancellatus　　　　　*R. spruneranus*
Dracunculus vulgaris　　　　*Tulipa australis*

The most interesting of these is the iris, which resembles *I. pumila* but with green (not papery) strongly-keeled spathes. It is essentially a Balkan species found also in Bulgaria, Romania, 'Yugoslavia', and flowers in April or May. Most of the others listed above will also have finished flowering by June or July when it is the best time to climb to the summit. The crocus, which also occurs in the Peloponnese, produces its white or mauve flowers before the leaves in autumn. *Anacamptys pyramidalis* is the most abundant orchid here and the flowers are regularly dark reddish-mauve as in the British pyramidal orchid and distinct from the pale-flowered forms of the western Mediterranean. Other orchids here include *Orchis morio picta, O. pauciflora, O. quadripunctata, O. simia.*

From the upper end of Litochoro a road leads up the mountain to Prionia 1100m which is a popular picnic site during the heat of summer and there is ample car-parking space. It is on the border of the area around the summit declared a National Reserve in 1938. This road is easily negotiable by car though only the lower part is at present asphalted. At first one passes through maquis; at about 700m the vegetation changes to forest consisting esentially of *Pinus nigra* but with specimens of *Castanea sativa, Celtis australis, Juglans regia* and *Taxus baccata*. There are also patches of beech forest. Along the roadside and under the trees a large number of interesting plants may be seen. Shrubby undergrowth includes box, holly, ivy, *Acer campestre, Cornus mas* and *Daphne laureola*. Interesting herbaceous species include:

Achillea nobilis	*Dorycnium pentaphyllum*
Astragalus monspessulanus	*Genista sakellariadis*
Campanula lingulata	*Geranium sanguineum*
C. spatulata	*Helianthemum nummularium*
Centaurea alba	*Salvia ringens*
C. grbavacensis	*Scutellaria rubicunda*
C. pindicola	*Teucrium chamaedrys*
Centranthus junceus	*Thalictrum saxatile*
Dorycnium hirsutum	

The achillea resembles a rather robust form of the common yarrow with small leaves and large heads of white, or pale yellow, flowers. *Campanula lingulata* grows to about 20cm, and has a head of purple-blue flowers and leaves with wavy margins, covered with bristly hairs. It is not uncommon throughout the mountains of northern Greece. *Centaurea alba* is a variable and fairly tall-growing species characterised by the papery bracts surrounding the heads of purple flowers. *Centaurea grbavacensis* is a rare, handsome, purple-flowered species which is also found in southern 'Yugoslavia', but I hesitate to suggest how its name should be pronounced. *Centaurea pindicola* resembles *C. triumfetti* but has blunt, slighlty-lobed leaves and white flowers with black anthers. The centranthus (= *C. longiflorus junceus*) is a somewhat 'weedy-looking' spur valerian with narrow leaves and is endemic to Greece but seems to be the same as one from Albania (15.7). The genista is a local endemic usually growing on rocky surfaces and forms neat low bushes with yellow flowers. The salvia, which is fairly abundant here, has impressive large 5cm long dark mauve-blue flowers.

Many butterflies and other insects can be seen on the way up to Prionia in summer including some strange members of the Nemopteridae which somewhat resemble large lacewing flies but with the hind wings long and ribbon-like. They seem to be particularly attracted to the flowers of achillea species.

At Prionia there is a waterfall and on the wet rocks and by the sides of the mountain stream *Geranium macrorrhizum* is especially abundant, flowering in June and July. With careful searching in wet areas one may find the butterwort *Pinguicula hirtifolia* producing its white, yellow-throated flowers. An even greater treasure to be sought here is *Jankaea heldreichii* - the beautiful and well-known endemic gesneriad of Mount Olympus. It grows mainly on shady rock faces with silky rosettes and bell-shaped pale lilac flowers from June to August. It is locally abundant in the Papa Rema and Xerolakki Rema (rema means a gorge) and also to be seen along the way to the Ag. Dionyssios monastery which is reached by a track to the north leading off the road to Prionia.

From Prionia, a three hour walk on a well-marked footpath through the woods takes one to the Spilios Agapitos refuge at 2100m. If one intends to stay the night there it is advisable to book beforehand at Litochoro for accomodation is limited, especially in July. By arrangement one may be able to hire a mule at Prionia to climb to the refuge and this is quicker than walking but it is questionable whether it is more comfortable. The woodland at the beginning of the walk is largely deciduous but there are scattered conifers *Pinus nigra pallasiana* and *Abies boris-regis* - the only abies on Olympus. Above about 1600m these give way to the panzer pine *Pinus heldrechii*. A large number of woodland species may be seen on the journey to the refuge, including:

Cardamine bulbifera	*Lilium chalcedonicum*
Cephalanthera damasonium	*Listera ovata*
C.longifolia	*Mercurialis ovata*
Clematis vitalba	*Orchis spitzelii*
Convallaria majalis	*Polygonatum pruniosum*
Coronilla emerus	*Primula veris columnae*

Daphne laureola *Sanicula europaea*
Doronicum columnae *Verbascum phlomoides*
Drypis spinosa *Vincetoxicum hirundinaria*

The drypis is a member of carophyllaceae - a much-branced perennial with spine-like leaves and mauve or white flowers with five deeply notched petals. It flowers in July, rather later than most of the other species listed here. The lilium is the celebrated turk's cap lily with beautiful red flowers, also in July. The form here is sometimes referred to as *L. heldrechii*, named after Von Heldreich who climbed Mount Olympus and described the plants there in 1851. It is generally agreed now that it is simply a robust form of the type species. The mercurialis is similar to the common dog's mercury which it replaces here. *Orchis spitzelii* is, of course, an uncommon species which usually flowers here in June and July. This list is just a selection of the many plants to be seen here.

Higher up the path as one approaches the refuge there are montane species such as *Asperula muscosa, Gentiana asclepiadea, Gentianella crispata, Prenanthes purpurea*. The white-flowered asperula, which is abundant in places, is endemic to Olympus. The gentianella is an uncommon annual species which flowers from August to October and has small, violet, ciliate flowers. On reaching the refuge hut Spilios Agapitos one is above the tree line and many interesting plant species can be seen, including:

Achillea grandiflora *Lamium striatum*
Aquilegia amaliae *Ranunculus sartorianus*
Cardamine carnosa *Saxifraga glabella*
Campanula albanica *Viola delphinantha*
Dianthus integer minutiflorus

The aquilegia is now considered to be a sub-species *A. ottonis amaliae*. It has white or pale blue flowers and is probably endemic to Olympus. The campanula resembles our harebell. The dianthus has white flowers with a very circular outline in July and August. The viola is a distinct and attractive species with long-spurred violet-pink flowers from May to August. It is not confined to Olympus, however, and found as far away as Chelmos and Mount Athos. Other species predominately growing in rock crevices are:

Aethionema saxatile *Saxifraga sempervivum*
Arabis bryoides *S. scardica*
Jovibaba heuffelii *S. spruneri*
Saxifraga griesbachii

The jovibaba is a kind of houseleek with pale yellow flowers. The four saxifragas are all well dealt with in the excellent Alpine Garden Society's guide 'Saxifrages' (Harding 1970).

It takes about three hours from the refuge to climb to Mitikas (2917m). Some alpine species to be seen here flowering around the edges of the melting snow patches in June and July are:

Corydalis densiflora *Iberis sempervirens*
C. parnassica *Scilla nivalis*
Edraianthus graminifolius *Viola heterophylla graeca*
Gentiana verna balcanica

The edraianthus (*O = Wahlenbergia graminifolia*) is an attractive member of the Campanulaceae with heads of blue flowers and grass-like leaves. The viola is a mountain pansy with yellow, purple, or bicoloured flowers. Like most violas the nomenclature is complicated and it is sometimes referred to simply as *Viola graeca*. It occurrs on many other Greek mountains.

It is hardly possible to do justice to Mount Olympus as a botanising centre. It is so easily accesible and has such a wealth of interesting plants. There are several ways to get to the peaks and ravines and interested readers would do well to consult the Greek Tourist Office or Greek Mountaineering Association.

16.5 Mount Ossa and Mount Pilion: If one is staying for a few days at Litochoro to climb Mount Olympus it is worth making an excursion to the much lower Mount Ossa 1978m that lies about

45km to the south east. One continues south from Litochoro along the road to Athens, crossing over the Tenbi river and turning sharp left along the route signposted to Omolio and Stomio. Shortly before reaching Karitsa a partly-surfaced, forest road to the right takes one across the northern flank of Mount Ossa. At first there are plantations of fine old sweet chestnut trees where it is lush with bracken and many roadside plants including:

Acanthus spinosus   *Lychnis coronaria*
Anthyllis barba-jovis   *Scutellaria altissima*
Cercis siliquastrum   *Verbena officinalis*
Galega officinalis

Further along, one passes through beech woodland where the flora has a distinct northern European character with the following species growing on banks and in clearings:

Asperula odorata   *Melittis melisophyllum*
Cephalanthera longifolia   *Neottia nidus-avis*
Dactylorhiza saccifera   *Orthilla secunda*
Daphne laureola   *Platanthera chlorantha*
Lathyrus grandiflorus   *Polygonatum pruniosum*
Lilium martagon   *Sambucus elebus*
Lysimachia vulgaris

The polygonatum is sometimes included with *P. odoratum*. The platanthera here seems to be an unusual form with long inflorescences carrying as many as twenty-five completely green flowers. Similar plants were seen near Metsovo by the Katara Pass (16.7) and they are reminiscent of *P. holmboei* from Cyprus (22.10). The orthilla is the serrated wintergreen (= *Ramischia secunda* = *Pyrola secunda*).

After traversing a region where the beech trees are mixed with pines and there is an undergrowth of *Erica arborea* one comes to open treeless countryside with fine views of Mount Olympus and further on the conical peak of Ossa. In the short-grazed turf here grow *Asphodeline lutea, Astragalus thracicus, Centaurea attica, Dianthus haematocalyx, Eryngium amethystinum, Inula oculus-christi, Salvia verticillata.* The last of these is an attractive species with dense heads of small, very dark purple flowers and bracts. It is not uncommon in other parts of northern Greece. On rocks near here grow *Aubretia thessalica* and *Campanula rupicola.* There is a foot track to the summit of Ossa where one may be fortunate to find *Corydalis solida, Crocus veluchensis, Erodium absinthoides, Tulipa australis.* The erodium is an uncommon and attractive species with relatively large violet flowers. The road continues down the west flank to Spili from where one can make ones way back to Litochoro. In places this route is flanked with many specimens of *Verbascum macrurum.* If ever Salvador Dali had designed a plant this would probably be the one. It is quite large with more or less heart-shaped basal leaves and long 1-2m rat's tail-like inflorescences which are rarely branched except at the base. These flowering stalks tend to hang over towards the ground in a random manner and sometimes straighten out. The whole effect with them scattered with yellow blossoms, and mixed with last years blackened specimens, produces a weird surrealistic effect.

Pilion lies further south and rises to only 1651m. It is mainly covered with forest but there are open places and a speciality here is *Campanula incurva* described by Polunin as "the finest of all Balkan campanulas". It is a low-growing species with large upright-facing flowers resembling pale violet Canterbury bells.

16.6 **Parnassos:** This is the highest of the southern mountains of northern Greece and lies close to the eastern end of the Gulf of Corinth. It is easily accessible and is best approached from the village of Arahova (Arakhova) which is near the base of its southern flank and on the road between Amfissa, through Levadia and on to Athens. Most travellers to the region will want to visit the classical site of Delphi some 12km to the west of Arahova where a number of interesting plants are to be seen amongst the ruins:

Alkanna graeca   *Onobrychis ebenoides*

Asphodeline lutea *Onosma frutescens*
Asphodelus fistulosus *Phlomis fruticosus*
Campanula topaliana *Salvia triloba*
Centranthus ruber *Salvia viridis*
Daphne jasminea *Silene gigantea*
Euphorbia acanthothamnos *Smyrnium orphanidis*
Ferulago nodosa *Stachys swainsonii*
Linaria chalepensis *Vicia dasycarpa*

The daphne is not a common species. It produces a very low-growing small bush with typical white star-shaped flowers that are tinged lilac on the under surface. The ferulago is an 'umbellifer' in which the nodes are swollen to the size of an acorn. It occurs on other parts of the Greek mainland and also on Crete (20.8) The silene resembles a robust form of the more widespread *S. italica*.

Travelling along the road from Delphi to Arahova one will note bushes of *Euphorbia dendroides* growing on the rocks and large patches of the woad *Isatis lusitanica* which closely resembles our native species and is found as far east as Israel (24.6). The impressive wild hollyhock *Althea pallida* (= *Alcea pallida*) grows in places by the roadside and at the Marmara site amongst the foundations of the old Temple of Athena. It is found throughout northern Greece but nearly always as scattered specimens rather than in large groups.

From Arahova an asphalted road rises up Parnassos to the ski site at about 1900m and then down the northern flank to Gravia. Much of the northern slope is covered with dense forest in which *Abies cephalonica* predominates at the higher levels, sometimes parasitised by mistletoe *Viscum album*, but on the southern flank there are open spaces. Although Parnassos was made a National Reserve as long ago as 1938, this has done little for the plant life and grazing is everywhere excessive. However, on open ground and in clearings on the way to the ski centre one may see may interesting plants including:

Ajuga orientalis *Euphorbia myrsinites*

Anemone blanda
Asphodeline lutea
Aubretia deltoidea
Centaurea triumfetti
Crocus sieberi
Digitalis laevigata

Iris pumila
Lamium garganicum
Lathyrus grandiflorus
Marrubium velutinum
Verbascum delphicum
Viola heterophylla graeca

The crocus can be a beautiful sight in April and early May. It is sometimes accompanied by *C. veluchensis* and the two species are not always easy to distinguish. The last has rather less rounded perianth segments and a white, rather than yellow, throat to the flower. The iris (= *I. pumila attica*) is typical of some Greek mountains and the sub-species is especially short-growing, with curved foliage and either purple or yellow flowers. The verbascum is one of several species to be found here. It has an impressive branched inflorescencs and is quoted here because of its specific name referring to Delphi, though it is found in other parts of Attica and in Euboea.

The forests are mostly very dense and contain relatively little undergrowth apart from *Fritillaria graeca, Helleborus cyclophyllus, Lilium chalcedonicum* and *Orchis pallens*. In April there are still some snow patches left near the ski lift and few plants in flower except for crocus, corydalis and draba but one may identify the following

Arum maculatum
Astragalus angustifolius
Cerastium candidissimum
Colchicum triphyllum
Corydalis bulbosa
Crocus sieberi
Crocus veluchensis

Daphne oleoides
Draba parnassica
Gagea fistulosa
Ornithogalum oligophyllum
Potentilla speciosa
Scilla bifolia

When it is not in flower, one could be forgiven for identifying the potentilla as *P. nitida* from the Alps, but it is a far less attractive species with white flowers during summer. It was once named

P. parnassica but is now considered as a small-leaved alpine form of *P. speciosa*.

From the ski lift it takes about 1½ hours to reach the highest point Liakoura 2457m where snow lies until well into June. Here one may encounter the following though they do not all flower at the same time:

Aethionema saxatile	*Hypericum rumeliacum*
Campanula aizoon	*Iberis sempervirens*
Colchicum variegatum	*Ranunculus brevifolius*
Edrianthus graminifolius	*Sternbergia lutea sicula*
Erigeron alpinus	*Viola poetica*
Erysimum pusillum	

The viola has a limited range but is not a very conspicuous plant - a perennial pansy with small, pale violet flowers.

16.7 **The Katara Pass:** The town of Larissa lies some 40km south of Mount Olympus. From here the E92 road crosses the country to the west coast at Igoumenitsa where there is a regular ferry to the island of Corfu. After leaving Larissa the route travels through the Thessalian plain which is important for its cereal crops. From Trikala the road rises to Kalambaka near to the extraordinary monasteries of the Meteora, perched high on rock pinnacles. It crosses the western Pindus range at the Katara Pass 1690m and the local authorities are proud of their efforts to keep this important road free from snow throughout the winter. In June there is little snow left at the pass, which is windswept, more or less free of trees and heavily grazed but of considerable botanical interest. At and near the summit the following were observed amongst a few scattered pines, abies and box bushes:

Bornmuellera tymphaea	*Iberis pruitii*
Campanula spatulata	*Leptoplax emarginata*
Cynoglossum creticum	*Minuartia baldaccii*
Daphne oleoides	*Ptilostemon afer*

Dianthus haematocalyx *Rosa pulverulenta*
Euphorbia myrsinites *Scrophularia canina*
Helleborus cyclophyllus *Viola magellensis*

The bornmuellera and leptoplax (O = *Peltaria emarginata*) are both white flowered 'crucifers' that thrive on serpentine. The rosa is a charming, low-growing and very spiny briar that produces white dog roses in May and June. These plants all have some protection against the severe grazing. They are mostly either prickly or distasteful to sheep. However the minuartia (= *Alsine baldaccii*) escapes by being so minute and produces thousands of tiny white stars in May and June. The viola is a small pansy, usually with purple flowers, but the genus is a very complicated one and I may not have the right name for it. It reminds one of the enigmatic species seen in Albania (15.7). Two other plants one may encounter near here are the rather tall sea-pink *Armeria canescens* and, growing in disturbed ground by the roadside, *Campanula hawkinsiana*. The last has a tap root with spreading procumbent stems and chalice-shaped flowers that are dark purple on the upper surface but paler below. The stems produce a milky latex when broken. This is another species which is tolerant of serpentine.

The road goes down fairly steeply to the west of the pass and there are woods of stunted beech trees and some other deciduous species. Under the trees grow *Cardamine bulbifera* (O = *Dentaria bulbifera*), *Corydalis bulbosa*, *Symphytum bulbosum* and *Veratrum album*. One may also see *Orchis mascula* and a green-flowered form of *Platanthera chlorantha*. Some bulbous species such as *Fritillaria pontica*, colchicums and ornithogalums are also to be seen. As one approaches the charming small town of Metsovo there are meadows with *Narcissus poeticus* in spring which quickly give way to such plants as the oxe-eye daisy *Leucanthemum vulgare*, yellow rattle, gladiolus and scorzonera. By the roadside here *Chamaecistus supinus* with its yellow, red-streaked flowers, makes a fine show in June accompanied by such species as *Morina persica* and *Inula oculus-christi*.

The road from Metsovo to Ioanina passes through mixed

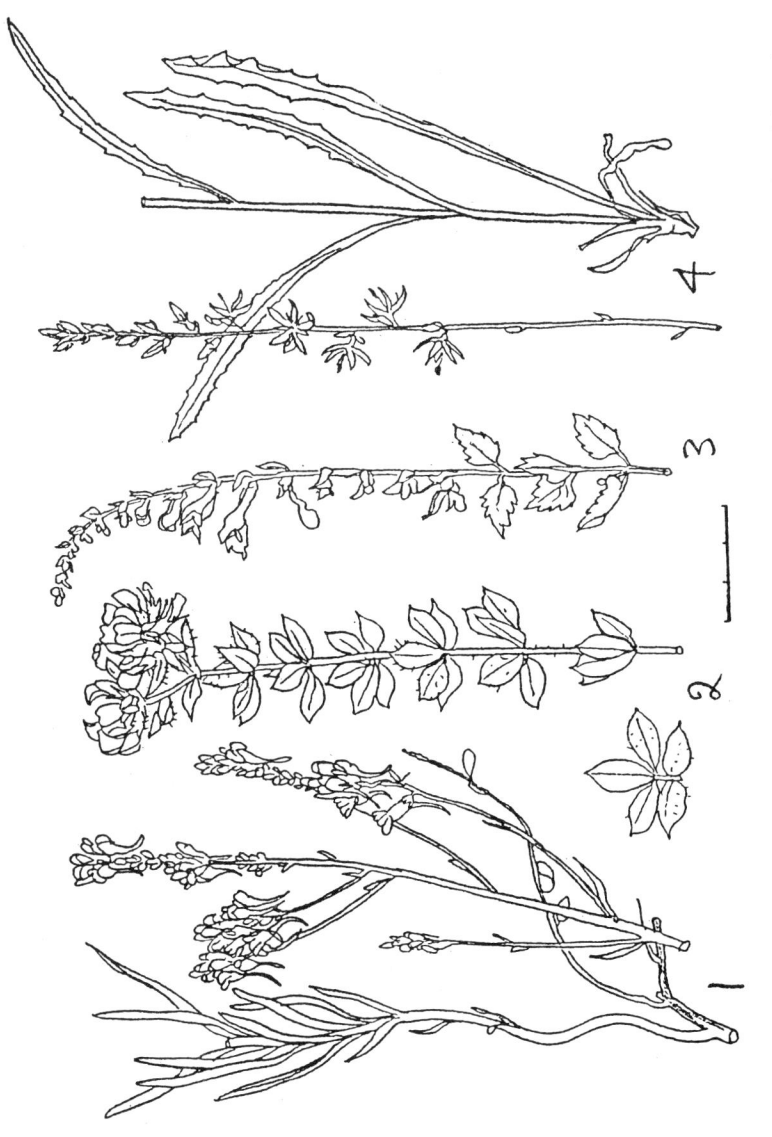

N. Greece 1. Linaria peloponnesiaca 2. Dorycnium hirsutum 3. Scutellaria rubicunda 4. Asyneuma limonifolium

countryside where *Putoria calabrica* is common on the rocks. The uncommon *Alkanna pindicola* with yellow flowers, grows here with *Acanthus spinosus*, *Helianthemum canum*, *Petrorhagia thessala* and *Salvia viridis*.

16.8 Zagoria and the Vicos Gorge: Zagoria is the name given to a region of wild, uncultivatable, limestone country north of Metsovo and Ioanina, bounded on the west by the E90 road and on the east by the Aoös river. It is known for the architecture of its village houses which belie the poverty of the region. Good hard limestone is readily available for walls and roofing tiles and this fact, coupled with the quality of the local joinery and wrought-iron work, has produced fine solid buildings. The countryside is under populated, rugged and with forests, ravines, open spaces and all that goes with good plant hunting terrain.

The region will probably be approached by driving northwards along the E90 out of the bustling town of Ioanina which was once Ali Baba's 'capital'. Shortly after leaving the town, vast groups of *Phlomis fruticosus* grow on banks by the roadside. Some 30km along this road there is a turning eastwards to Monadendri which is a typical Zagoria village of stone houses and paved streets and a convenient place to find accomodation. The road continues above the village and then gives way to a negotiable unsurfaced track which is a good region for botanising with fine views of the snow covered Pindus mountains in May to July. There are patches of woodland which include beech, hawthorn and *Acer monspessulanum*, *Corylus colourna*, *Juniperus oxycedrus* and *Ostrya carpinifolia*. The corylus is the Turkish hazel which is grown on a large scale for its nuts in Trabizon. Unlike the common hazel, it is a small tree rather than a large shrub. A large number of other species are to be found in the woods, clearings and by the roadside:

Armeria rumelica *Leopoldia weissii*
Asphodeline lutea *Linaria peloponnesiaca*
Campanula spatulata *Micromeria juliae*
Coronilla varia *Orlaya grandiflora*

Dianthus haematocalyx *Petrorhagia illyrica*
Echium italicum *Ranunculus sumhomophyllus*
Eryngium campestre *Silene italica*
Helleborus cyclophyllus *Stipa pennata pulcherrima*
Hieracium pannosum *Teucrium polium*
Inula oculus-christi *Verbascum macrurum*
Legousia pentagonica *Vicia dalmatica*

The hieracium has basal leaves covered with greyish hairs as in *Hieracium lanatum* of the Alps and our rock gardens. The stipa is a grass that has extraordinarily long feathery awns up to 20cm. It is most attractive and is prized in France where it is collected for decoration as 'plumet de Vaucluse' (10.10). It is distinctly uncommon in Greece but occurs in fair quantity here. The species is sometimes split into a number of sub-species and this one is almost certainly *pulcherrima* but some taxonomists may class it as *S. rechingeri*.

From Monadendri one can enter the impressive Vikos gorge through which the Voidomatis river runs to join up with the Aoös river. It has walls nearly 1,000m high in places. The gorge itself and part of the mountain Tymphi to the north were declared a National Park in 1973. It is certainly worth a visit to see the flora but should not be taken lightly. Flash floods may occur, especially from snow melt in spring and stones brought down by storms are a danger; it is also possible to get lost there. It takes about five hours to walk through to the village of Vikos at the northern end and another two hours to the village of Papingo (Papingon) where there is accommodation. Plant species to be seen in the gorge include:

Acanthus balcanicus *Nepeta spruneri*
Achillea abrotanoides *Ramonda serbica*
Centaurea pawlowskii *Staehelina unifloculosa*
Lilium carniolicum *Symphytum ottomanum*
Moltkia petraea *Valeriana crini epirotica*

The centaurea is a rare, but not particularly distinguished, species growing to about 40cm and with pale violet flowers. It is generally found in rock crevices. The lilium has small flowers and somewhat resembles the well-known *L. pyrenaicum*. It is a rather uncommon Balkan species which has also been recorded near the Katara pass. The valeriana has a very limited distribution but is similar to *V. montana* of the Alps and Pyrenees.

The whole of this area is of great botanical interest and would be very well worth while exploring by the adventurous traveller. Polunin (1980) describes some of the regions in the vicinity which are not dealt with here.

16.9 Around Mount Smolikas: The E90 road continues northwards past the turnings to Monadendri and Papingo to the small market town of Konitsa. and the broad flood basin where the Sarantaporos river mingles with the Aoös and Voidomatis. Around cultivated fields with much *Paliururs spina-christi* one may see a number of interesting 'weeds' such as the tall *Echinops spinosissimus* and curious *Scabiosa tenuis* - an annual with reddish mauve flowers and long black hairs on the calyx. Another curious, one might even say extraordinary, plant here is *Centaurea zuccariniana*. At first sight it is very difficult to decide what family it belongs to for the flower heads appear like tufts of small feathers and the sparse, mauve florets are rather inconspicuous. Centaurea is known for the variety of its bracts round the capitulae but here the genus seems to have gone to an extreme. By the roadside one may also encounter the yellow knapweed *Centaurea salonitana* which is a handsome perennial that grows to a metre tall and has 'normal' floral bracts.

As one goes up into Konitsa from the south there is a sharp turning to the right that is signposted to Eleftero and Distrato and this is a particularly good route for botanising. It is quiet, well-surfaced nearly to Eleftero and there is little difficulty stopping to make foreys into the woods. At first one passes through grazed land with bushes of *Juniperus oxycedrus* and some *Phlomis fruticosa*. There are patches of the wild form of the garden sage *Salvia officinalis* and some other species such as *Althaea cannabina, Linum aroanum* and *Nigella damascena*. There are fine views of the Albanian mountains to the west. Soon one passes through woods composed mainly of pine but with some deciduous species intermixed, and many interesting smaller plants to be seen including:

> *Asyneuma limonifolium* *Hypericum richeri*
> *Atropa beladonna* *Knautia drymeia*
> *Cerinthe minor* *Lathyrus grandiflorus*
> *Cistus incanus* *Legousia pentagonia*
> *Coronilla varia* *Linaria dalmatica*
> *Dictamnus fraxinella* *Lysimachia vulgaris*
> *Digitalis viridiflorus* *Salvia sclaria*
> *Dorycnium hirsutum* *Silene vulgaris*
> *Echium vulgare* *Trifolium purpureum*

The asyneuma has narrow, toothed leaves and a long unbranched inflorescence of blue flowers - it is a member of Campanulaceae and related to phyteumas. The digitalis is an unprepossessing foxglove with small greenish yellow flowers. The dorycnium, by contrast, is a fine form of the species with relatively large white flowers that are pink on the outer side of the standard. The knautia (O = *Scabiosa sylvatica*) is sometimes appropriately referred to as the wood scabious. It has typical lilac-pink flowers but characteristic broad, toothed and undivided leaves. I am not certain about the correct identity of the hypericum, which is a robust species with black dots and streaks on the petals and sepals - it is a difficult genus and produces hybrids.

Various orchids may be seen here including *Anacamptys pyramidalis, Cephalanthera rubra* and epipactis. Towards the highest part of the road there are some good specimens of abies and through clearings one gets a superb view of Mount Timfi 2497m with its jagged snow spattered peaks. By comparison, the higher, conical summit of Mount Smolikas 2637m looks less impressive from here.

Getting to the top of Mount Smolikas 2367m, the second highest in Greece, is no afternoon jaunt. It does not have a convenient road that will take one much of the way, as for Olympus and Parnassus. The journey has to be done largely on foot, probably carrying camping equipment, though it may be possible to hire a mule. The most usual way of making the ascent is to take the road from Konitsa towards Distrato described above and to stop at the village of Paleoselli. From here there is a track which goes up the mountain to lake Drakolimni 1600m in 4-5 hours. Further along the road towards Distrato a rough vehicle track from Padhes also leads to this lake. From here it will take about 3 hours to reach the summit. One can also make the journey by continuing past Distrato along a route to the village of Samarina from where there is a mule track to the summit of Smolikas.

There are mixed woodlands on the lower slopes which in places have wet areas where the water seeps from impervious lower rocks. Here one may find *Gentiana asclepiadea, Ranunculus platanifolius* and the curious *Rhynchocorys elephas*. The last is a semi-parasitic member of Scrophulariaceae which grows to about 50cm tall. The yellow flowers have a pronounced 'beak' and, as the specific name implies, look like the head of an African elephant with the ears outstretched. It is an uncommon plant, often found on serpentine and also occurs further east towards Iran. The Greek form is usually allocated to the sub-species *boisseri*. Higher up one encounters *Pinus heldrechii*, as on Olympus, and by the time one reaches the lake Drakolimni one is largely above the tree line. Plants one may encounter on the summit include:

Alyssum smolikanum *Fritillaria epirotica*
Aubretia glabrescens *Gentiana verna balcanica*
Barbarea sicula *Leptoplax emarginata*
Bornmuellera baldacci *Linum punctatum pycnophyllum*
Campanula tymphaea *Saxifraga adscendens*
Cerastium smolikanum *S. paniculata*
Dianthus deltoides degenii *Silene multicaulis*
D. haematocalyx pindicola *Viola albanica*
Draba lasiocarpa

Much of Smolikas is serpentine and most of the plants listed above are able to cope with its toxicity. The campanula somewhat resembles an edrianthus and, as its name implies, is also found on the neighbouring Mount Timfi. The fritillaria is an attractive short-growing species with rather large flowers like our snakeshead lily which are dark purplish and mottled but variable. It is special to the area but is said to grow also at the Katara pass. The flowers are produced in May and June. The linum is a prostrate-growing species with fine blue flowers in June and July. It occurs also in south east Turkey and another sub-species is found in Sicily (13.9). The viola was once included in *V. magellensis*. It is fairly common on Smolikas and produces its beautiful pale reddish-violet flowers from June to August.

16.10 **Prespes National Park:** This interesting region lies in the extreme north-west of Greece close to the group of lakes which include Lake Ohrid and those of Prespes. It is probably most easily reached from the small town of Florina.

Leaving Konitsa one travels in a northerly direction along the E90 to Kastoria and then the E86 to Florina. At first the route follows the Sarandaporos river and there are fine views of the mountains. By the roadside one can hardly fail to notice the 2m tall *Cirsium candelabrum* which is architecturally impressive but produces rather insignificant yellowish white flowers in July and August. Other plants of the grass verges are *Campanula lingulata, Centaurea salonitana* and *Linum aroanum*. On gravel slopes and sometimes in the gravel of the river bed one may see:

Putoria calabrica	*Teucrium montanum*
Salvia candidissima	*Tussilago farfara*
Scutellaria orientalis	

The salvia is a particularly attractive species with a rosette of heart-shaped, white-felted leaves and an inflorescence to 30cm of large white flowers that have bright blue stigmas. The scutellaria is a low-growing species with yellow flowers and the tussilago our

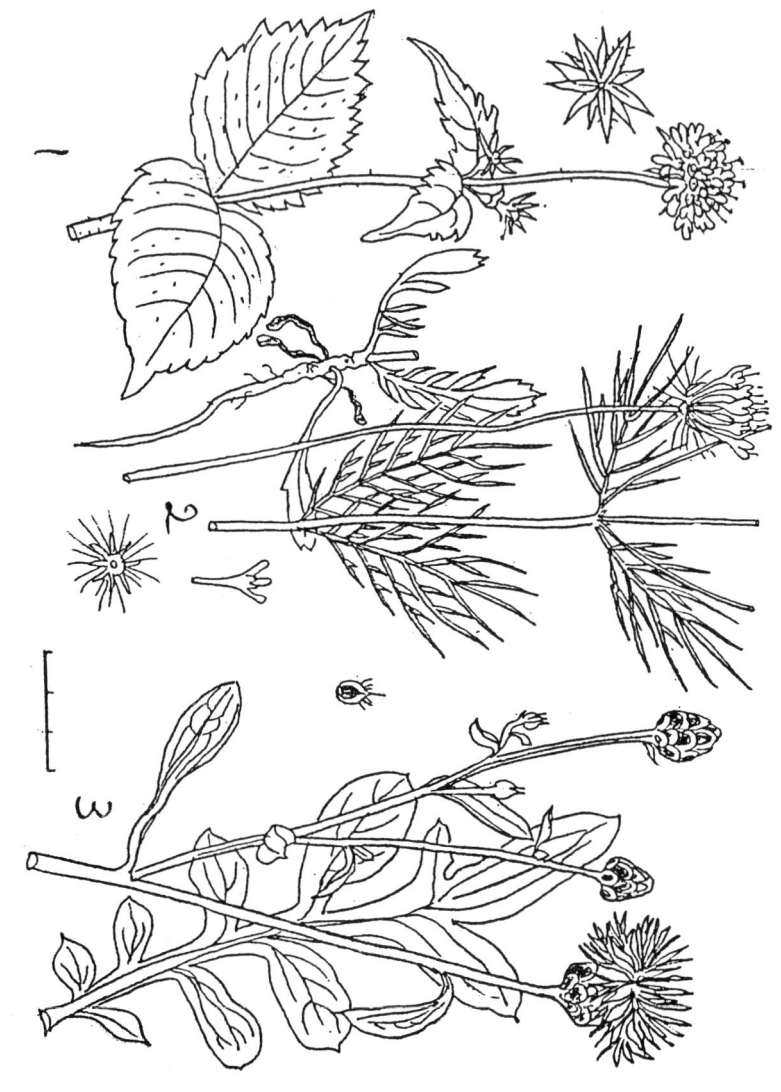

N. Greece 1. Knautia drymeia 2. Scabiosa tenuis 3. Centaurea salonitana

own coltsfoot that can be found growing under alpine conditions here. One may also be fortunate to see *Linum hirsutum* which at first sight resembles *L. narbonense* but the large flowers are paler blue streaked with darker blue veins and the leaves are hairy.

From Florina one drives out westwards for 30km to the turning to Prespa. The road rises through woodland and there are snow poles at the highest part. Much bracken grows in clearings and by the roadside one may see *Geum coccineum* with beautiful scarlet flowers. This is the true species; the so-called *G. coccineum* of gardens is nearly always *G. chiloense* from Chile. Amongst the plants to be found in woodland here is *Geranium versicolor* which is distinguished by its rather large white flowers with conspicuous dark purple veins.

The Prespes (or Prespa) National Park was founded in 1971. It embodies most of the lake Mikri Prespa (often referred to as Mikri Limni - literally the small lake) and part of Megali Prespi a larger expanse of water that is shared between Greece, Albania and 'Yugoslavia' and surrounding land which rises to 2120m. The lakes are at about 850m above sea level. There is a good asphalted road into the Reserve and about 2.5km after leaving the E86 one rises to a picnic point where there is a fine view of the lakes. Near here one may see the rosebay willow-herb *Epilobium angustifolium*, so common in Britain but rare in Greece. This beautiful 'weed' is an introduction from Alaska. On descending from here the road forks and it is most profitable to bear to the right signposted Megali Limni. One passes through grazed land with woodland and in summer past extensive cultivations mainly of climbing beans supported by canes from the local reeds tied up in wigwam fashion. The shores of the lakes are bordered with dense reedbeds in places.

The whole of the area has a somewhat mysterious back-of-beyond atmosphere and is designated as a reserve mainly to protect the interesting birdlife which includes breeding colonies of both white and Dalmatian pelicans and cormorants in large numbers. Other species such as ducks, egrets and the occasional sea eagle may be seen. Bear, wolf, fox, deer and wild boar are said to occur. Water-loving plants growing near the shores include the beautiful flowering rush *Butomus umbellatus, Caltha palustris laeta* which has rather smaller flowers than the usual form, and the waterlilies *Nuphar luteum* and *Nymphaea alba*. Plants noted in the drier areas include:

Anchusa hybrida
Campanula lingulata
Consolida ambigua
Echium italicum
Galega officinalis
Hypericum olympicum
Linaria genistifolia

Linaria peloponnesiaca
Onosma frutescens
Sambucus ebulus
Scutellaria galericulata
Stachys sp.
Veratrum album flavum
Verbascum macrurum

The two yellow-flowered and rather tall linarias occur together here. *L. peloponnesiaca* has numerous rather small flowers in somewhat tight inflorescences whereas *L. genistifolia* has whip-like branches and fewer, but larger flowers that give the impression of rows of small birds sitting on a branch. The stachys is a rather tall-growing plant with small, often trifed, leaves and pale mauve flowers. It is one of the *glutinosa* group but I have not been able to identify it with certainty. The veratrum is a green-flowered form sometimes referred to as *V. lobelianum*. The above list is based on a single visit and many other interesting plants may undoubtedly be found. The official hand-out for the park mentions an endemic *Centaurea prespana* and there were at least two species of this genus I was unable to identify. One was very tall and the other dwarf with finely-divided leaves and yellow flower heads.

16.11 **Mount Vermion:** This is an isolated mountain group some 60km west of Thessalonika and north-west of Mt. Olympus. It is somewhat drier and less interesting botanically than some of the others but has the advantage of a ski resort road, much used in winter by the inhabitants of Thessalonika. The nearest country town is Naoussa but it is also possible to obtain some accomodation at Kato Vermion, near the ski station of Seli at 1,400m. The road up from Naoussa passes through woodland which is well worth examining though the horseflies can be especially troublesome there in June. *Gentiana cruciata* and *Knautia drymea* (16.9) are two of the plants to be seen by the wayside.

The area around the ski centre is deserted in summer and very heavily grazed. Some of the few plants that escape the worst attention of sheep are *Astragalus angustifolius, Daphne oleoides, Juniperus communis, Stachys cretica* and *Potentilla recta*. Not far from the ski lift there is mixed woodland with a number of species common to the area but including *Geranium reflexum* which somewhat resembles *G. phaeum* with narrower and more-reflexed dark mauve petals and brown, hairy sepals. A speciality is *Isatis vermia* which is endemic to the mountain. It closely resembles the common woad *I. tinctoria* but the basal leaves are hairy on both sides and flowering takes place much later - in September and October. *Lilium chalcedonicum* is also found in some of the wooded areas. It takes several hours to get to the highest peaks where there is a chance of seeing some of the following:

Aethionema saxatile *Hypericum rumeliacum*
Aubretia deltoidea *Jovibaba hueffelii*
Corydalis densiflora *Primula veris columnae*
Crocus cvijicii *Ranunculus psilostachys*
C. sieberi sublimus *R. subhomophyllus*
Dianthus deltoides *Scutellaria alpina*
Draba lasiocarpa *Saxifraga paniculata*
Globularia bisnagarica *S. tridactylites*

Crocus cvijicii is a yellow, spring-flowering species which is probably at its best here in June. The problem is to know how to pronounce the name. Bowles (1952) in his book on crocus and colchicum aptly says " I have never discovered how this name should be pronounced, whether it is better to sneeze it or, as a witty friend of mine put it, "to play it on a violin". The species has very small corms and does not take kindly to cultivation.

N. Greece 1. Salvia candidissima 2. Geum coccineum 3. Geranium versicolor

17. SOUTHERN GREECE

17.1 The southern part of Greece consists essentially of the Peloponnese (Peloponesos peninsula or Morea) that is separated from the mainland by the Corinth Canal. The narrow isthmus of Corinth is no more than 5.5km wide so the region has somewhat the character of an island. It lies at about the same latitude as Sicily but is slightly smaller in area. The section dealt with in this chapter also includes the countryside around Athens since it is likely to be the starting point for most visitors to the region.

All plant enthusiasts should visit this part of the Mediterranean. In addition to plants common to Greece in general and a fair number of endemics; it is home for some species originating in Asia which are not found in northern Greece. A botanising excursion can be coupled with visits to the numerous ancient sites but an organised 'classical' tour is unlikely to give the best view of the extraordinary rich flora. Package holidays are available to resorts south of Athens and to seaside towns in the Peloponnese, such as Nauplia, but the area is too large for it to be covered adequately by staying at any one resort. Modest overnight accommodation is not expensive at the time of writing and, although the use of a hired car is a great help, it is possible to travel from place to place by bus.

The majority of the lowland plants can be seen at their best between the middle of March and the middle of May though some interesting bulbous species flower earlier. However, the higher mountains have a snow cover at this time of the year and most alpine species flower there in June and July. Autumn-flowering bulbous plants are to be seen in mountains and lowlands in September to November and a few winter-flowering plants are most likely to be at their best from November to February.

Several good 1:400,000 scale maps are available but, as for the rest of Greece, a few are cluttered up with names given in both Greek and the Roman lettering equivalents and there is no standardised method of transliteration - German maps often differ markedly in this respect from those published in Britain. The keen traveller who does not know Greek would find it well worth while mastering the alphabet.

17.2 **The terrain:** The Peloponnese is roughly like an outstretched hand with a thumb and only three fingers pointing southwards. All of these 'digits' have interesting plants but the middle finger which stretches to the Mani peninsular goes to the most southerly part of mainland Greece is the most exciting for the botanist. Most of the region is hilly and there are substantial mountains in the north, especially the Chelmos (Aronia) and Killini (Kyllene, Ziria) groups that rise to over 2,000m. Further south the Taiyetos (Taïgetos, Taygetos) range, running north to south and down to the Mani peninsular, also reaches over 2,000m and is often impressively snow-clad when viewed from the Plain of Sparta. The greater part of these mountains are composed of hard limestone, though a few areas of softer limestone, and sandstone occur. Igneous rocks are rare in the Peloponnese.

17.3 **Literature:** The most useful work for this area is 'Flowers of Greece and the Balkans' (Polunin 1980). It is relatively inexpensive and of a suitable size for a suitcase but it deals primarily with plants of the mountains. Flowers of Greece and the Aegean (Huxley & Taylor 1977) is a useful small handbook dealing with the more common species and is still available. An up-to-date treatise on the mountain flora is Mountain Flora of Greece, Volume 1 (Strid 1986) and Volume 2 (Strid & Kit Tan 1991); essentially a work for the specialist and considerably more concise than Flora Europaea about the mountain species of this area, though it describes only those species recorded above 1800m.

17.4 **Attiki - Athens and surroundings:** The visitor who has only a day in Athens can still see many typical native Grecian plants whilst visiting the ancient sites there. The Acropolis, which is probably the most impressive of all classical sites, is home to many wild plants. Around the lower part of the path leading to the top there are many bushes of *Medicago arborea* - a shrub growing to 3m, with yellow flowers and pods in a single flat spiral. Some of the

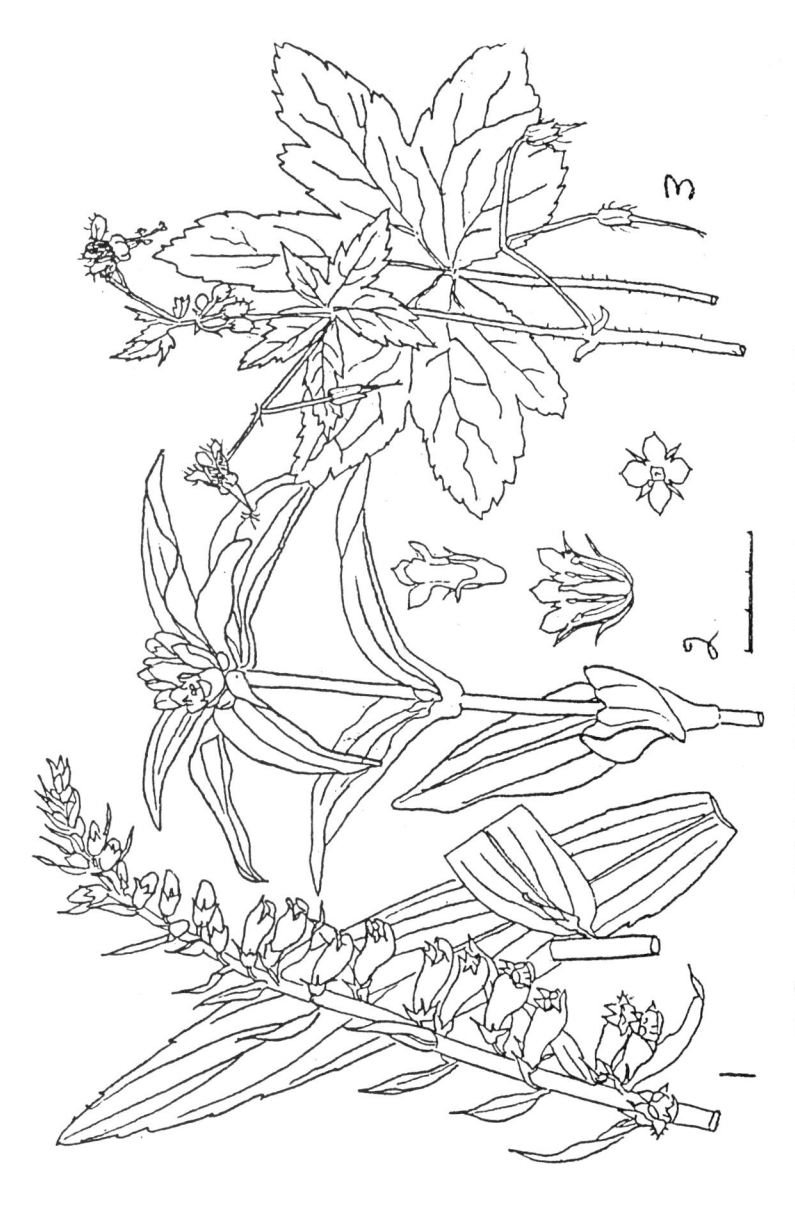

N. Greece 1. Digitalis viridiflora 2. Gentiana cruciata 3. Geranium reflexum

specimens here may have been planted, but it is certainly a native species rather typical of Greece. With it grows the smaller shrub *Teucrium fruticans* that is found throughout the Mediterranean. One may also see the 'round headed leek' *Allium sphaerocephalum* growing to ½-1m tall with dusky purple flowers; it is a rare British native sometimes grown for ornament in gardens. Species which occur higher on the hill, within and around the Parthenon site, include:

Alyssum saxatile   *Papaver rhoeas*
Asphodelus fistulosus   *Parietaria officinalis*
Calendula arvensis   *Reseda alba*
Diplotaxis erucoides   *Salvia verbenaca*
Erodium gruinum   *Scrophularia canina*
Ferula communis   *Tragopogon porrifolius*
Lamium moschatum   *Urtica pilulifera*
Malva cretica

No doubt a keen observer can add to these, all of which are widespread species found by roadsides and in waste places. The lamium often has 'variegated' foliage and the papaver is the large dark-red poppy that has black blotches at the base of the petals and is so greatly admired by visitors, even those with no special interest in plants. It is very different from the relatively pale form of the species we see wild at home and some authorities recognise a similar species *Papaver apulum*.

Another site popular during fine weather with tourists and Athenians alike during fine weather is the Lykavitos. This is also on a hill overlooking the city and it is possible to get to the top by cable car. However, it is not a very arduous climb along the path and by tackling it on foot one may be rewarded by seeing some interesting plants. The most notable is *Campanula rupestris* (*C. celsii* in Polunin (1980)) - a variable, and beautiful monocarpic species that has suffered name changes. It grows in rock crevices and forms rosettes of softly hairy leaves followed by branched stems lying flat against the rocks and carrying mauve-blue flowers. It is found at low altitudes as well as on some of the high mountains. A selection of the plants seen at the Acropolis can be found here and, in addition, the following have been noted:

Biscutella didyma   *Phagnalon rupestre*
Euphorbia characias   *Psoralea bituminosa*
Leopoldia comosa   *Silene colorata*
Malcomia maritima   *Verbascum undulatum*
Misopates orontium   *Vicia dasycarpa*

The vicia, which is sometimes classed as a sub-species of *V. villosa*, usually has bicolour mauve/white or mauve/pale mauve flowers. It is widespread and common throughout southern Greece, sometimes covering wide areas.

About 50km south of Athens lies the Cape of Sunion which is easily reached by bus to make a pleasant day out from the capital. Many of the plants seen in Argolis (17.5) can be found here and a few which are especially worth looking for in summer include; *Althea cannabina*, *Convolvulus oleifolius* and *Hypericum empetrifolium*. The althea is a hollyhock-like plant which grows to 2m tall and has deeply cut leaves and relatively small pink flowers and the hypericum is a small shrub with heath-like leaves. Visitors there in the autumn may enjoy the bulbous and tuberous-rooted species common to this part of the world - *Cyclamen hederifolium, C. graecum, Narcissus serotinus, Spiranthes spiralis* and *Sternbergia lutea*.

Mount Hymettos (Ymettos), which rises to 1026m, is even nearer than Cape Sunion to Athens and a road leads to the summit. The top is treeless and from here one may have an introduction to the mountain flora of Greece though it is by no means so rich in species as the higher peaks. A visit in autumn and early winter is well worth while for several interesting bulbous species are to be found here including:

Colchicum cupanii   *Crocus laevigatus*
Crocus boryi   *Scilla autumnalis*

C. cartwrightianus   *Sternbergia sicula*

It is not easy to identify the crocus species without digging up some corms to examine their covering. However, *C. boryi* has creamy white flowers and *C. cartwrightianus* is usually lilac purple. *C. laevigatus* may vary from white to lilac.

Mount Parnes (Parnis) lies some 30km to the north of Athens and there is a road that leads nearly to the top. At 1413m this has more of an alpine feeling than Hymettos and carries quite an interesting flora. Around the melting snow in spring one should look for *Crocus sieberi var atticus* with pale lavender to deep violet blue flowers and a yellow throat. It is locally common here but restricted to this area and one or two places in southern Euboa and the island of Andros. Other species here include:

Aethionema saxatile	*Gagea reticulata*
Anemone blanda	*Hermodactylus tuberosus*
Aubretia deltoides	*Iris pumila attica*
Corydalis solida	*Orchis quadripunctata*
Doronicum caucasicum	*Tulipa orphanidea*

The aethionema is a sub-species graeca which is, not surprisingly, endemic to Greece. The aubretia is the parent species from which our garden varieties have been developed. The tulipa has cup-shaped flowers on a bronze red colour which are paler on the outside. It used to be common as a cornfield weed throughout Greece but it is declining with the increased use of herbicides, though still common enough in some places to be picked and sold in bunches by the roadside and in markets throughout Greece.

17.5 Argolis - Corinth, Mycenea and Epidavros: Most visitors leaving Athens for the Peloponnese will travel along the road to Corinth. The two mountains of Gerania 1351m and Pateras 1132m just north of here may be of botanical interest but they are not particularly accessible. If one digresses along the road that leads north-westwards to Thebes one can turn off westwards to Egosthena

(Porto Germano). This is a small seaside resort on the east end of the Gulf of Corinth and has the interesting ancient ruins of Aigosthena; it is also a convenient place to study the flora of the region. In the rocky area around the castle one can expect to see:

Acanthus spinosus　　　　*Lagoecia cuminoides*
Ballota acetabulosa　　　　*Micromeria graeca*
Campanula drabifolia　　　*Nigella damascena*
C. rupestris　　　　　　　*Stachys spruneri*

Other species in this area, especially along the approach road to the resort, include:

Anacamptis pyramidalis　　*Orchis italica*
Anthyllis vulneraria　　　*O. quadripunctata*
Centaurea mixta　　　　　*Thymelaea tatonraira*
Helianthemum hymettium　　*Trifolium uniflorum*
Ophrys lutea murbeckii

The centaurea has flat rosettes of leaves and stemless purple flowers. The helianthemum, found only in the Peloponnese and Crete, carries somewhat flattened heads of rather small yellow flowers and the leaves are covered with bristly hairs. Sometimes it is classed as a form of the variable *H. canum*.

The ruins of ancient Mycenae lie some 30km south-west of Corinth, just off the main road to Argos, and it is worth stopping on the way there in April to look for orchids. The specialities of the region are *Ophrys spruneri, Op. argolica* and *Op. reinholdii*. The first of these can sometimes be seen in large stands. It resembles a robust form of *Op. sphegodes* and is sometimes regarded as a sub-species but seems fairly distinct when seen in the wild. *Ophrys argolica* is on its home ground here, though it is found sparingly throughout the Peloponnese, at Delphi, Crete and Carpathos - the rounded, bee orchid flowers have a red-brown or maroon lip, usually with two lighter spots - hence the name 'Eyed bee orchid'. *Ophrys reinholdii* also has a dark lip with two spots though these

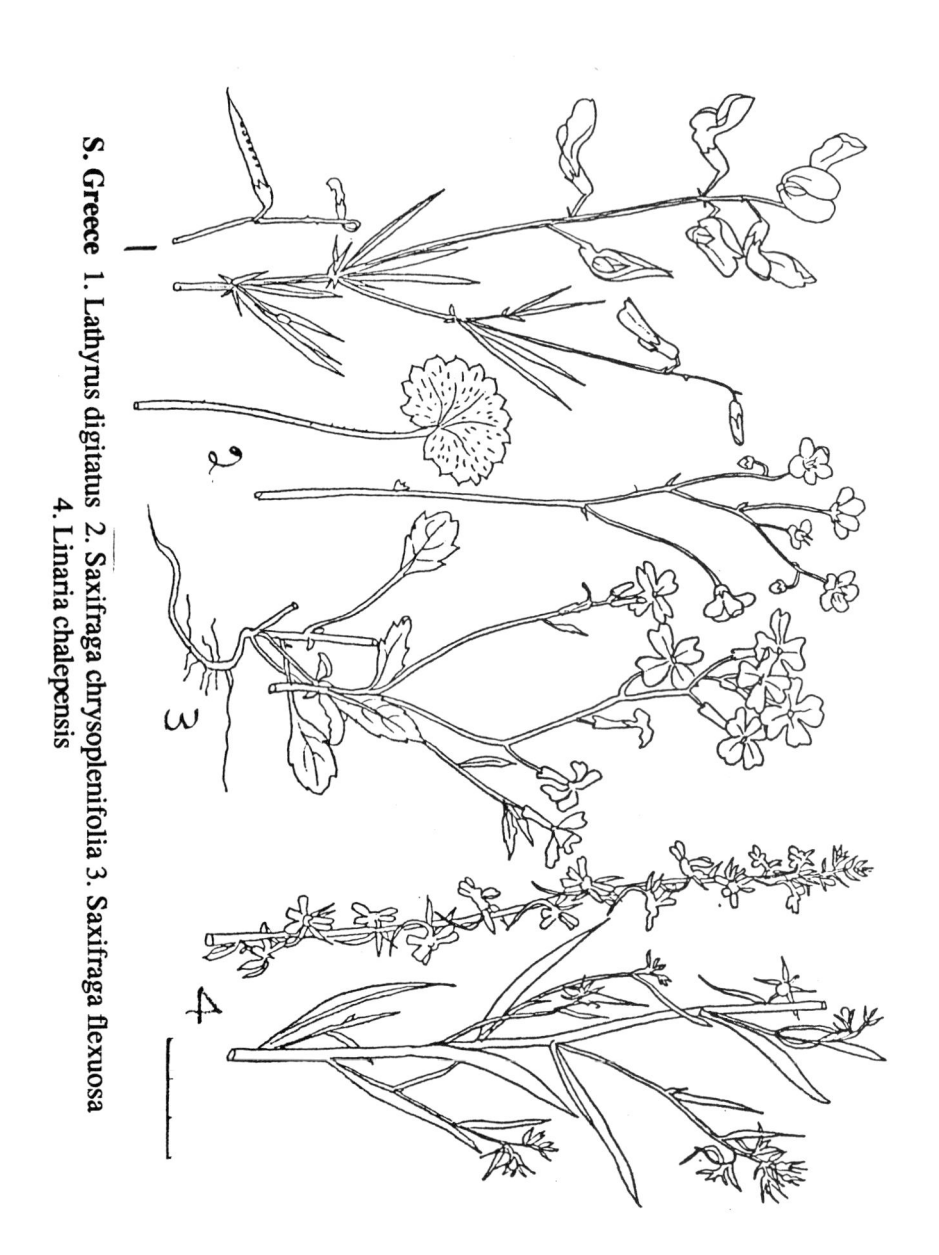

S. Greece 1. Lathyrus digitatus 2. Saxifraga chrysoplenifolia 3. Saxifraga flexuosa 4. Linaria chalepensis

are frequently joined to give an horizontal bar but considerable variation occurs and it can be confused with *Op. argolica.* It is found in some of the Greek islands and in south-west Turkey (23.9).

Weedkillers are used along the paths of the ancient site but not over the whole area and there is a restriction on grazing animals, so the vegetation can seem very lush in places. Here one can find:

Alyssum saxatile	*Phagnalon rupestre*
Anagallis foemina	*Psoralea bituminosa*
Asphodelus fistulosus	*Reseda alba*
A. microcarpus	*Scorzonera cana*
Lamium moschatum	*Trigonella balansae*
Lathyrus cicera	*Verbascum undulatum*
Papaver rhoeas	*Vicia dasycarpa*

Several specimens of the lamium, though not all, have variegated foliage. The lathyrus climbs amongst the grasses with large, single flowers of an unmistakable brick-red colour. The scorzonera is a somewhat unusual, though widespread species sometimes referred to as *Podospermum canum* or *Scorzonera jacquiniana.* It has yellow flower heads and leaves with narrow pinnate segments. The trigonella is a legume which grows to 50cm tall, has tri-partate leaves and heads of yellow flowers - it is especially common here in early summer. The vicia also grows in quantity, sometimes covering large areas as it does throughout the Peloponnese and the flowers are usually bicolour with dark and light mauve but here there are stands where they are wholly mauve coloured. It may be classed as a sub-species of the fodder vetch Vivia villosa and could be a survivor from cultivation.

On rocks one may see the beautiful *Campanula rupestris, Cotyledon horizontalis* and the yellow-flowered *Onosma frutescens* which is somewhat of a feature of the Peloponnese in early summer.

Apart from the well-photographed Lion Gate at the main site of Mycene, there is little to be seen except fallen stones but the nearby burial chamber is well worth a visit. There are patches of garigue near here with much *Phlomis fruticosa* and *Sarcopoterium spinosum*. Amongst these shrubs grow *Lathyrus digitatus* and *Petroraghia velutina*. The first of these is a vetch without tendrils that has very narrow, pointed leaves and inflorescences carrying up to seven flowers of reddish mauve colour with bluish coloured wings. The petrorhagia is a small dianthus-like plant with mauve-pink flowers about 1.5cm diameter and bifed petals. It appears to have a large inflated calyx but this is a series of bracts enclosing several flower buds, only one of which is usually open at a time. Huxley & Taylor (1977) say that *Fritillaria graeca* may be seen between boulders here in April. It occurs throughout the Peloponnese but is not easy to find.

To get from Corinth to the ancient site of Epidavros with its great ampitheatre, one can either take the road that passes near the east coast or go via Mycene and Nafplio (Nauplion). Both routes provide good opportunities for botanising. Furthermore, Nafplio is a rather charming resort which was once the capital of Greece and the seat of government and is well worth seeing and has good accommodation. Along the coast road one passes hills covered with pines and by the roadside are many of the plants already mentioned in this section. Additional shrubs include: *Calicotome spinosa, Cistus incanus, C. monspeliensis, Colutea arborescens, Euphorbia dendroides* and *Spartium junceum*. Both *Anemone pavonina* and *A. coronaria* can be seen here; the first of them is usually the form with scarlet flowers. They are often accompanied by *Muscari commutatum* and the autumn-flowering cyclamen *C. graecum*, and *C. hederifolium*.

The road from Nafplio to Epidavros passes through hilly country with mixed light garigue which includes the following shrubs:

Cistus salvifolius   *Pistacia lentiscus*
Euphorbia acanthothamnos   *Quercus coccifera*
Helianthemum lavandulifolium

Amongst these grow:

Anchusa variegata
Bellardia trixago
Centaurea mixta
Gagea graeca

Gynandriris sisyrinchium
Muscari commutatum
Onobrychis alba
Ophrys ciliata

Whilst at Epidavros it is worth taking a tour round the end of the Argolis peninsula to the coast at Drioli, then past Galatas, where one has a close view of the island of Poros, and along the coast to Ermioni and Kranidi to cross the Didimo hills back to Epidavros. Approaching the coast at Driopi there is maquis type vegetation with *Acer sempervirens, Arbutus unedo, Quercus coccifera* and *Lavandula stoechas*, all of which suggest that there is probably a pocket of acid soil. By the roadside one may see patches of the annual *Saponaria calabrica* producing a carpet of bright pink flowers. The higher parts of the road over the Didimo hills pass through patches of garigue with much *Juniperus sabina* and *Quercus ilex*. There are interesting plants here including the following:

Anemone blanda
Aethionema saxatile
Anchusa variegata
Centaurea mixta
Convolvulus althaeoides
Crocus chrysanthus
Cyclamen graecum

Gagea graeca
Iris pumila attica
Lathyrus digitatus
Malcomia maritima
Muscari commutatum
Parentucellia latifolia
Ranunculus subhomophyllus

Several orchids can be seen amongst these, especially *Aceras anthropophorum, Orphrys scolopax heldreichii, Op. lutea, Op. tenthredinifera* and *Orchis quadripunctata*. The aethionema has flowers like shepherd's purse which are followed by heart-shaped fruits held horizontally and with brown side wings. The crocus produces its yellow flowers as early as January or February and is not visible when the rest of the above plants flower in April to May. The ranunculus has parsley-like leaves and yellow flowers with a strongly reflexed calyx; it belongs to a complicated group of species that have a cluster of small oval tubers close to the base of the stem. The cyclamen flowers in autumn.

17.6 Lakonia - Mystra, Taiyetos: This sector embraces the central of the three fingers and stretches down to Cape Tenaro (also called Tainaron or Matapan), the most southerly part of mainland Greece and as far south as Tunis. In ancient times the people of this region, which included the Spartans, were said to be brief and concise in their speech and hence the word 'laconic'.

Travelling from Nafplio to Tripolis in early summer one is bound to see the attractive yellow-flowered *Alkanna graeca* growing on rocks. Other species not previously mentioned here include the small-flowered herb robert *Geranium purpureum*, the shrubby *Globularia apulum* with 2cm diameter mauve-blue flowers and *Vicia pannonica*. The last of these is a vetch with pale straw-yellow flowers though in some areas they may be purple or clear yellow. Another interesting species that one may be surprised to see growing in garigue here is *Vinca herbacea*. It closely resembles the common periwinkle but, as its name suggests it is herbaceous; the stems die back to the rootstock at the end of the growing season.

Present day Sparta is not a particularly attractive town and shows nothing of its former glory. However, the open market is always interesting and sometimes one may see bunches of flowers of *Tulipa orphanidea* and bulbs of the tassel hyacinth *Leopoldia comosa* for sale. The bulbs have been regarded as a kitchen delicacy since ancient times together with those of various species of muscari. Ruins of the ancient city are close by but little is left standing. The vegetation around is very lush in places and I once noted a plant that seemed to be a species of lamium but on close examination was probably *Stachys germanica* but it had white, instead of the usual pale pink, flowers.

From Sparta, one of the first places to visit is the old Byzantine city of Mistras (Mystra) which is beautifully situated in the foothills of the Taiyetos mountains some 5km from the modern city. Steep paved roads rise among the churches and other buildings to the Convent at the top. It is a flowery site and, although most of the

Campanula topaliana Delphi ruins, Greece. (p.259)

Verbascum macrurum Mt. Ossa, Greece. (p.258)

Viola graeca Mt. Parnassos, Greece. (p.260)

Quercus alnifolia Platres, Cyprus. (p.390)

plants to be seen there are common to the surrounding countryside, so many people visit the place that a rather full list of plants is given here. Perhaps the most noticeable in spring and early summer are the Judas tree *Ceris siliquastrum* which is wreathed in mauve flowers and the giant fennel *Ferula communis* that rises to 2 or 3m. Others include:

Alyssum saxatile	*Lunaria annua*
Anchusa hybrida	*Malcomia flexuosa*
Anemone pavonina	*Micromeria graeca*
Arisarum vulgare	*Onosma frutescens*
Asphodelus microcarpus	*Saxifraga chrysoplenifolia*
Bellevalia dubia	*Senecio squalidus*
Campanula rupestris	*Silene colorata*
Cymbalaria microcalyx	*Tragopogon porrifolius*
Euphorbia characias	*Trifolium stellatum*
Lathyrus aphaca	*Umbellicus erectus*
Leopoldia comosa	*Vicia melanops*

The cymbalaria and saxifrage are both uncommon species initially described from Mistra and Taiyetos. The first of these closely resembles the ivy-leaved toadflax but has a longer spur and is not unlike the Sicilian endemic *Cymbalaria pubescens* (13.12). The saxifrage resembles the common *S. rotundifolia* and grows on wet rocks especially near the newly restored Senate House.

The magnificent Taiyetos mountain range, which stretches approximately in a north-south direction for some 40km, is composed of hard limestone and dolomite. Its highest peak, Agios Ilias 2404m, is not difficult to climb, apart from the physical effort required. The eastern face of the range is cut by six deep gorges that carry away the snow melt to the river Evrotas whose water is used to irrigate the Plain of Sparta. All of these are of considerable interest to botanists and the one most commonly visited is the Langada Gorge for it is of easy access from the main Sparta to Kalamata road. On rock faces and steep rocky slopes by the roadside and in the gorge, sometimes dripping with water, one may find:

Arabis verna
Aubretia deltoidea
Bupleurum fruticosum
Colchicum boissieri
Heptaptera colladonoides
Iris unguicularis
Lamium garganicum
Malabaila aurea
Malcomia flexuosa
Minuartia pichleri

Onosma erecta
O. frutescens
O. montana
Scabiosa crenata breviscapa
Scilla messeniaca
Silene echinosperma
S. goulimi
Stachys candida
Thalictrum orientale

Several of these are uncommon or endemic species and may be difficult to identify. The colchicum produces its mauve flowers in autumn before the leaves appear and is distinguished by having curious long stolons instead of the usual bulbs of the genus. The heptaptera and malabaila are uncommon but rather insignificant umbellifers. The minuartia is a small cushion-forming species with white flowers - a member of the caryophyllaceae and resembling alsine. It is endemic to this area. The scabious is a neat, mat-forming plant with heads of mauve-pink flowers on 3-10cm stems. Although the species is widespread in the Mediterranean, it is very variable and the form *breviscapa* is endemic to southern Greece. *Scilla messeniaca* resembles a robust form of the much more common *S. bifolia*. The stachys is an attractive perennial species limited to this area, with rounded, white-woolly leaves and white flowers with small purple dots.

Much of the gorge is wooded with the oriental plane *Platanus orientalis* which is a typical feature of many parts of Greece, *Pinus nigra pallasiana* and the Greek fir *Abies cephalonica*. Around these grow the yellow-flowered shrubs *Phlomis fruticosus* and *Spartium junceum*. Under the canopy of trees there are carpets of *Cyclamen repandum* with its sweet-scented, rose red flowers in spring accompanied in places by primroses *Primula vulgaris*, the small white-flowered comfrey *Symphytum ottomanum*, *Anemone blanda*, yellow *Doronicum orientale* and a violet - probably *Viola riviniana*. Here and there one may come across a plant of a truly wild madonna lily *Lilium candidum*. In clearings one can also find *Hermodactylus tuberosus* with its green iris flowers in March.

The easiest way to get up Taiyetos is to take the road to the mountain shelter. About 10km south of Sparta on the main E961 road there is a turning to the west for the village of Paleopanagia from where a unsurfaced road leads to the hamlet of Poliana at 1650m. Near here is a rather obscure signpost that points to the refuge - a journey on foot which takes some 1½-2 hours. A further 2 hours' climb is needed to reach the summit at 2,404m and part of the way may be blocked by snow in April.

The unsurfaced road is rough but negotiable by car. At the lower part in spring there is a fine show of scarlet *Anemone pavonina* with *Aubretia deltoides* on the rocks and large groups of *Doronicum orientale*. As one climbs higher the road passes through black pine woods with more doronicum, primroses, *Anemone blanda*, *Symphytum ottomanum* and *Orchis quadripunctata*. Eventually one arrives at a charming area by a stream with plane trees and woods of the native *Abies cephalonica*. In these woods the winter-flowering snowdrop *Galanthus regina-olgae* is fairly plentiful and by the stream grow *Lamium bifidum*, *L. garganicum* and *Ranunculus ficarioides* together with a dactylorhiza species. In a flat area by the stream there are thousands of plants of the interesting little aroid *Biarum tenuifolium* and bushes of the very prickly wild gooseberry species *Ribes uva-crispa*. The road continues upwards to the edge of the tree line and in places there are plants of a particularly fine form of *Ornithogalum nutans*, with its racemes of greenish-white flowers. The rocky turf above here is home to several interesting species including:

Alkanna graeca	*Hermodactylis tuberosus*
Anchusa hybrida	*Lathyrus digitatus*
Anemone pavonina	*Lilium chalcedonicum*
Anthemis chia	*Ranunculus ficarioides*
Aristolochia pallida	*R. psilostachys*
Colchicum boisseri	*Scilla bifolia*
Corydalis densiflora	*Stachys candida*
Erysimum pectinatum	*Viola parvula*

The alkanna is probably the mountain sub-species *baeotica*. The colchicum, which flowers in autumn, has already been mentioned in connection with the Langarda Gorge. It usually produces only two leaves in spring. The corydalis closely resembles *C. solida* and is sometimes classed as one of its sub-species but has a more compact inflorescence of pink flowers. The erysimum, with small yellow flowers, is a Greek endemic. The beautiful scarlet turk's cap lily flowers mainly in July. The viola is a tiny annual pansy with yellow flowers and hairy leaves.

Near the top of the unsurfaced road there is a track that leads off to the katafigian or refuge at about 1600m and the journey there takes about 1½ hours on foot. The flora is not so different here to that lower down but one may see other species such as *Morina persica* which produces its white, pink-tinged flowers on 1m tall stems in June and the unusual foxglove *Digitalis laevigata* that has small rusty-coloured flowers. Some orchids are here including *Anacamptis pyramidalis* and the beautiful red helleborine *Cephalanthera rubra*. *Nigritella rubra* also has been recorded from Taiyetos.

It takes about 2-2½ hours from the refuge to climb along a fairly well marked track to the summit where there is a small chapel. Snow covers most of this area in the spring so it is not worth attempting the journey until most of this has melted and the plants are coming into flower in June. There are numerous treasures to be searched for here, including the less common and endemic species:

Acantholimon echinus	*Colchicum graecum*
Achillea nobilis	*Jurinea taygetea*
A. taygetea	*Micromeria taygetea*
Alyssum taygeteum	*Nepeta camphorata*
Armeria canescens	*Rindera graeca*
Astragalus taygeteus	*Saxifraga marginata*
Campanula papillosa	*Scutellaria rupestris*
Centaurea athoa	*Trifolium parnassi*
Cerastium candidum	*Veronica erinoides*

It is not possible to describe all of these specialised species in detail and the enthusiastic reader should refer to The Mountain Flora of Greece Vol 1 (Strid 1986) and Vol 2 (Strid & Kit Tan 1991). The campanula has solitary, stalkless blue flowers set in rosettes of hairy leaves and is unlike any other Greek member of the genus. It flowers later than most other plants there - in August and September. The rindera is a member of the Boraginaceae somewhat resembling a moltkia. From a woody rootstock with grey bristly leaves it produces a stem up to 20cm tall with heads of small, dark, reddish-purple flowers in June.

17.7 **The Mani:** This is the name given to the peninsula due south of Sparta. It is a distinct region with its own special history. The people who lived there were a tough independent group never fully conquered by the Ottoman Turks. Not only did they hold their enemies at bay, they also feuded amongst themselves, carrying out family vendettas, and for this purpose they built many tall square towers which are specially noticeable at Vathia in the south. Now, the buildings are falling into ruin and most of the inhabitants have left, taking their livestock with them. The combination of the rocky terrain and relative lack of grazing in places makes the area of very special interest to plant enthusiasts.

Many visitors to the area will arrive at the port of Yithion (Yithio, Githion) in the north east and perhaps come to see the caves at Pirgos in the north west. *Euphorbia dendroides, Phlomis fruticosa* and *Salvia triloba* are frequently seen here and in other parts of the peninsula. A number of fairly common species grow in great profusion in places, including:

Allium neapolitanum	*Parentucellia viscosa*
Centranthus ruber	*Psoralea bituminosa*
Convolvulus althaeoides	*Verbascum undulatum*
Crepis rubra	*Vicia dalmatica*
Erodium gruinum	*V. melanops*
Malcomia maritima	

Some other, rather less widespread, species to be seen in flower from March to May include:

Asphodeline lutea	*Lupinus micranthus*
Campanula spatulata	*Onosma fruticosa*
Cerinthe major	*Petrorhagia glumacea*
Consolida orientalis	*P. velutina*
Echium italicum	*Scabiosa argentea*
Euphorbia characias	*Silene gigantea*
Isatis lustianica	*Stachys germanica*
Leopoldia comosa	*Trifolium aurantiacum*

The campanula is a noticeable feature of the Peloponnese with thin, sparsely-leaved stems carrying large mauve-blue flowers which are held upright. It grows in grass and between shrubs. The consolida is a larkspur - an annual with delphinium-like flowers. *Petrorhagia glumacea* is a dianthus-like plant with pink flowers about 2cm diameter. *P. velutina* is similar but has smaller flowers with bifed petals. The silene is a robust species which usually has white flowers. The scabiosa is an attractive and variable species with pinnate lower leaves and the outer florets much larger than the inner ones. The plants here differ from the norm by having mauve flowers, whereas they are usually described as yellowish white or pinkish white. It somewhat resembles *S. caucasica*, which at one time enjoyed a great vogue as a cut flower and it would be worth an introduction to British gardens if it proves to be hardy in our climate. The trifolium resembles a very large hop trefoil with heads of yellow flowers that turn orange brown as they age. A fortunate visitor may see the rare *Fritillaria davisii* amongst these in February and March. It closely resembles *F. graeca* but is much shorter, growing with the lowest leaves at ground level and dull purple flowers. It is well worth visiting the Mani in September and October with several tuberous and bulbous-rooted species in flower:

Allium callimischon	*Mandragora autumnalis*
Campanula versicolor	*Narcissus serotinus*

Colchicum boissieri *Scilla autumnalis*
C. variegatum *Spiranthes spiralis*
Crocus goulimyi *Sternbergia lutea*
Cyclamen graecum *Urginea martima*
C. hederifolium

The allium has thread-like leaves which appear in autumn; the flower stalk grows in spring but the small, whitish, bell-shaped flowers do not open until the autumn. The campanula has a rosette of basal leaves and a tall unbranched 1.5m stem carrying flowers that are pale blue with dark blue centres. *Colchicum boisseri* has yellow anthers whereas *C. variegatum* has black anthers and larger, strongly tessellated flowers. The crocus is a distinct species, only recently discovered and is probably confined to the Mani. Flora Europaea describes it as from near Areopoli - on the north-west coast of the Mani. It has lavender-coloured flowers with a white throat and yellow stigma and stamens.

Towards the southern end of the peninsula there are patches of garigue with the shrubs *Calicotome villosa, Euphorbia acanthothamnos* and *Sarcopoterium spinosum* among which grow:

Centaurea cyanus *Gagea graeca*
Cerinthe major *Gynandriris sisyrinchium*
Cyclamen graecum *Linum pubescens*
Erica manipuliflora *Onosma fruticosa*

The centaurea is the common blue cornflower and this contrasts with the linum which is a short-growing annual with relatively large pink flowers. It grows also in Israel (24.5). The erica resembles a short form of *E. arborea* and produces its pink flowers in autumn, not in spring.

Near the very end of the peninsula there are clumps of *Matthiola incana* - the wild form of our garden stock which usually grows in sight of the sea and has scented purple or white flowers. Another interesting plant that may be seen here but is most common on the west side of the peninsula is *Astragalus lusitanicus*

- a strong growing species rather resembling a broad bean but with large inflated pods. It has an unusual distribution, being found in southern Portugal (4.4), southern Greece and Cyprus (22.10) though the eastern form is slightly different and is referred to the sub-species *orientalis*.

Orchids abound in the Mani and other parts of the Peloponnese, including:

Anacamptis pyramidalis	*Ophrys lutea*
Barlia longibracteata	*O. reinholdii*
Neotinea maculata	*O. scolopax cornuta*
Ophrys argolica	*O. sphegodes*
O. carmelii	*O. spruneri*
O. ferrum-equinum	*O. tenthredinifera*
O. fusca	*O. ciliata*

17.8 Achaia, the northern mountains: Travelling north from the Mani, many visitors will wish to visit Ancient Olympus near the north-west coast. In their book on The Flowers of Greece, Huxley and Taylor (1977) describe this site as having 'flowers in plenty' including orchids, notably the monkey orchid *Orchis simia*. However herbicides are now used there and it can be disappointing as a botanical site though the Judas trees make a good show in spring. Nevertheless, *Anemone pavonina* in both the red and mauve forms may be seen in plenty at places beside the road and in woods the oak trees are sometimes infested with mistletoe and the somewhat similar *Loranthus europaeus* which has yellow fruits that drop off and litter the ground below in spring. Under the trees grow the orchid *Limadorum abortivum* and the small comfrey with yellowish flowers *Symphytum bulbosum*.

There are four main mountain groups in Achaia and the neighbouring region of Korinthia. They are Erimanthos (Lambia 2224m), Panahaiko (Voïdia 1926m), Aronia (Helmos 2341m) and Killini (2376m). Of these the Aronia group, often referred to as simply Chelmos or Helmos, is by far the most easily accessible and will be described in most detail. However, the others are well worth exploring by the enthusiast and all except Erimanthos have mountain huts.

The best way to approach Chelmos is from the small town of Kalavrita. It lies at the southern end of the spectacular Vouraikos gorge, which is itself well worth traversing, though it is not accessible to motor vehicles but has a charming small-gauge railway leading to the coast on the Gulf of Corinth. There is a good new road that takes one from Kalavrita, up past a war memorial to the ski-lift on Chelmos. Amongst the weathered junipers and prickly hummocks of *Astragalus creticus rumelicus* and melting show in April and early May one may see:

Corydalis densiflora	*Ranunculus subhomophyllus*
Crocus sieberi	*Scilla bifolia*
Gagea fistulosa	*Viola aetolica*
Ranunculus brevifolius	

The gagea is a rather fine species with relatively large flowers and somewhat hairy stems. Both ranunculi are dwarf buttercups, usually with only one flower per stem and spreading sepals in *R. brevifolius* which has entire, or shallowly, three-lobed basal leaves and those of *R. subhomophyllus* are divided into narrow lobes. The viola is an annual, or short-lived perennial, pansy with pale yellow flowers 1-1.5cm in diameter.

It is necessary to wait until mid June when most of the snow has cleared before proceeding to the refuge hut at 2100m and beyond to find some of the rarer plant treasures of this mountain. There are so many interesting species here that an enthusiast may find it worth while spending a week or more in the area. In addition to those cited above, it is possible to see the following:

Aquilegia ottonis	*Rindera graeca*
Arabis bryoides	*Saxifraga exarata*
Arnebia densiflora	*S. rotundifolia*
Asperula boissieri	*S. scardica*
A. oetaea	*S. sempervivum*

Aster alpinus
Asyneuma limonifolium
Gentiana verna
Linum punctatum
Potentilla speciosa

Silene auriculata
Solenanthus stamineus
Verbascum acaule
V. delphinantha

Some of these have already been mentioned but to be sure of the identifications it will be necessary to consult the new 'Mountain Flora of Greece' (Strid 1986, Strid & Kit Tan 1991). The arnebia is an especially interesting member of the Boraginaceae with a rosette of hairy leaves which tend to wither at flowering time and produce domed inflorescences of rather large yellow flowers. It is found only in this part of Greece but is widespread in Turkey and is well illustrated in Polunin (1980) - though under the name *Macrotomia densiflora*. The linum is a perennial with prostrate stems carrying rather large blue flowers. It also grows in the Madone and Nebrodi mountains of Sicily (13.9) and in northern Greece (16.9). *Viola delphinantha* has narrow leaves and mauve-pink flowers with narrow petals and a very long spur. It is a distinct species, also found in northern Greece and Bulgaria. There is an Asiatic component in the flora here but also a north European one with the presence of *Gentiana verna*.

In grassy places one may find *Tulipa australis*, *Fritillaria graeca* and *Dactylorhiza sambucina* and an expedition to the valley of the Styx can be very rewarding botanically. It is advisable to obtain local advice on the best and safest way to get there.

The other four mountain groups of the region are well worth visiting for their flora though if time is limited, Chelmos would be the first choice. Killini can be reached by taking the road south out of the seaside resort of Xilocastron on the shore of the Gulf of Corinth, some 35km north-west of Corinth town. This leads to Trikala and from there a track winds up to the refuge at 1650m. The flora here resembles that of Chelmos. In June and July one can expect to see *Aster cylleneus* - named after the mountain - which is, in fact, a form of the widespread and variable *A. alpinus* that is common in the Alps and in northern Greece and also found on Chelmos.

Panahaiko is fairly easily accessible from the city of Patras. One should take the road southwards to the village of Kastritsi and from there on a track leads to the summit (Voïda 1926m) and the refuge at 1500m. One special advantage of visiting this mountain is that the staff of the Botanical Institute of the University of Patras is actively involved in the study of the native flora and can be of considerable help to serious enthusiasts. Polunin (1980) lists some of the species to be seen here; the best time to go there is June or July. Erimanthos lies south of Panahiko and is less accessible though a road south from Patras goes through Kritharakia from where there is a road eastwards to the village of Agios Dimitrios and thence a track up the mountain. A thorough study of the flora of this area would be an interesting challenge for an enthusiast.

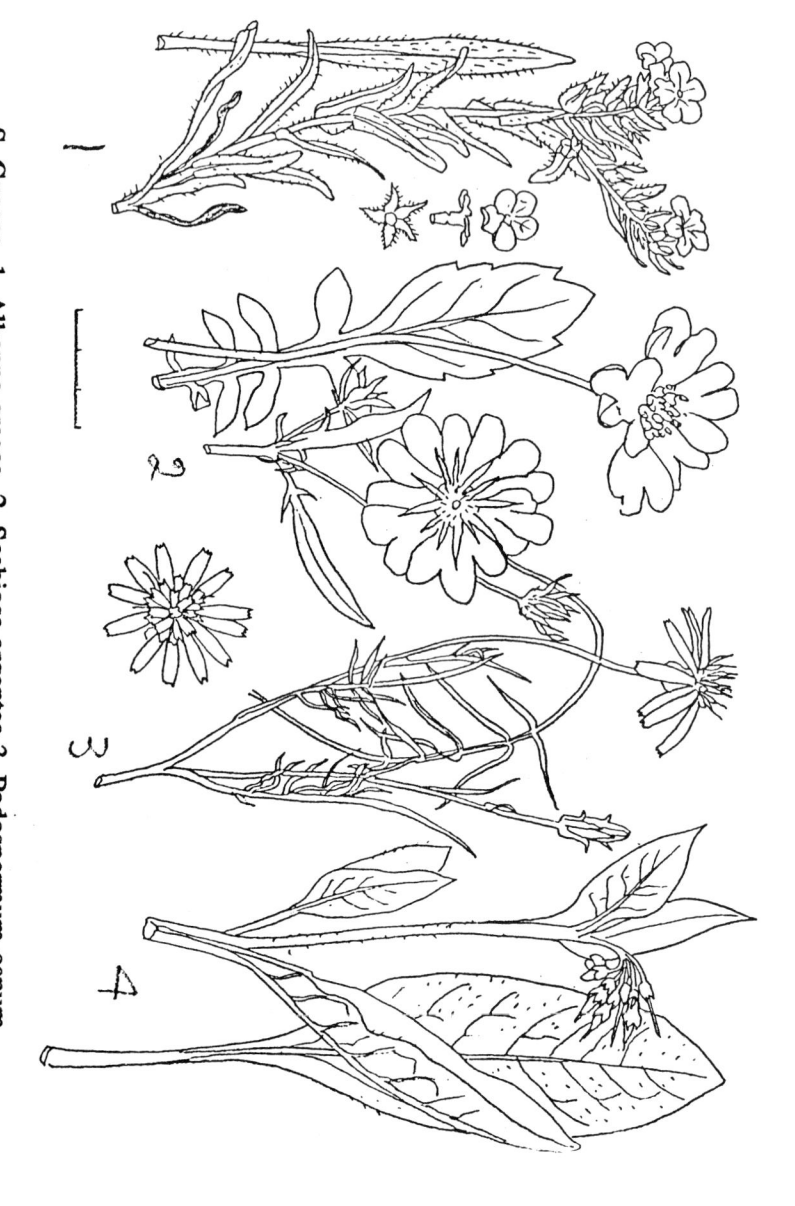

S. Greece 1. Alkanna graeca 2. Scabiosa argentea 3. Podospermum canum 4. Symphytum bulbosum

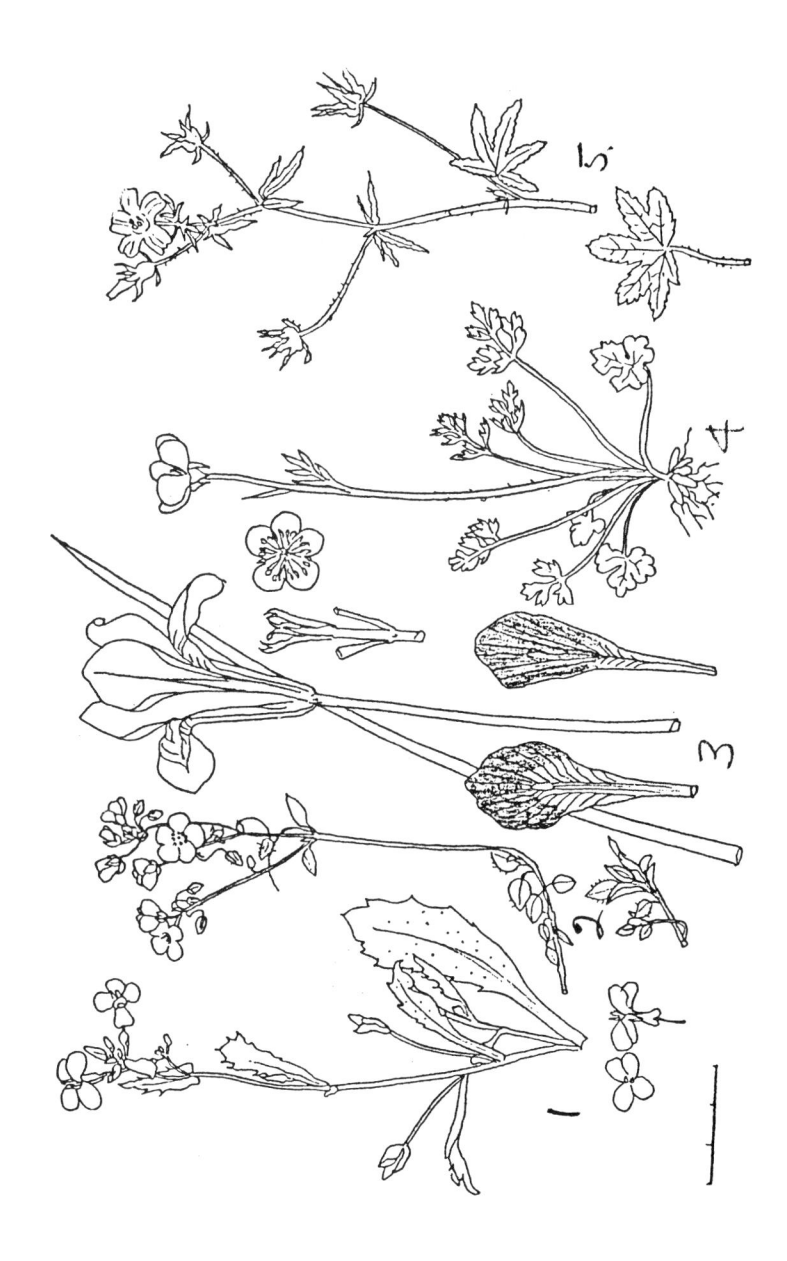

S. Greece 1. Arabis alpina 2. Helianthemum hymettum 3. Iris unguicularis 4. Ranunculus subhomophyllus 5. Malva cretica

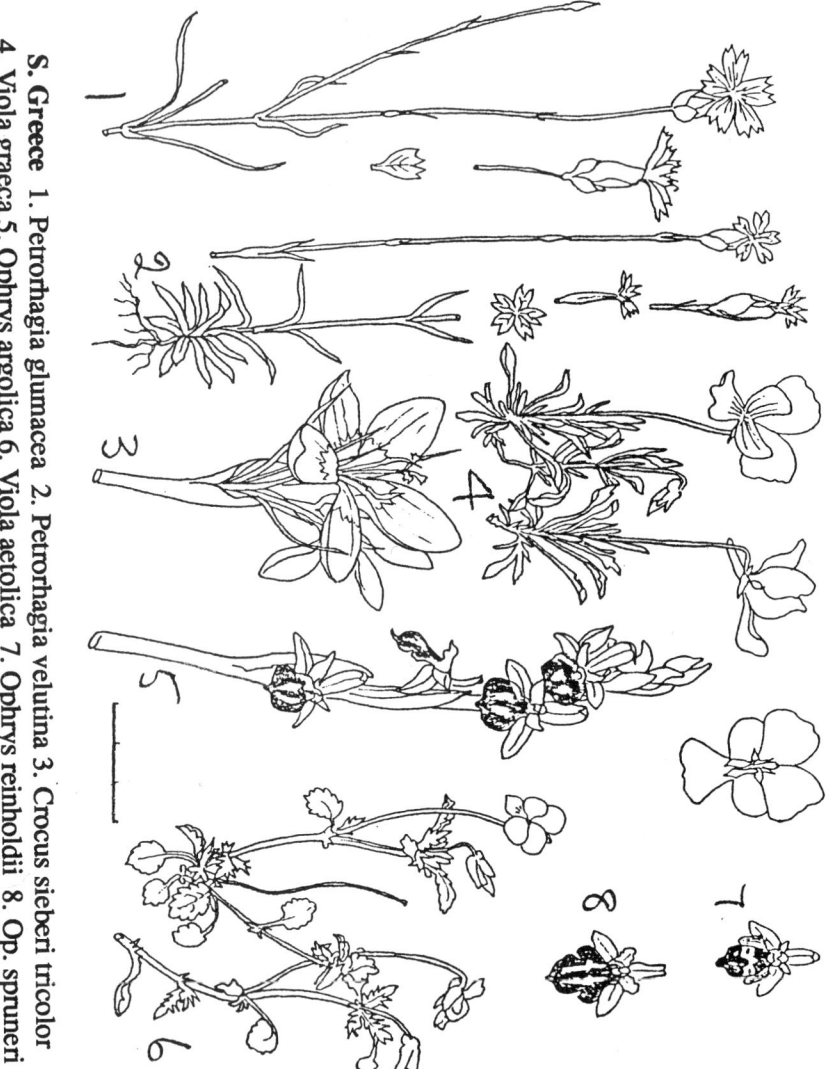

S. Greece 1. Petrorhagia glumacea 2. Petrorhagia velutina 3. Crocus sieberi tricolor 4. Viola graeca 5. Ophrys argolica 6. Viola aetolica 7. Ophrys reinholdii 8. Op. spruneri

18. CORFU AND CEPHALONIA

18.1 Corfu and Cephalonia, like the other Ionian islands which lie off the west coast of Greece, have a somewhat higher rainfall than those to the east of the mainland. For this reason they are more clothed with taller shrubs and trees than the Aegean islands and appear lusher and greener, especially in spring. In 1953 Cephalonia suffered a severe earthquake that destroyed most of the habitable buildings and caused many of the inhabitants to emigrate. As a result farming was considerably reduced leading to a low level of grazing - another factor favouring the lush appearance of the vegetation.

The island of Corfu or Corfou (Kerkira or Kerkyra in Greek) vies with the Costa del Sol for position in the league table as the most popular Mediterranean resort with British tourists. It becomes very crowded in Summer but from mid-March to mid-May, when the wild flowers are at their best, it is less populated and an excellent place for botanising. Indeed it might well be the first choice for the complete beginner to start looking at the Mediterranean flora. By contrast, Cephalonia (often called Kefallinia in Greece) has avoided mass tourism but is now being opened up to visitors though it is still a place for a 'quiet' holiday.

Travelling to Corfu is straightforward. It has an international airport served from London and from several regional airports in Britain. Corfu is also a good starting point for travel to the northern part of the Greek mainland with frequent boat services to Igoumenitsa and to Albania for the more venturesome. The distances on the island are small and one can botanise the whole area from a single venue with the use of a car or motor scooter or motor cycle. A hired pedal cycle can be useful for the energetic visitor though it is a stiff climb up Mount Pantokrator. Most of the resorts and accommodation are on the east coast, including the town of Corfu itself. There is, however, a sizeable resort at Paleokastritsa on the north west coast and smaller centres along the north coast at Ag. Georgios, Roda, and Archaravi. These northern resorts are probably the best places to stay for botanising as this part of the

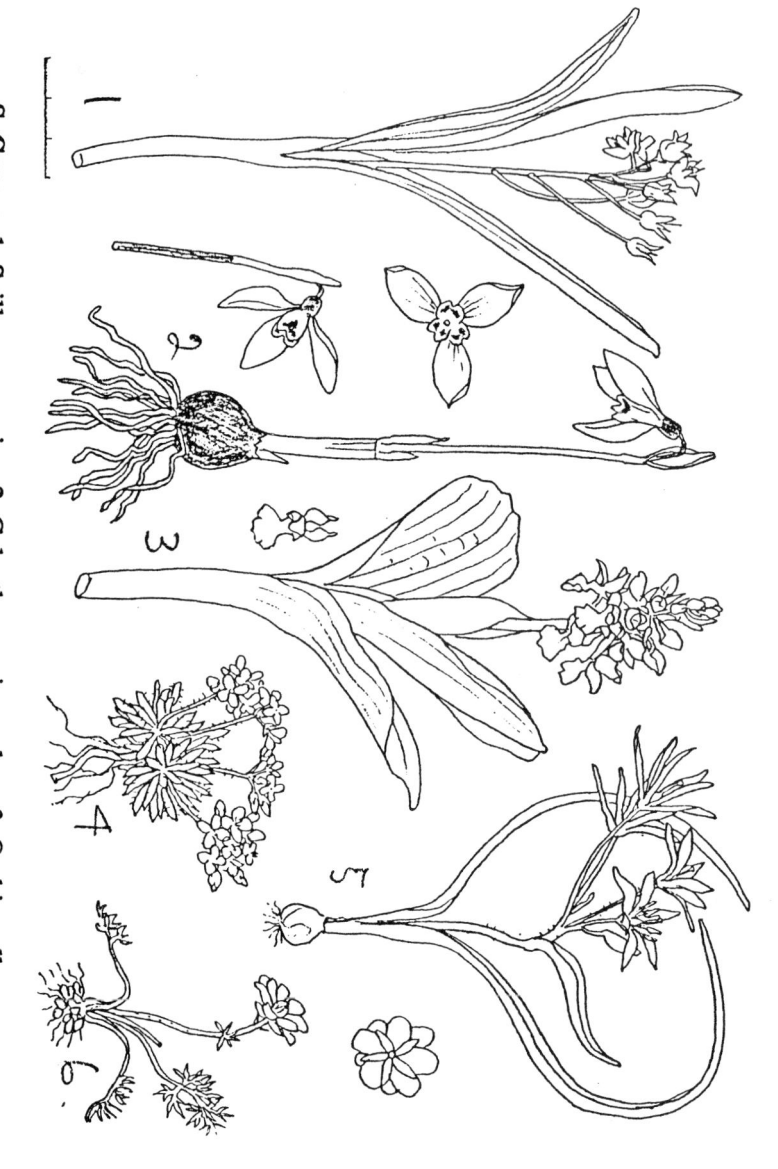

S. Greece 1. Scilla messenaica 2. Galanthus reginae-olgae 3. Orchis pallens 4. Draba parnassica 5. Gagea fistulosa 6. Ranunculus sartorianus

island is both the most interesting for its wild flowers and the least 'developed'. Maps of 1:100,000 scale are readily available on the island even though they may be difficult to find in Britain.

During the last few years an international airport has been built on Cephalonia so it also is easily accessible from Britain and one no longer needs to travel there via Athens. Most of the new resorts situated towards the south-east of the island are suitable for excursions to study the flora. However, Cephalonia is the largest of the Ionian islands and it is almost essential to have the use of a hired car. Many of the roads are still unsurfaced and hard on smaller vehicles and those with poor tyres. All plant enthusiasts visiting the island will wish to visit Mount Enos 1628m and this is not difficult for a road leads to the top though it is rough and unsurfaced near the summit. A 1:100,000 Efstathiadis map is available.

18.2 **Literature:** Books on the flora of Greece, such as Huxley & Taylor (1977), Polunin (1980) and the more general works on the Mediterranean flora as that by Blamey & Grey-Wilson (1993), Schönfelder (1990) are a great help but there is no popular literature which deals specifically with the flora of either of the islands. If one has a copy of 'The wild flowers of Crete' by Sfikas (1987) then it is worth taking it to Corfu for a number of the species described in it are found on both islands. Two recent papers that may be of help to serious enthusiasts are those by Georgiou (1988) for Corfou and Phitos & Damboldt (1985) for Cephalonia in English and German respectively.

18.3 **The terrain:** The northern part of Corfu is hilly, rising to 906m at Mount Pantokrator in the north east. Further south there are only isolated lower hills such as Ag. Deka 576m, not far south of Corfu town. Substantial cliffs rise from the sea in the north of the island, and interesting coloured sandstone cliffs occur at Sidari (Sidar) in the north-west where there are marshy areas. The extreme south of the island is mainly rather flat and with salt lakes at Alikes and Korission that are interesting for bird watchers and have maritime species of plants. The greater part of the island, especially in the central area, is covered with olive orchards which provide an important export commodity. Lemons and other citrus fruits are also grown.

Cephalonia is larger and considerably more mountainous than Corfu with several summits of 800-1000m rising as a central backbone to the island. Many of them are covered with thick maquis, without roads and not easily accessible. Fortunately it is easy to get to the top of the highest part Mount Ainos which has a cap of abies surrounded by treeless stony pastures. As in Corfu, olives are an important crop and there are substantial vineyards in the south-west which produce wines that are appreciated throughout Greece for their quality.

18.4 Corfou. The north-west, Paleokastritsa and The Troumpeta Pass: Paleokastritsa (Palaiocastritsa) lies on a headland with rugged sea cliffs partly clothed with shrubs. A feature here, and indeed in many parts of the island, is the number of the tall narrow cypress trees. The wild *Cupressus sempervirens* is a native of the eastern Mediterranean area but the narrow fastigiate form which originated in cultivation is often seen in Italy and grows frequently in the Ionian Islands but is rarely seen in other parts of Greece. In the quiet, pleasant corner of Corfu around Paleokastritsa fireflies can be seen on a warm spring evening and a few years ago the eerie call of a jackal could occasionally be heard during the night. Some of the plants that flower in fields and by the roadside here in spring are:

Allium subhirsutum	*Centaurium pulchellum*
Alyssum saxatile	*Fumaria capreolata*
Anagallis arvensis	*Geranium purpureum*
Aristolochia rotunda	*Helianthemum nummularium*
Bellis perennis	*Malope malacoides*
Calendula arvensis	*Smyrnium rotundifolium*
Cerinthe major	*Stachys cretica*
Convolvulus althaeoides	*Scrophularia peregrina*
Coronilla scorpioides	

The alyssum is the yellow-flowered species so common in rock gardens at home. The convolvulus here is the sub-species *elegantissimus* with more finely divided leaves than the type. The geranium is very similar to our herb Robert *G. robertianum* but has smaller and generally darker purple flowers. The smyrnium is an interesting umbellifer with yellowish-green foliage - of a kind to interest flower arrangers. Other plants that one might see when walking down to the village of Lipades, south of Paleokastritsa, include:

Crepis rubra   *Malope malacoides*
Dipsacus fullonum   *Ornithagalum narbonense*
Leopoldia comosa   *O.umbellatum*
Lunaria annua   *Symphytum bulbosum*

The lunaria is the annual species of 'honesty' and fairly common on the island. The symphytum is a rather small-growing comfrey with yellow flowers from which the stamens and stigmas project. Several orchids grow here, especially *Orchis morio* and *Orchis italica*. The giant orchid *Barlia longibracteata* has also been found but flowers much earlier than the others in February to March. Less common are the monkey orchid *Orchis simia* and *Serapias ionica*. The last of these is a particularly attractive plant with large orange-brown flowers similar to those of *S. neglecta* of which it is sometimes classed as a sub-species. It is only found on the Greek Ionian islands and far away in 'Yugoslavia'. There is a description and illustration of it in Buttler (1991). Another beautiful plant one may encounter in the grass near here is *Scilla hyacinthoides*. It produces flowering stems up to 80cm tall bearing many small, star-shaped blue flowers on long mauve pedicels. It is a Mediterranean native and often seen in Israel (24.8), but frequently grown for ornament and the plants here may be escapees - there is a record of it being found in the park of the Villa Monrepos near Corfu town in 1885.

An interesting short tour from Paleokastritsa is to cross over the northern range of hills by the Troumpeta pass. This can be done by taking the road towards Corfu town and turning northwards at Iatri to Tsolou and Skripero. In the grassy areas here grow many plants of *Gynandriris sisyrinchium* and fine specimens of *Ophrys ferrum-equinum*. As one climbs up the winding road of the pass there are patches of typical garigue. On the island the main shrubby species of this type of association include:

Calicotome villosa	*Phillyrea angustifolia*
Cistus incanus	*Phlomis fruticosa*
C. monspeliensis	*Pistacia lentiscus*
C. salvifolius	*Salvia triloba*
Erica arborea	*Sarcopoterium spinosum*
Genista acanthoclada	*Thymus capitatus*
Teucrium polium	

In places there are scattered bushes of *Arbutus unedo* with *Anemone pavonina* and *Ophrys lutea* growing below. The anemone seems to have narrower petals than the usual form and reminds one of *A. hortensis* (= *A. stellata*). A tall bugle *Ajuga genevensis* is also seen here together with the autumn-flowering *Cyclamen hederifolium* and ubiquitous *Asphodelus aestivus*.

At the end of the pass there is a road junction and the turning to the west takes one to Kastellani and Pagi and from there a winding mountainous road climbs to Prinilas from where one can return to Paleokastritsa. Here is fine scenery and a chance of finding more interesting plant species.

18.5 Corfou. Mount Pantokrator and the north east: A visit to Mount Pantokrator 906m is a highlight for the plant enthusiast. The usual approach is from Pyrgi (Piryi) on the east coast, some 15km north of Corfu town. Coming from Paleokastritsa, on the other side of the island, one will probably take the road towards Corfu town that joins the coastal road near Gouvia. A few kilometres before reaching the village of Sgombou there was a marshy area with thousands of tall specimens of *Orchis laxiflora* accompanied by the yellow flag iris *Iris pseudacorus*. Unfortunately this wet area may now be drained.

The road climbs steeply from Pyrgi to the village of Spartilas. In the grass verges and fields there are thousands of *Gynandiris sisyrinchium* in spring but they may not be noticeable in the morning for the flowers only open after midday and the foliage rather resembles the grasses amongst which they grow. There are also good forms of the attractive *Ophrys ferrum-equinum* here and plants of the 'modest' *Cruciata levipes* (O = *Galium cruciata*) - the crosswort that is fairly widespread in Britain.

At the next village of Strinilas one has reached a height where the winter is too cold for olive trees and the flora changes. In spring there are colourful carpets of the pink *Geranium brutium* studded with the white flowers of the Greek chamomile *Anthemis chia* and stars of *Ornithogalum umbellatum*. There is also a cerastium species which is difficult to identify with certainty.

Above Strinilas a good road continues through rocky country to the summit with scattered patches of *Quercus coccifera* garigue. Species to be seen here in the spring include:

Aceras anthropophorum	*Leopoldia comosa*
Anemone blanda	*Matthiola fruticulosa*
A. pavonina	*Muscari neglectum*
Ajuga orientalis	*Orchis provincialis*
Cynoglossum creticum	*Parentucellia latifolia*
Euphorbia myrsinites	*Ranunculus gracilis*
Fritillaria graeca	*Saponaria calabrica*
Geranium lucidum	*Senecio vernalis*
Lamium garganicum	*Veronica persica*

The ajuga is a rather tall bugle covered with woolly hairs and having small, very dark purple flowers. The fritillaria is the subspecies *thessala* with flowers that are mainly green on the outside but brownish within. The matthiola is a form with peculiarly brown flowers and the senecio is a dwarf ragworth with few, rather large bright yellow flowers.

There are also scattered plants of a yellow-flowered

erysimum which may be *E. raulinii* and a white-flowered arabis - perhaps *Arabis alpina*. Visitors in the autumn here may see *Crocus boryi, C. biflorus, Galanthus reginae-olgae* and *Sternbergia sicula*. The galanthus is sometimes referred to as *G. corcyrensis* and was claimed once as an endemic of Corfu, but it differs so little that it is no longer considered to be distinct from the above.

The summit of Pantokrator has a small monastery and is straddled by an ugly broadcasting mast, but on a fine day affords an excellent view to the east over the straits to the Albanian mountains that are usually snow-capped in spring. Early in the year, when the flowers are at their best, this is a quiet spot, untouched for centuries but with a hint of modernity. On one occasion I saw a monk in full robes repairing the road with a barrowload of smoking asphalt and in the distance an old shepherd was tending his sheep with a transistor radio playing very loud pop music. Presumably the signal from the mast on the summit was very strong.

18.6 Corfou. The north coast region: On the coast due north of Pantokrator there is a saltwater lagoon called Lake Antionoti with salt pans and on the seaward side of it lies the most northerly tip of the island Cape Agios Ekaterinis. The lake is more or less ringed with reeds and this is a fine place for bird watching where one can frequently see herons, spoonbills and sometimes glossy ibis. To the south of the lake there is woodland and to the north patches of turf, garigue and sand dunes next to the sea. The turf here, and indeed all the way eastwards to Kassiopi, is rich with orchids, especially:

Anacamptis pyramidalis	*Orchis morio*
Ophrys ferrum-equinum	*O. papilionacea*
O. heldreichii	*Serapias. cordigera*
Orchis coriophora frangrans	*S. lingua*

The *Ophrys heldreichii* is the horned orchid of the region and generally considered as a sub-species of *O. scolopax* - it differs only slightly from the other 'horned' sub-species *cornuta*. The

Orchis morio here often has pale coloured flowers and seems somewhat untypical of the species (18.7). In addition to these one may occasionally see *Ophrys bombyliflora* and *Orchis lactea*. Other plants associated with the orchids include *Echium plantagineum, Gladiolus communis, Petrorhagia velutina* and *Verbascum phoeniceum*. There is some doubt about the identity of the last, for the species is said to have purple flowers and the plants here have flowers of a distinct coppery colour and with the filaments of the stamens carrying numerous outstanding purple hairs. It is very like the plants seen in Albania (15.7) and would seem to resemble *Verbascum arcturus* that is said to be a Cretan endemic. It is hoped that some reader will solve this problem.

The composition of the garigue is similar to that already described (18.4), but there are other interesting species one should look for here, including:

Alkanna tinctoria	*Romulea bulbocodium*
Echium italicum	*Spartium junceum*
Narcissus tazetta	*Vitex agnus-castus*
Pyrus amgdaliformis	

The alkanna (=*A. lehmanii*) is the dyer's alkanet with beautiful blue flowers and was used at one time for the red dye produced from its roots. It is not the only member of the genus found on the island. *Alkanna corcyrensis* also occurs, especially in rocky places. It has whitish flowers and, in spite of its name, is not confined to Corfu but grows also on the western part of the Greek mainland and in Albania. The pyrus is a small spiny tree with white flowers and rounded fruits - it is especially common in Crete (20.8), on some of the other Greek islands, and also found in the south of France (10.4).

The north-west corner of Corfu is somewhat low lying and has extensive olive orchards with some very old trees. Because of its situation it receives the brunt of the rainfall and supports a river which flows into the sea on the north coast at Sidari. Here is a marshy area with water buttercups and other moisture-loving plants.

In April and May it is home to thousands of frogs which are often hunted by grass snakes. On the coast there are interesting cliffs of compressed sand layered in various colours from pale yellowish colour to a dark purple-brown. Amongst the plants here grow a form of *Anthyllis vulneraria* with dark red flowers, *Buglossoides pupurocaerulea* (= *Lithospermum pupurocaerulea*), *Trifolium stellatum*, *Hedysarum coronarium* and *Polygala nicaensis*.

On the north-west coast at the bay Ormos Ag. Georgiou amongst the *Cistus villosus* bushes I once saw a very robust specimen of the butterfly orchid *Platanthera chlorantha*, with a stem 30cm tall carrying 50 flowers. One would not expect to see this species so far south, except in the mountains, but it probably thrived here because of the relatively high rainfall. Growing with it were large numbers of the horned orchid *Ophrys scolopax heldreichii* and isolated specimens of the bee orchid *Ophrys apifera* together with large stands of a lupin that has blue flowers with a white mark on the keel - probably *Lupinus varius*.

18.7 Corfou. The South: The southern part of the island is well worth visiting, especially the coast, even though it is not so rich in plant species as the area further north. One place particularly worth seeing is the lake Limni Korission near the south west coast. Along the relatively narrow strip of sandy soil on the seaward side grow many typical shore plants:

Dittrichia viscosa	*Parentucellia viscosa*
Euphorbia parialis	*Plantago coronopus*
Juniperus oxycedrus	*Pseudorlaya maritima*
Medicago littoralis	*Silene colorata*
M. marina	*Thymus capitatus*

The lake is fringed in places with tall reeds and in wet areas one can expect to see *Alisma plantago, Lythrum junceum, Orchis laxiflora* and *Ranunculus ophioglossifolius*. There is an interesting form of *Orchis morio* here with rather pale and largely unspotted flowers. It is reminiscent of what is now called *O. morio ssp. syriaca*

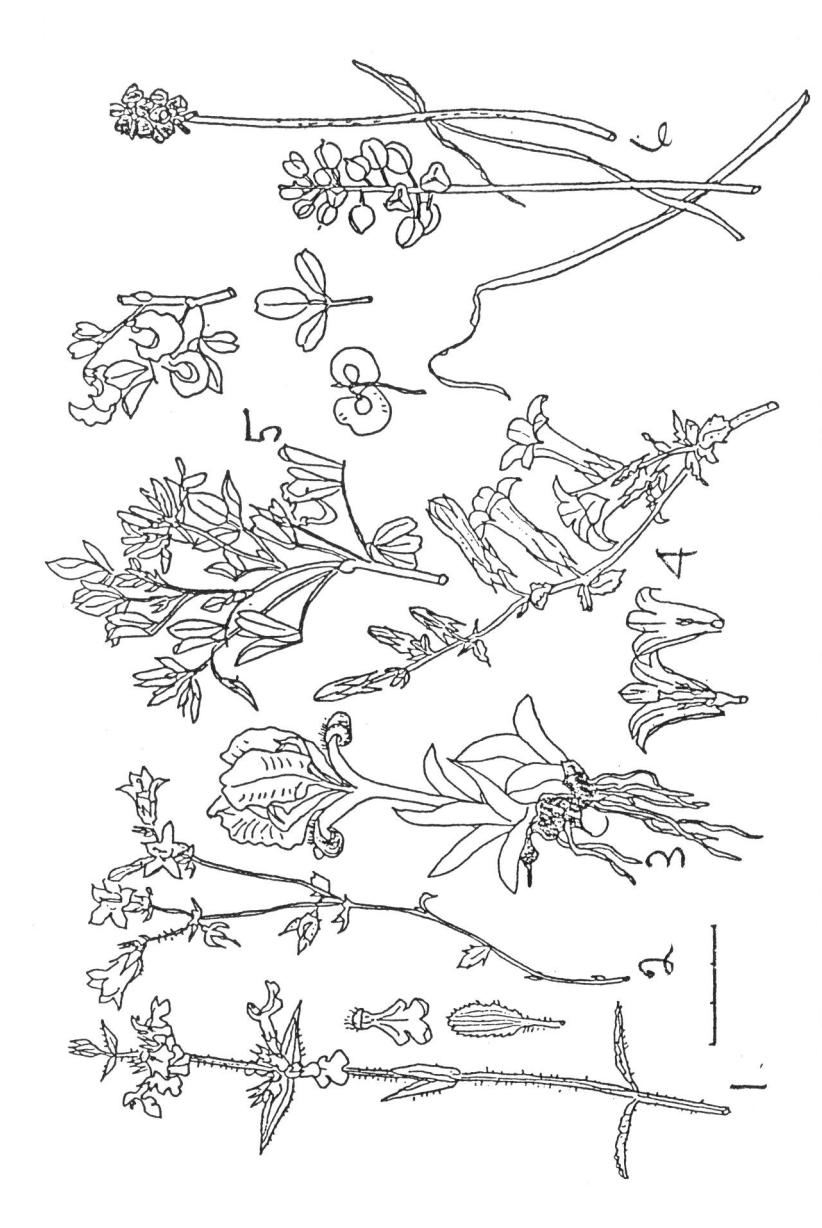

S. Greece 1. Stachys spruneri 2. Campanula drabifolia 3. Iris pumila attica
4. Campanula rupicola 5. Medicago arborea 6. Muscari commutatum

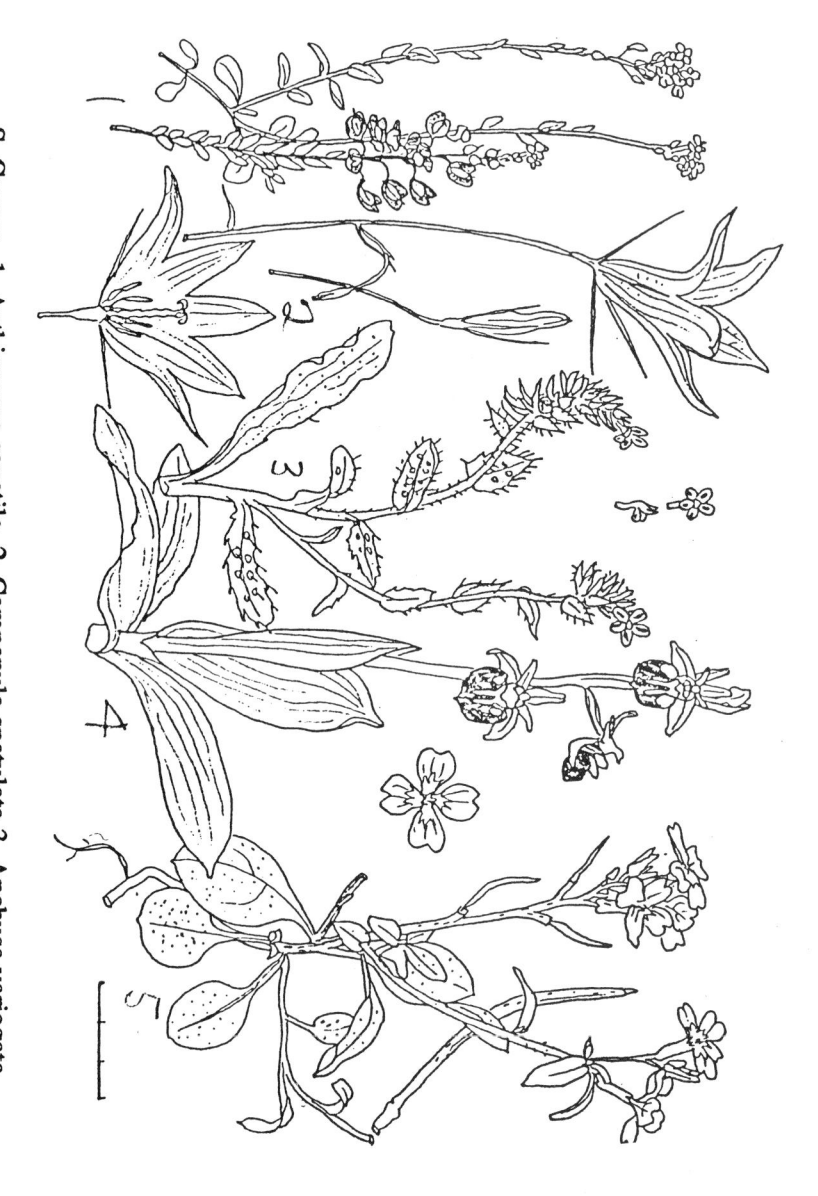

S. Greece 1. Aethionema saxatile 2. Campanula spatulata 3. Anchusa variegata
4. Ophrys sphegodes mammosa 5. Malcomia flexuosa

but it has some features of the newly-discovered *O. albanica* that is also included as a sub-species of *O. morio*. A similarity would not be surprising, for the Albanian orchid grows mainly along the west coast which is no more than 3km from Corfu at the nearest point in the north. In any case the green-winged orchid in Corfu is variable and unusual and would make a good subject for further examination.

    18.8 **Cephalonia. The South Coast Region:** Some of the coastal resorts have pleasant sandy beaches and Skala is a typical example. Here, at the small square where the buses stop, there is a row of *Poplus nigra afghanica* (= *thevestina*) - like a Lombardy poplar but with a whitish trunk (22.9). Growing in the sand by the shore one may see:

Anthemis maritimus	*Matthiola tricuspidata*
Cyperus kalli	*Pancratium maritimum*
Eryngium maritimum	*Papaver nigrotinctum*
Glaucium flavum	*Valantia hispida*
Malcomia maritima	

    All these common shore species can make a colourful show, especially the matthiola. Unfortunately they are vulnerable to mechanised cleaning of the beaches for the 'benefit' of the tourists and are often swept up in their prime. They may be seen also at most of the other coastal resorts accompanied, at times, by the rather tall annual *Silene dichotoma*, which often has pink rather than the usual white flowers in this region. In places the sand is fouled at times by the debris of *Posedonia oceanica*. I have watched it being harvested at Kato Katelios for use as a top-dressing on the fields nearby. Although this marine flowering plant can be troublesome at resorts, its presence is an indication that the water is relatively clean, for it is not able to tolerate high levels of pollution. It also provides a convenient nursery for fry and can help considerably in boosting local fish stocks.

    Where there is a rocky shoreline, as at the charming small

port of Poros, a number of interesting plants may be seen growing on the cliffs, including *Capparis ovata, Ptilostemon chamaepeuce* and *Lavatera arborea.*

By roadsides inland and around regions which have once been cultivated there are many interesting and colourful plants to be seen in spring. During a visit to the abandoned ruins of the old town of Skala destroyed by the 1953 earthquake, the following species were noted:

Aristolochia rotunda	*Orobanche ramosa*
Bellardia trixago	*Oxalis pes-caprae*
Campanula spatulata	*Parentucellia viscosa*
Foeniculum vulgare	*Plantago afra*
Gladiolus italicus	*Psoralea bituminosa*
Galactites tomentosa	*Reseda alba*
Knautia integrifolia	*Trifolium resupinatum*
Lythrum junceum	*T. stellatum*
Orobanche lavendulacea	*Vicia dalmatica*

The bellardia and parentucellia often grow together in large groups and at first sight they are sufficiently similar to be taken as pink and yellow-flowered forms of a single species. The campanula is the sub-species *spruneriana* not the large-flowered form seen in the Peloponnese. It resembles a rather tall harebell with upright facing flowers in spring. *Orobanche lavendulacea* is one of the less common species of the genus. It has very dark purple flowers and is usually parasititc on psoralea. *Orobanche ramosa* is a very widespread and abundant species with a wide range of different host species - here it seems to specialise in parasitising the oxalis (Bermuda buttercup).

In the hedgerows *Spartium junceum* abounds with *Rosa sempervirens* and *Lonicera etrusca.* Several orchids are to be found in grassy places including *Ophrys scolopax, Orchis coriophora fragrans, Orchis laxiflora* in places and *Serapias vomeracea.*

18.9 Cephalonia. Garigue: This type of vegetation is well represented on rocky areas inland. There is relatively little grazing in most places and the shrubs are somewhat taller than usual and sometimes merge into dense maquis. The shrubby species include:

Anthyllis hermanniae	*Fumana thymifolia*
Calicotome villosa	*Myrtus communis*
Ceratonia siliqua	*Pistacia lentiscus*
Cistus incanus	*Sarcopoteium spinosum*
C. salvifolius	*Teucrium polium*
Euphorbia acanthothamnos	*Thymus capitatus*

The anthyllis is especially abundant in places. It is a small or medium-sized dense prickly shrub covered with elongated yellow, broom-like flowers in May. Amongst these shrubs one may find:

Anthemis chia	*Leopoldia weisii*
Anthyllis tetraphyllia	*Linaria pelisseriana*
Bellardia trixago	*Linum pubescens*
Dorycnium hirsutum	*Petrorhagia velutina*
D. pentaphyllum	

Several orchids are to be found but they do not seem plentiful. *Anacamptys pyramidalis* and *Ophrys scolopax* occur in moderate numbers and an unusual form of *Serapias lingua* with an almost white lip has been found. One may also have the good fortune to see *Ophrys ferrum-equinum gottfriediana* and *Ophrys sphegodes cephalonica* both of which are special to the island. The last of these is sometimes raised to specific rank as *Ophrys cephalonica*.

18.10 **Cephalonia. Mount Enos:** The best surfaced route to the summit of Mount Enos starts from Drapano on the east side of the narrow gulf by Agostoli. The road from here through Rozato towards Sami climbs to a pass and after some 10km there is a turning southwards signposted to Enos. This road can also be reached from Ag. Georgios (6km SE of Argostoli) climbing via the monastry of Ag. Gerasimou. Yet another way is from the east at Digaleto (see 18.11) where an unsurfaced road leads to join the Ainos summit road at the church of Ag. Eleftherios.

On the road from Agostoli to the Enos turn one passes in places dense stands, which might be described as woods, of the tall columnar Italian Cyprus *Cupressus sempervirens var. pyramidalis* (= *fastigiata* = *stricta*). Although this is said not to occur in the wild it nevertheless seeds itself here and covers considerable areas. The wild spreading form is found further east, especially in Crete.

After the Enos turn the landscape becomes stony and treeless. There are many plants of the pink *Crepis incana* with a yellow-flowered helianthemum, *Thymus capitatus* and *Erysimum cephalonicum*. The last of these is a modest, typical, yellow-flowered treacle mustard that is endemic to the region and also grows at lower levels. Before reaching the radio station, the dishes of which are clearly visible on the skyline, one may see *Putoria calabrica* growing on rock faces and by the roadside in the shelter of boulders *Ajuga orientalis* and *Lamium garganicum striatum*. The last of these is a particularly handsome sub-species with extra large, well-striated flowers.

At about 1300m the road enters the fenced Nature Reserve forest of *Abies cephalonica* where the gate is usually open and one can drive in without restraint. Before entering it is most certainly worth examining extensively the rocky treeless pastures that surround the forest. Much of the land by the roadside near the entrance to the reserve is fenced and inaccessible but it is possible to get to this type of habitat a little lower down near the small roadside church of Ag. Eleftherios. In May here the ground is covered with myriads of yellow or near white flowers of *Orchis pauciflora*, dotted here and there with purple *Orchis quadripunctata*. Other species to be seen here are:

Anchusa variegata   *Helianthemum nummularium*
Anemone hostensis   *Ophrys lutea*
Crepis incana   *Ornithogalum montanum*
Crocus sieberi sublimis *Parentucellia latifolia*

Cyclamen sp.   *Quercus coccifera*
Euphorbia rigida *Ranunculus millefoliatus*

The crocus flowers very early in spring and the cyclamen in the autumn - possibly *C. graecum* though it may be *C. hederifolium*.

In the nature reserve the trees are almost entirely *Abies cephalonica* - a fine species which is found on many parts of mainland Greece stretching as far north as Albania. *Anemone blanda* grows in profusion under the trees and here and there one may see *Corydalis densiflora* and *Scilla bifolia*. *Saxifraga adscendens parnassica* and *S. tridactylites* grow on moss-covered rocks. The non-prickly *Astragalus depressus* forms mats with white flowers and one occasionally comes across plants of *Arabis alpina*. In the higher parts there are still snow drifts under the trees in May and in surrounding wet patches grow the crocus, scilla and *Ranunculus ficarioides*. A feature of rocky areas in clearings is *Veronica sartoriana* that may be confused with the common wall speedwell *V. arvensis*. It forms magnificent mats of gentian blue in May. Other species which should be searched for on the mountain include: *Draba lasiocarpa, Fritillaria mutabilis, Lathyrus digitatus, Scutellaria rupestris cephalonica, Viola cephalonica*. The last of these is an endemic small pansy with mauve flowers in May-July. It grows in the abies forest and resembles *V. calcarata* of the Alps. The fritillaria is said to be very similar to *F. graeca thessala* - could it be the same as the one that grows on Corfu's Mount Pantokrator?

18.11 **Cephalonia. Poros to Sami Road:** This is an attractive, well-watered route worth taking for its plant life. About 3km south-west of Poros, at Dimisianata, a road leaves for Sami. It is somewhat winding and the southern part is not fully asphalted at the time of writing but work is in progress to improve it. In places there are fairly extensive patches of sarcopoterium garigue, perhaps resulting from ground fires. Other plants to be seen here include *Campanula spatulata, Gagea graeca, Silene cretica* and several orchids including *Ophrys ferrum-equinum gottfriediana* and *Ophrys scolopax*.

In places the road runs alongside a small river lined with plane trees and moist margins where one may see maidenhair fern *Adiantum capillus-veneris* and *Selaginella denticulata* accompanied by cyclamen, probably *C. hederifolium*. Other plants which grow in grassy places here are *Allium nigrum, Crepis incana, Echium italicum, Hypericum perforatum, Linum pubescens, Pulicaria disenterica*, and *Securigera securidaca* (= *Coronilla securidaca*) with its curious seed pods (15.7).

Further north along the road, just past Digaleto, a road leads off westwards to climb up Mount Enos and join the summit road at the small church of Ag. Eleftherios (18.10). It is unsurfaced and rough but the countryside is quiet and charming. In the lower area there is woodland comprised mainly of *Quercus coccifera* that has suffered no grazing and grows to 3m or taller and resembles *Q. calliprinos*. However there are flocks of goats and it is not uncommon to see individuals climbing these trees and causing some interesting sculptured specimens. The ground beneath these trees is worth exploring and in spring it is colourful in places with the flowers of *Anemone hortensis*.

18.12 **Cephalonia. The West Coast Route:** Using the roads near the west coast it is quite possible to travel from near Kato Katelios in the south, via Argostoli to Fiskardo in the extreme north and back comfortably in one day allowing time for botanising - or of course in the opposite direction. Some 10km after leaving Kato Katelios there is a sharp turning to the right that leads to Poros. By the roadside, a short distance from the turn, I saw *Silene viridiflora*. This is one of the less common members of the genus and, unlike some of the others, one which is easy to identify - a rather tall perennial growing to 1m or more with distinctly green flowers and with the usual sticky stems.

Back to the main road one passes interesting garigue (18.9) before coming to a series of villages on the way to Argostoli. Here, beside the roadside and on stone walls one may see fine strands of the pale yellow-flowered snapdragon *Antirrhinum siculum*. This is said to be endemic to Sicily and south-west Italy but it has become naturalised in parts of Spain, France and grows extensively on the old buildings of Malta. Here, in Cephalonia, it may have been inadvertently brought over from Italy or perhaps by immigrants from Malta.

The road from Drapano (near Argostoli) northwards to Katohori follows the coast with rock cliffs in places and fine views over the Bay of Argostoli to Lixouri. *Euphorbia dendroides, Ptilostemon chamaepeuce* and *Putoria calabrica* grow on the rocks and by the roadside one may see *Salvia pomifera*, sometimes bearing the round insect galls that give it its name. At Katahouri junction with the road to Lixouri there is a convenient stopping place to examine the grazed garigue but the plants are more or less as described for other parts of the island though one may see *Helianthemum salicifolium* and *Orlaya grandiflora*.

A diversion to the north end of the bay at Argostoli may be worthwhile. Near the coast here it is marshy with frogs, herons and coarse reeds. Further along the coast there is an unsurfaced road that leads northwards through wild, uninhabited country via Artheras to the coast at Ag. Spiridonas. Some of the plants to be seen include:

Campanula drabifolia   *Micromeria graeca*
C. ramosissima   *Notobasis syriaca*
Centaurium pulchellum   *Petrorhagia prolifera*
Linum hirsutum   *Prasium majus*

Back on the road from Katohari northwards one passes a turning westwards to the interesting small peninsular of Asos. On the steep road down there are more woods of the narrow Italian cyprus. The road continues northwards to the interesting small port of Fiscado (Fiskardho, Phiscardo) said to be named after the Norman king Robert Guiscard, who died here in 1085. It escaped the ravages of the 1953 earthquake and has some interesting Venetian houses, so it is well worth seeing; but it has few special plants, though *Lavatera arborea* decorates the buildings in places.

Cephalonia 1. Parentucellia viscosa 2. Erysimum cephalonicum 3. Linaria pelisseriana

19. Lesvos

19.1 This is the third largest of the Greek islands, situated in the north-east of the Aegean Sea and close to the Turkish mainland - about 15km at the nearest point. It is sometimes referred to as Lesbos but this is a less accurate transliteration than Lesvos, for beta which looks like our B, is pronounced as a V in Greek. To complicate matters, on some maps the island is called Mytilini or Mytilene, which is also the name of the largest town there.

Lesvos is comparatively undeveloped for tourism and a good place for a quiet holiday. It has several small coastal resorts, as at Methymna in the north, Plomari in the south, Eressos in the west, Kalloni on the bay of that name and the region south of the capital Mytilene. A few package holidays are arranged, especially to Methymna but it is not difficult to find accommodation on arrival there in spring without booking in advance. The airport near Mytilene now takes international flights so it is no longer necessary to fly via Athens or travel there by sea.

Although Lesvos has a few special plants, it is of only moderate botanical interest, except perhaps for the orchids which are rather plentiful. However, no serious appraisal of the flora seems to have been made since 1898 so botanising there could be an exciting challenge for the enthusiast. One needs some kind of motorised transport for the island is fairly large. Most of the roads are narrow and winding and it often takes longer than expected to get from one place to another. A useful 1:140,000 scale map is available on the island and includes such important details as the siting of petrol pumps.

19.2 **The terrain:** The rocks of the north and west are mainly volcanic and have given rise to rather dry infertile soils. Those of the south east are older granites with scattered calcareous outcrops with support more luxuriant plant growth.

Land around the coast is largely cultivated with many olive orchards and some less common crops such as pistachio (*Pistacia vera*) but there are many areas inland with rocky pasture land. The highest parts are the Lepetimnos mountains in the north and Olympos in the south west, both claimed to rise to exactly 968m. The southern mountains are more wooded and verdant than those in the north and this area is best for botanising.

Two large bays stretch well inland, with relatively narrow entrances to the sea. These are the Kolpos Kallonis, which is more or less central, and the smaller Kolpos Geras near Mytilini. Salt marshes and salt pans are found on the northern coast of the first of these.

19.3 **Literature:** There is no up-to-date literature specifically on the flora of Lesvos, the last being that by Candargy (1898) which is in French, available only in academic libraries, and virtually a list of species found there. Apart from a general work on Mediterranean plants it is worth taking Huxley & Taylor (1977) and Polunin (1980) on a visit there if they are readily available.

A fairly recent publication in German (Baumann et al. 1981) maps the orchid species which occur on Lesvos, Chios, Samos, Kos and is potentially very useful to the enthusiast but not easily available. The flora of Lesvos is included with that of Turkey in Davis et al. (1965-76) and not in Flora Europaea (1964-76).

19.4 **The North around Methymna:** Methymna, or Mithimna, a small village on the north coast is also known as Molyvos from the Turkish occupation that ended in 1912. It is a quiet holiday resort with hotels, a harbour and a large impressive Genoese fort. The steep main street of the old Turkish village centre has wires stretched across it supporting wisteria plants which bloom in April. It is a real pleasure to sit in a restaurant for an evening meal here with the scent of the wisteria and the sound of scop's owls calling to one another. These attractive small birds are considered to be sinister by some of the local inhabitants for according to Greek legend they are lost souls who weep drops of blood whilst pining for their lovers. The sound may be realistic but they are, in fact, rather sociable little birds that like human habitation.

Although the cultivated land here is considered to be fertile with olives, potatoes and several fruits including pistachio nuts, as major products, most of the higher areas tend to be dry and treeless with stony acid soil and a rather restricted flora. The Lepetimnos hills near Methymna, which rise to 968m at Karakas, show clear evidence of tertiary or quaternary volcanic activity with eroded lava flows and occasional hot springs as at Loutra Eftalous near the north coast about 5km east of Methymna. The typical garigue here consists mainly of *Sarcopoterium spinosum, Calicotome villosa, Quercus coccifera* and scattered small stocky trees of *Pyrus spinosa* which are covered with snowy white flowers in spring. The ground is carpeted in places with the prostrate *Trifolium uniflorum* which produces its small pale pink flowers singly rather than in heads like a typical clover. It is widespread in the east of the Mediterranean and occurs as far west as southern Italy. Associated with these one may find:

Anemone pavonina　　　*Muscari neglectum*
Asphodelus aestivus　　　*Ornithogalum nutans*
Gynandriris sisyrinchium

The anemone here is usually the beautiful scarlet form. On the east side of the Lepetimnos there are areas covered by bracken, some plants of the impressive *Ferula communis* and autumn-flowering cyclamen, mainly *C. graecum* but possibly also *C. hederifolium*.

What the area lacks in variety of plant life it makes up for in the insects and other animals. The beautiful eastern festoon butterfly (*Alancastria cerisyi*) the scarce swallowtail and the hummingbird hawk moth are common here. In April one may be fortunate also to see thousands of migrating painted lady butterflies which rise like a swarm of locusts from the fragrant carpets of *Anthemis cretica* near the shore. Tortoises and lizards, including geckos, are frequently seen and near streams one may find terrapins, frogs and tree frogs.

By the coast near Petra, some 4km south of Methymna, there is more typical garigue with *Cistus monspeliensis, C. salvifolius, Lavandula stoechas* and *Legousia speculum-veneris*. By the shore grow spectacular stands of *Malcomia flexuosa, Matthiola sinuata* and on the rocks by the sea species *Campanula rupestris* or what could be *C. lyrata* - a similar Turkish species (23.7). In moist areas behind the shore one may find the orchids *Ophrys lutea minor, Ophrys mammosa*. The last of these seems to be a particularly robust and fine form here and is found in other parts of the island though it is by no means common.

19.5 Methymna to Kalloni: The road south from Petra passes through Aleppo pine woods where *Orchis provincialis* and *Limodorum abortivum* are common. By the roadside grow patches of the Greek pink dandelion *Crepis incana* and the curious dark-purple flowered *Lysimachia atropurpurea*, which is widespread but often overlooked. After a short time one reaches Kalloni with its marshes and salt pans by the bay called Kolpos Kalloni. This is a very good centre for birds, including flamingos, avocets, glossy ibis, stilts, various herons and spur-winged plover. Little owls and red-legged falcons often nest nearby. In one marshy area there are large numbers of *Orchis laxiflora*.

From Kalloni a winding road of some 35km takes one to the western coastal villages of Sigrion (Sigri) and Eressos. The vegetation is not particularly interesting here but in the extreme west there is a chance to see several orchid species which do not seem to occur in other parts of the island, including *Ophrys argolica, O. tenthredinifera, Orchis collina, O. sancta* and *Serapias orientalis*.

19.6 Kalloni south towards Mytilini: Travelling south-east from Kalloni towards Mytilene the road passes through more extensive forests of *Pinus halepensis* with remnants of garigue as undergrowth. The sparse undershrubs here include *Cistus salvifolius, Erica manipuliflora, Fumana thymifolia, Pistacia lentiscus* and *Sarcopoterium spinosum*. Under the light shade of the trees are large colonies of *Orchis morio picta, O. provincialis* and *Limondorum abortivum*. The Provence orchid is especially fine here, with beautifully spotted leaves and sometimes ivory-coloured or near white flowers. Other orchids found dotted here and there under the pines include *Neotinea maculata, Ophrys fusca, Orchis papilionacea* and *Orchis tridentata*. Autumn-flowering cyclamen *C. graecum, C. hederifolium* abound and there is also a yellow-flowered gagea.

19.7 **Mount Olymbos area:** Travelling south through the pine woods described above one comes to lower ground near Lambu Mili and shortly after there is a turning westwards to the village of Keramia and leading further to Ajiassos. Around Keramia are old olive orchards and the roadside verges and turf between the trees here are particularly rich in orchids. Other plants include the scarlet form of *Anemone pavonina* in abundance and scattered plants of *Anemone coronaria*, mostly the mauve-flowered form which tends to flower in March and is mostly over by the time the scarlets are at their best. To set off and liven up the colour scheme there are thousands of the glistening white flowers of the Greek chamomile *Anthemis chia*. The main orchids are *Orchis morio picta, O. papilionacea, O. italica* and *Serapias vomeracea*. The first two of these often hybridise here to give intermediate forms. Other orchids which may be found here with careful searching include:

| *Anacamptis pyramidalis* | *Ophrys lutea* |
| *Ophrys fusca* | *Orchis quadripunctata* |

The Olympos area has two peaks of which Profitus Ilias 968m is the highest, though not easily accessible, for there is no readily negotiable road to the summit. There are, incidentally, at least two other hills called Profitis Ilias on the island. The second peak of Olympos called Xilokodouno 914m, lies some 3km to the southwest and is much more easily reached for the road leading to Plomari on the south coast traverses the top of it.

Continuing along the road from Keramia one comes to the village of Ajiassos (Hagiasos on some maps) where there is accommodation making it a good place to stay to explore the region in detail. Just to the south are woods with some magnificent old specimens of Aleppo pine many of which are tapped for resin used in the production of retsina wine. Moderately dense undergrowth here includes such species as *Cistus, Euphorbia rigida* and *Phillyrea latifolia. Orchis quadripunctata* is fairly plentiful and one may see *Ophrys ferrum-equinum* and *Ophrys scolopax*, but the real prize to be found is *Comperia comperiana* (= *C. taurica*). It was first discovered here by Candargy in 1897 on a site described as Mt. Buro, Olimbos. In 1981 my wife and I found 21 robust plants within an area of some ten square metres a short distance south of Ajiassos. It flowers rather later than most orchid species and is at its best in late May or early June. It has also been discovered on Rhodes (21.6) but its main sites are in Turkey (23.9).

Progressing further along the road the Aleppo pines give way to a fine sweet chestnut forest. Along the roadside here and amongst the trees in places are a number of interesting plants:

Anemone coronaria *Fritillaria pontica*
A. pavonina *Galanthus elwesii*
Colchicum variegatum *Orchis quadripunctata*
Cyclamen graecum *Ornithogalum montanum*
C. hederifolium *Tulipa orphanidea*
Doronicum orientale

The colchicum is a fine autumn-flowering species with very checkerboard marked flowers and producing usually four wavy-edged leaves in spring. It grows in many of the Aegean islands. The identify of the doronicum is not absolutely certain - it might possibly be *D. columnae*. The fritillaria is the variety *substipelata* and a fairly tall and robust form with broad bluish green leaves. Its flowers, produced in May and June are plain green and not chequered. The galanthus is a rather broad-leaved snowdrop that flowers during the winter until March. It is fairly easily grown and not uncommon in home gardens. The tulip has rather narrow, untwisted leaves and orange-red flowers with narrow segments that are flushed with green on the outside. It is sometimes called *T. hageri*.

The road continues over the mountain down to the south coast at Plomari and near here one may find *Orchis collina*, which is

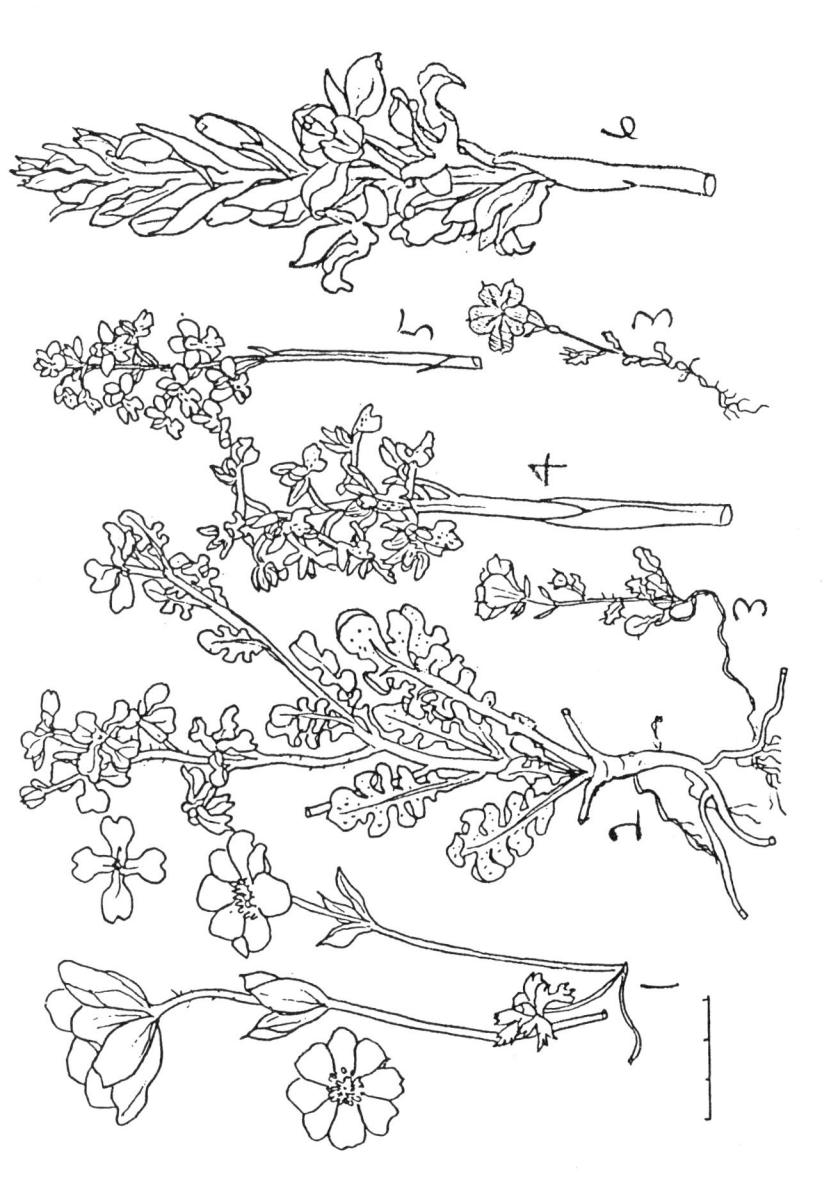

Lesbos 1. Anemone pavonina 2. Matthiola tricuspidata 3. Legousia speculum-veneris 4. Orchis morio 5. Orchis quadripunctata 6. Limodorum abortivum

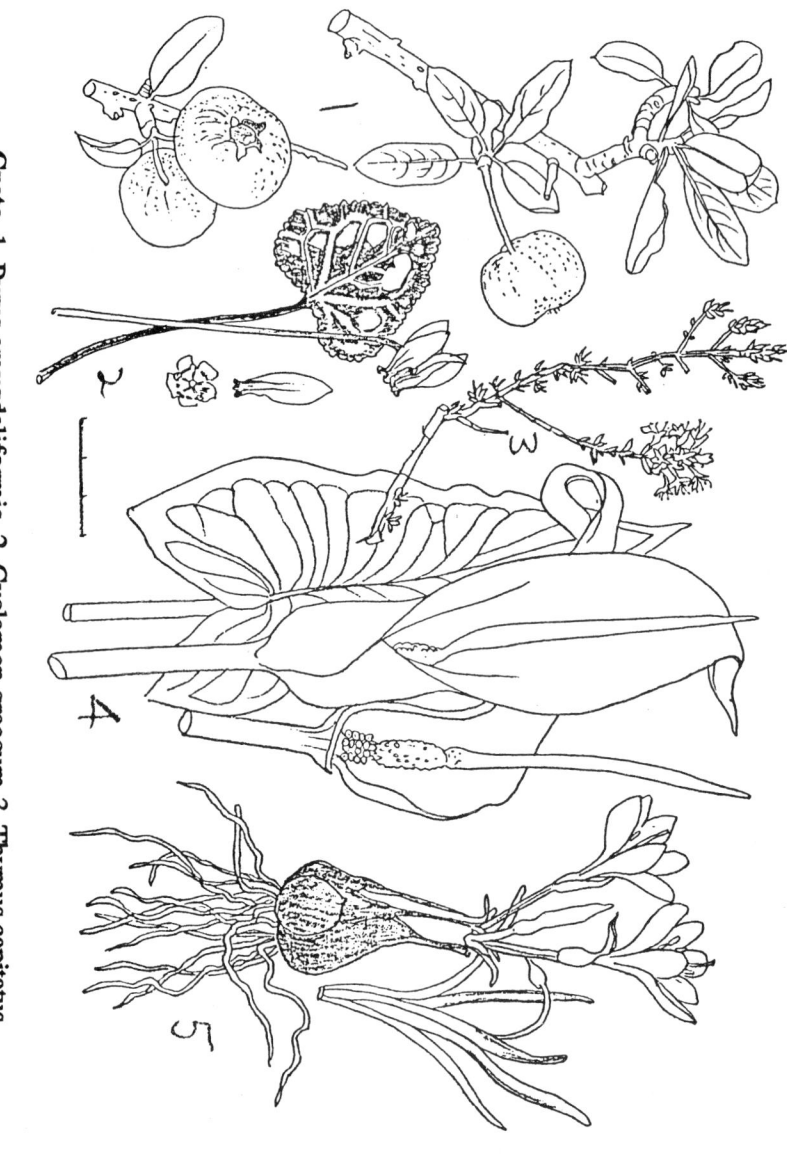

Crete 1. Pyrus amygdaliformis 2. Cyclamen graecum 3. Thymus capitatus 4. Arum creticum 5. Sternbergia lutea sicula

somewhat rare in other parts of the island. It flowers earlier than most other species and is at its best in early April. In the lusher valleys near the sea grows *Rhododendron luteum* which is a beautiful yellow-flowered 'azalea' frequently seen in home gardens. It is mainly a plant of the Pontus region of Turkey and this is its only site in Greece. It is occasionally found in the Balkans and has established itself in part of Scotland near Aberfeldy, where it makes a magnificent sight with its yellow, scented flowers in spring.

19.8 Addendum: South of Lesvos lie the other eastern Aegean islands of Chios, Samos and Kos which are near to the Turkish coast. All of them are of interest botanically and can be reached by boat from Lesvos. They are generally greener and more wooded than the smaller islands to the west. Chios which is the nearest to Lesvos, and fairly easily reached from it by sea, has mountains over 1,000m with fine woodland including the Greek fir *Abies cephalonica*. Some of the most interesting plants of this island are:

Crocus nubigenus   *Muscari parviflorum*
Crocus olivieri   *Ptilostemon chamaepeuce*
Fritillaria bithynica *Sternbergia colchiflora*
Hyacinthus orientalis *S. lutea*
Muscari macrocarpum

Crocus nubigenus is a mauve-striped, spring-flowering species similar to the variable *C. biflorus*. *Crocus olivieri* has orange yellow flowers in January to March and, like the species just mentioned, is more frequent in Turkey. The fritillaria is a small campanulate type with flowers that are green outside but yellow within. It also grows in Turkey and is rather like *F. carica* (23.10) also found on Chios and Samos. *Muscari macrocarpon* is a rare species with yellow flowers that are purplish in bud, and grows mainly in Turkey and Crete.

Crete 1. Scrophularia lucida 2. Anemone coronaria 3. Heliotropium hirsutissimum 4. Crithmum maritimum

20. CRETE

20.1 Of all the Greek islands, Crete is the largest (approx. 256km long by 56km at the widest part) and the most southerly with Cape Lithinon on a lower latitude than Malta though not as far south as part of Cyprus. It has a variety of habitats and substantial mountains often carrying snow to the end of May. With some two hundred endemics found nowhere else in the world, it is undoubtedly one of the most exciting places in the Mediterranean for botanising. The flora embraces species that have a reference to Asia Minor dating back some five million years when it was joined to what is now Turkey, as well as some North African species. In addition to the plants there are exciting Minoan remains and Crete is a place where the larger raptors, such as eagles and vultures can frequently be seen - including the rare and very large lammergeier.

Crete has an international airport near the main town Heraklion (Herakleion, Iraklion) which is situated about mid-way along the north coast. Most of the holiday resorts are in the north and if one chooses a single venue then somewhere here is advisable, as the main road that serves the island, referred to as 'Nea Odos' (New Road), provides the best access to most parts. Travelling from some of the south coast towns may be more difficult. Because the island is relatively long, and many roads are very winding, it is not possible to cover it from one resort and several visits to different parts would no doubt be rewarding. The best time to see most plants is from March to May but some interesting species flower in the autumn, though not as many as for example in the Peloponnese.

The weather can become very hot and dry during the summer, as in most other Mediterranean islands, and it can change suddenly with torrential rains and strong winds, especially in the west. The south coast enjoys most of the sunshine. From time to time the warm, oppresive scirocco wind blows from Africa bringing sand from the Sahara and giving the mountain snow an orange brown covering.

20.2 **The terrain:** Most of the island is hilly and there are three distinct high mountain groups over 2000m which are, from west to east, Levka Ori or white mountains (Pakhnes or Pahnes 2453m), Idhi or Ida (Psiloritis 2456m) and Dhikti or Dikti Ori (Dikti 2148m). The surface rocks of Crete are mostly limestones easily penetrated by water so dry ground conditions often prevail. As a result many of the plants are adapted to cope with water shortages. There is an underlying base of metamorphic rocks which come to the surface at some lower levels, especially in the extreme west.

One special feature of the Cretan landscape is the number of limestone gorges. They occur throughout the island but are most common in the west. The Samaria is said to be the deepest and longest of its kind in Europe and has become an important tourist attraction. These gorges, with their steep rock faces, provide a home for many interesting species that have escaped the agricultural activities of cultivation and grazing. Such, rock inhabitants are referred to as chasmophytes.

The island also has sizeable plains, such as the relatively large Messara in the south and higher ones like the Lassiti and Omalos, near to the Dikti and Levka mountains respectively. These regions are cultivated with crops suited to the respective altitudes.

20.3 **Literature:** The most useful introduction to the plants of the island for the amateur is 'Wild Flowers of Crete' (Sfikas 1987) which is available in the main towns there and a few specialist bookshops in Britain. However, it has the disadvantage of rapidly falling to pieces after a little use. 'Flowers of Greece and the Balkans (Polunin 1980) is an authoritative book on the subject and has a summary of the Cretan flora and 'Flowers of Greece and the Aegean (Huxley and Taylor 1977) is a useful introduction. The most up-to-date, and extremely useful, publication is 'Flora of the Cretan Area' (Turland, Chiltern & Press 1993). However, the title is somewhat misleading as it is essentially a check list of species with distribution maps and it needs to be used along with some publication describing species such as Sfikas (1987) and Strid (see below). It includes the neighbouring Karpathos group of islands which have a somewhat similar flora.

A new and helpful book published in Heraklion and entitled 'The Wild Orchids of Crete (Alibertis 1989) is available on the island. It has fine illustrations and, in keeping with other modern works on orchids, it includes several new or 'invented' species according to ones views on the subject. Flora Europaea covers Crete but is now somewhat out-of-date and is in process of revision. There is no modern treatise on the flora of Greece as a whole but the recently published 'Mountain Flora of Greece' (Strid 1989 and Strid & Kit Tan 1991) is the definitive work on the subject and, of course, embraces the mountains of Crete. However, as the title implies, it deals only with those species that grow above 1,800m.

20.4 **The extreme West:** Near the western end of the 'New Road', some 20km past Chania (Hania), a road crosses the mountains southwards through the towns of Voukolies and Kandanos to reach the south coast at Paleohora. There are interesting sites for botanising along here with patches of garigue typical of the island as a whole and to be discussed in further detail later. Some of the more attractive and interesting species to be seen here include:

Anemone coronaria   *Ranunculus creticus*
A. heldreichii   *Hermodactylus tuberosus*
Cyclamen creticum   *Rosularia serrata*
Petromarula pinnata

Anemone heldreichii is a Cretan endemic like a small version of *A. hortensis* and sometimes considered as a sub-species of it. The flowers have many narrow petals, usually of a pale mauve colour. It occurs scattered throughout the island, usually in the hills. The cyclamen is also an endemic, rather like a white-flowered form of *C. repandum*. The petromarula is yet another Cretan endemic, belonging to the Campanulaceae and has pinnately lobed leaves and flowering stems to 1m carrying numerous mauve-blue flowers in which the petals are swept back as in a cyclamen. It grows mainly on rock faces and is classed as one of the chasmophytes also found in gorges. The ranunculus is an attractive small buttercup with large yellow flowers up to 3cm diameter and hairy spreading calyxes. It is common throughout the island and also found in Karpathos, Rhodes and parts of south-west Turkey. The rosularia resembles a houseleek with blunt, leathery-margined leaves and stems carrying numerous small purplish-red flowers. It is a genus more widespread in Turkey. A number of orchids can be seen along this road including: *Ophrys lutea, Op. sphegodes mammosa, Op. scolopax heldreichii, Orchis italica, O. lactea, O. papilionacea* and *O. quadripunctata.*

The village of Paleohora and the coastline here and westwards towards the small settlement at Gialos is a good place to study the flora of the turf and rocks by the seashore. Here one may find:

Allium rubrovitatum	*Romulea ramiflora*
Asphodelus aestivus	*Muscari macrocarpum*
Bupleurum semicompositum	*M. speitzenhoferi*
Gynandriris sisyrinchium	*Nigella doerfleri*
Linum strictum	*Petromarula pinnata*
Malcomia flexuosa	*Silene colorata*

The allium is a small-growing species, up to 20cm tall, with campanulate flowers in April and May that are reddish with white edges. It is a rare endemic of Crete and Carpathos described by Sfikas (1987). Both of the muscari species are rare. The first has flowers that are purple in bud but open to become a dull yellow colour. The second is similar to a leopoldia with violet-coloured sterile florets and fertile flowers that are greenish with a yellow base. It was once thought to be a Cretan endemic but now known to occur in North Africa. The nigella is a small love-in-a-mist with greenish flowers and another Cretan endemic.

A track leaves Gialos westwards towards the monastery of Chrissoskalitissa but it is a long walk there, though probably of special botanical interest. Offshore here is a small island called Elafonissi which would be a worth visiting if one could find a local fisherman to take one there. This is the only place in Europe where one can

Crete 1. Globularia alypum 2. Bellis longifolia 3. Quercus aegilops (macrolepis)
4. Taraxacum gymnanthum 5. Crocus laevigatus 6. Colchicum pusillum

Crete and Cyprus 1. Iberis sempervirens 2. Otanthus maritimus 3. Crocus veneris 4. Alnus orientalis 5. Spiranthes spiralis 6. Limoniastrum monopetalum

see *Androcymbium rechingeri* that grows between the rocks. It somewhat resembles an ornithogalum with broad bracts surrounding the inflorescence and white flowers having a touch of purple or pink. Its nearest relative in Europe is *A. europaeum* that grows on the Cabo de Gata of south-west Spain (7.8).

The area west of the road from the north coast down to Paleohora is largely unexamined botanically and would be a worthwhile site for an expedition. Along the north coast of the region there are three peninsulas - the Korikos, Rodopou and Akrotiri from west to east respectively. These carry interesting regions of garigue and are well worth a visit.

20.5 **The Omalos plateau:** This is approached from Chania by taking a road southwards to Alikianos where there are orchards of oranges and some other fruits such as avocados. Shortly after passing the village of Fournes, the road begins to climb and the hills in places have a covering of typical maquis including the following shrubs:

Arbutus andrachne   *Erica arborea*
A. unedo   *Juniperus phoenicea*
Calicotome villosa   *Phlomis fruticosa*
Cistus incanus creticus *Rhamnus alaternus*
C. salvifolius

Some 30km from Chania one arrives at the village of Omalos which twenty years ago was practically deserted but now has a few restaurants to nourish the thousands of tourists that come to walk the Samaria Gorge. The Omalos plateau is a high flat polje used for growing cereals and in the spring before the season's cultivation begins there are sometimes carpets of *Anemone coronaria*, mostly of the purple form but dotted here and there with scarlet-flowered specimens. Amongst them grows the pink tulip *Tulipa bakeri* which generally flowers a week or so later than the anemone. The precise status of this species is still to be determined. It is very similar to another Cretan species *T. saxatilis* and there is some evidence to suggest that it may be a stable hybrid between it and the white-flowered *T. cretica*. All three of these tulips are Cretan endemics. Growing amongst this delightful carpet of flowers one may also find *Gagea bohemica, Hermodactylus tuberosus, Leopoldia comosa, Ranunculus ficariiformis. Arum creticum* grows in places here and after late summer grazing, which can be severe, the very prickly *Centaurea idea* is one of the few plants to survive intact and produce its yellow flowers in July and August. It is a montane species endemic to Crete and closely resembles the more familiar *C. solstitialis*. By September or October the pale mauve flowers of *Crocus laevigatus* appear by the roadside. They have feathery orange stigmas and white or cream-coloured anthers.

The road ends at the entrance to the Samaria Gorge, a place generally referred to as Xyloscalo - the wooden staircase - alluding to the 'stairs' down to the bed of the gorge. There is a convenient stone-built tourist pavilion here where one may spend the night. The view is truly magnificent, especially in spring when the mountains are capped with snow, particularly the impressive peaks of Volakias 2116m, Psilafi 1983m, and the more distant Pahnes 2453m. This is the home of the bearded vulture or lammergaier - a huge majestic bird that generally nests on the slopes of Volakias and can be seen frequently flying with its irritating flocks of mobbing choughs. There is, or used to be, a rather pathetic-looking stuffed specimen in the tourist pavilion.

A track leads up from the pavilion to Strifomadi 1942m and there are scattered and picturesque old trees of *Cupressus sempervirens*. Many interesting smaller plants can be seen as the snow melts here in April. They include:

Aubretia deltoidea	*Crocus sieberi*
Berberis cretica	*Onosma erecta*
Centaurea raphanina	*Erysimum raulinii*
Cerastium scaposum	*Prunus prostrata*
Chionodoxa cretica	*Verbascum spinosum*
C. nana	

The aubretia is the sub-species *deltoidea* and parent of the many garden varieties we grow at home. The berberis is a low, mat-forming bush that is extremely prickly and a hazard for walkers who do not wear boots. It is not uncommon in Greece and Cyprus but often overlooked, for it is one of the least attractive species of its genus, though the yellow flowers sometimes make a show in summer. The centaurea is a neat small plant with a basal rosette of leaves and almost stemless capitulae of mauve flowers. It is closely related to *C. mixta* from southern Greece (17.5) and *C. aegiolophila* from Cyprus (22.5) - two species that grow at low altitudes. Of the two chionodoxas; nana has small nearly white flowers whilst those of *cretica* are blue. The two species seem to merge but the one seen here is most likely *C. nana*. This genus, or 'has been genus' is mainly Asiatic and Crete is its only European station, both 'species' are endemic. Unfortunately the latest thinking by taxonomists does away with the genus chionodoxa and classes them all as species of scilla so those here become *Scilla cretica* and *S. nana*. To the layman this is a pity for 'chionodoxa' appropriately means - Glory of the Snow, the plants that flower amongst the melting snows. The crocus is an especially handsome species with large flowers that are mainly white with an orange throat, and the form here is sometimes referred to the sub-species *versicolor*. The onosma is typical of its genus with yellow flowers and rough hairy foliage; it is also found in mainland Greece. The verbascum is a small spiny shrub with yellow flowers and another Cretan endemic.

Other species which may be seen here but flower later in the summer include; the prickly *Astragalus angustifolius* with white flowers and the two hypericums *H. trichocaulon* and *H. kelleri*. These last two look alike but are not closely related; they are procumbent, mat-forming plants with somewhat heath-like leaves and small yellow flowers. Later in September and October one may see *Colchicum pusillum, Crocus laevigatus* and *Taraxacum gymnanthum*. The first is a small species with pale mauve flowers, usually no more than 2cm diameter and some botanists recognise another similar species, C. cretense, distinguished by the fact that the foliage is not evident until after flowering but this seems to be a doubtful border line. The crocus also has small lilac or white flowers, darker striped outside and with orange stigmas and creamy-white anthers. The taraxacum is a neat, dwarf, autumn-flowering dandelion.

20.6 **The Samaria Gorge:** The northern entrance to the gorge is near the end of the road by the Tourist Pavilion and one can also enter it from the southern end at Agia Roumeli. There is a charge for entrance which goes towards the maintenance of the pathway and the provision of first aid in case of accidents. It is not possible to make the traverse in spring and late autumn because of the amount of water flowing through the narrow parts and at other times flash floods may occur. The 5-6 hour walk is easy for those who are fit and young but elderly and unfit travellers are advised to think carefully before committing themselves. It is usual to make the journey from north to south and catch the ferry at the coast which takes one eastwards to Chora Sfakion where there is a good road northwards. The route along the gorge may become congested in summer with hundreds of tourists - some keen to prove their fitness by making the journey as quickly as possible.

This gorge is famous amongst botanists for the number of unique plants growing there, especially the 'chasmophytes' which are mostly relict species that have survived man's interference from cultivation and grazing. The Samaria gorge is not the best place to see many of them for one can not often get close and, because of the distance to be covered may not leave sufficient time to search around. However, the flora here will be described as it is more or less typical of all similar habitats on the island. From the north entrance the path goes down steeply at first by the so-called 'xyloscalo' built originally by the drovers to take their herds for grazing. Here there is patchy woodland with *Cupressus sempervirens, Pinus brutia* and *Platanus orientalis.* The last of these resembles the London plane and has similar mottled bark and may indeed be one of the parents of that hybrid whose origin is uncertain. It is a very long-lived tree with some specimens thought to survive for two thousand years. By the side of the path in spring one may see carpets of the white-flowered *Cyclamen creticum*, clumps of *Onosma erecta* and the white-flowered endemic *Paeonia clusii* closely resembling *P. rhodia* of Rhodes (21.6). There are also plants of *Ranunculus creticus* and several orchids, especially *Ophrys fuciflora*, *Orchis pauciflora* and *O. quadripunctata*. Along the floor of the gorge by the stream in autumn one may see:

Arisarum vulgare	*Narcissus serotinus*
Colchicum pusillum	*Ranunculus bullatus*
Crocus veluchensis	*Taraxacum gymnanthum*

Several orchids grow here and from mid-May to mid-June one may be lucky to find the newly-described *Himantoglossum samariensis* (Albertis and Albertis 1989) recorded only from here. It has typical lizard orchid flowers tinged with mauve-pink. At the collection of buildings of the now deserted village of Samaria, where most travellers stop for a rest and refreshment, there are trees of the elm-like *Zelkova abeliaca* which also grows amongst the surrounding rocks. This is an Asian genus and here is its only representative in Europe.

There are so many interesting plants thriving on the rock faces that a selection only of what may be found is included here. Many of these are illustrated by line drawings and coloured plates in Polunin (1980).

Brassica cretica	*Ptilostemon chamaepeuce*
Campanula laciniata	*Putoria calabrica*
C. tubulosa	*Rosularia serrata*
Ebenus cretica	*Scutellaria sieberi*
Eryngium ternatum	*Staehelina arborea*
Helichrysum orientale	*S. fruticosa*
Linum arboreum	*Symphyandra cretica*
Petromarula pinnata	*Verbascum arcturus*
Precopiana cretica	

The brassica is a cabbage-like species similar to *B. insularis* of Corsica (11.9) and *B. hilaronis* of Cyprus (22.12). It usually has large white flowers though there is a form with pale yellow petals. Both of the campanulas are beautiful plants with mauve-blue flowers that are particularly large in *C. laciniata*, the most common of the two species. The ebenus is a Cretan endemic of some renown; a shrubby sainfoin-like plant with leaves of 3-5 leaflets covered with silky hairs and heads of mauve red flowers in April and May. It forms large hanging tufts at the north entrance to the gorge and throughout its entire length. The linum is also a small shrub and has bright yellow flowers. The precopiana is a small borage-like plant with mauve flowers. Ptilostemon is also a small shrub with narrow leaves and rounded heads of mauve, centaurea-like flowers in May and June; it is fairly widespread in Turkey. The scutellaria is a downy Cretan endemic with dense heads of cream-coloured flowers tinged with red in April and May. The staehelinas are evergreen shrubs belonging to the Asteraceae (Compositae). The first of them is another Cretan endemic, growing to 1m tall with ovate dark green leaves that are silvery below and heads of pinkish, centaurea-like flowers in summer and autumn. The second grows taller, has smaller leaves and heads of white flowers. The symphyandra is a campanula-like plant and somewhat of a speciality of the Samaria gorge though it is found in other places on Crete. It produces huge hanging mats with blue or occasionally white flowers in late summer with a few still out late into the autumn. It is especially plentiful towards the southern end of the gorge.

Although the symphyandra is found mainly in the Samaria gorge, most of the other species can be observed and photographed easier in other gorges within the area such as the Imbros, Imbriotico and the Aradhena. These are best approached from the road south from Vrisses to Chora Sfakion and the first is reasonably accessible. A long trip from Chora Sfakion westwards to Anapoli is necessary for acces to the Aradhena (Aradena). Further details of these two sites is given in Polunin (1980).

20.7 **The Levka Mountains:** These mountains with some fourteen peaks over 2,000m are home to an extraordinary number of interesting plant species. Access to the summits can only be obtained by a considerable amount of rough walking though mountaineering experience is not generally necessary. There are several different ways of approaching them but a start from the village of Ammouda on the Vrisses - Chora Sfakion road is probably one of the most direct routes. Even by this way it is advisable to leave adequately equipped for spending a night on the mountainside. The time to go is late May or June when much of the snow has melted but some species are in flower during July. A 'taste' of the mountain flora can be had by continuing up the slope behind the tourist pavilion by the entrance to the Samaria Gorge or by crossing the road and climbing up the slope there. Here one may be rewarded in May and June by seeing *Anchusa caespitosa*. It is a tufted plant with long, narrow bristly leaves and large blue forget-me-not like flowers and, of course, a Cretan endemic. An impressive number of other endemics to be found in the Levka mountains include:

Acantholimon echinus creticum
Alyssum fragillimum
A. idaeum
A. sphacioticum
Arabis cretica
Centaurea baldacci
C. idaea
Corydalis rutifolia uniflora
Dianthus sphacioticus
Draba cretica
Erysimum mutabile
Erysimum raulinii
Galium incanum creticum
Hypericum empetrifolium tortuosum
Nepeta sphaciotica
Paracaryum lithospermifolium
Scabiosa sphaciotica
Senecio fruticulosus
Silene sieberi
Tulipa saxatilis
Viola cretica
Viola fragrans

These are described in Sfikas (1987) but for fuller details one needs to consult Strid (1986) and Strid and Tan (1991). The acantholimon (= *A. androsaceum* in Sfikas) forms prickly clumps with purple or pink flowers in June and is one of the species especially typical of the region. The corydalis is a small plant with very pale purple flowers - another form of this species grows in Cyprus (22.10) and Turkey (23.9). The dianthus is woody-based with pink flowers and, like other species with 'sphaciotica' is named after the Levka Mountains. The paracaryum is a small forget-me-not like plant with dark bluish-violet flowers. It is referred to as *Mattiastrum* in Sfikas (1987). The tulipa is very similar to *T. bakeri* found on the Omalos plateau and not all botanists agree on the identity of these two species. *Viola cretica* is a violet with mauve flowers in April and May whereas *V. fragrans* is a pansy with yellow, or occasionally mauve, flowers later in May and June.

Of course, not all the plants of the Levka mountains are endemics and amongst them one may find some of the following:

Aethionema saxatile	*Ranunculus laterifolius*
Aubretia deltoidea	*Rosa heckeliana*
Campanula aizoon aizoides	*Satureja spinosa*
Euphorbia herniaria	*Saxifraga chrysoplenifolia*
Galium rotundifolium	*S. tridactylites*
Iberis sempervirens	*Sedum album*
Lamium garganicum	*S. rubens*
Linum arboreum	*Trifolium psysodes*
Prunus prostrata	*Veronica thymifolia*

The euphorbia is a tiny mat-forming species. The iberis is an attractive evergreen perennial with white flowers and is often grown in rock gardens at home and several cultivars exist such as 'Snow Queen' and 'Little Gem'. The ranunculus is a small annual buttercup that mainly grows in wet places. The rose produces compact bushes with rounded leaves and pink or white flowers in July. *Sedum album* is a widespread small stonecrop with white flowers that also grows in Britain, especially on the Malvern Hills.

20.8 **Around Rethymnon:** The town of Rethymnon (Rethimno) is the third largest on the island and has some interesting old buildings. Just to the east of the town there is a costal resort with a number of tourist hotels. In the sandy turf behind the beach one may see a number of shore plants that are typical of the island including *Atractylis gummifer, Cichorium spinosum, Euphorbia peplis, Otanthus maritimus* and *Pancratium maritimum*. The atractylis is a thistle with a rosette of prickly leaves and large, stemless heads of purple flowers in late summer and autumn. It is a widespread species occurring as far west as Portugal and growing in stony places and near the shore. The euphorbia is a strange prostrate-growing species whose leaves tend to turn reddish under drought conditions.

On a grassy area near the harbour at the centre of Rethymnon there are groups of the spiny cocklebur *Xanthium spinosum* with its prickly fruits in late summer - a native of North America. The extensive Venetian fortress is well worth a visit and the grassy areas inside are dotted in autumn with *Ranunculus bullatus* and the tiny blue *Scilla autumnalis*. There are groups of the 'tree of heaven' *Ailanthus altissima*, a suckering shrub with small greenish flowers and winged reddish fruits. It is a native of China sometimes planted to stabilise soil and has grown so vigorously in some areas of the Mediteranean as to oust the indigenous flora, as on the island of Monte Cristo east of Corsica.

A number of minor roads lead southwards out of the Rethymnon area to such villages as Maroulas, Gallos and Prassies. These will help the visitor who has recently arrived to get to know the local flora. One often passes through olive orchards at the lower levels, infested, as usual, with the Bermuda buttercup *Oxalis pescaprae*. In places there are specimens of *Pyrus amagdaliformis* - a small, somewhat spiny tree usually covered with white flowers in spring and bearing small, apple-shaped fuits in September and October. They are said to be edible when fully ripe. The oriental plane tree *Platanus orientalis* is fairly common here and occasionally one sees *Styrax officinalis*. A large number of smaller-growing plants occur by the wayside, including:

Acanthus spinosus	*Lavatera bryonifolia*
Arum concinnatum	*Leopoldia comosa*
Capparis ovata	*Mandragora officinalis*
Carlina corymbosa curetum	*Ornithogalum narbonense*

Cyprus 1. Ajuga chamaepitys palaestina 2. Anchusa aegyptica 3. Ranunculus asiaticus 4. Lamium moschatum

Cistus incanus creticus
Convolvulus althaeoides
Ecballium elaterium
Ferula communis
Ferulago nodosa
Heliotropium hirsutissimum
Hypericum triquetrifolium

Psoralea bituminosa
Rubus ulmifolius
Salvia triloba
Thymus capitatus
Urginea maritima
Verbena officinalis

The arum (= *A. nickelii*), sometimes considered as a subspecies of *A. italicum*, is not recognised by Flora Europaea. It produces its leaves in autumn and these persist throughout the winter. The large 'Lords-and-ladies' inflorescences develop in summer with yellow spathes and spadixes. It is widespread in Crete, especially near cultivation, and is also found in southern Turkey. The carlina differs very little from the widely-spread type species but is endemic to Crete and Carpathos. The ferulago is an umbellifer which is easily recognised by its curious, swollen nodes. The lavatera is a tree mallow with deeply-cleft leaves and typical mauve flowers. The rubus is a blackberry with rather small leaves and the flowers are usually pink coloured, often a dark pink. It frequently forms very prickly, inpenetrable thickets.

Southwards out of Rethymnon to the village of Galos one passes uncultivated stony ground with scattered bushes of species common to the area such as *Calicotome villosa, Genista acanthoclada, Sarcopoterium spinosum, Pistacia lentiscus* and *Thymus capitatus*. The calicotome is the only species of this genus found in Crete whereas the very similar *C. spinosa* is the common species occurring in the western Mediterranean. A number of autumn-flowering plants can be found amongst the stones including:

Atractylis gummifer
Colchicum pusillum
Dittrichia graveolens
D. viscosa
Erica manipuliflora

Narcissus serotinus
Ranunculus bullatus
Scilla autumnalis
Urginea maritima

Other autumn flowering species, growing in more wooded areas to the south of here, include *Cyclamen graecum* and the lady's tresses orchid *Spiranthes spiralis*.

Driving westwards along the new road out of Rethymnon, after a short time one passes rising rocky land to the south with most of the species described above and eventually comes to the mouth of a small gorge near the village of Gerani. On the rocks here one may see some chasmophytes typical of the island.

An alternative to the new road westwards is a part of the old road passing through Prines, Gonia, Episkopi and Vrisses. The countryside is a mixture of fields and patches of trees where the valonia oak *Quercus macrolepis* (= *Q. aegilops*) is common and easily recognised by its large scaly acorn cups. Amongst the plants already mentioned one may also see *Hypericum empetrifolium* and *Osyris alba* which superficially resembles an ephedera but has tiny yellow flowers - both have bright red berries in autumn. North of Vrisses there is a rocky peninsular that terminates in the north at the cape Akrotiri Drapano. This may also be reached by leaving the new road at the pleasant small resort of Georgiopolis. Around the villages of Sellia (Selia), Likotinaria and Kefalas there is an area of karst like limestone which cannot be cultivated. In cracks and pockets within the rock there are a few shrubs such as *Hypericum empetrifolium, Phlomis fruticosa, Sarcopoterium spinosum.* Amongst them one may see large numbers of *Cyclamen graecum* and other autumn and winter flowering species such as *Crocus laevigatus, Narcissus serotinus, N. tazetta*. In one site near the village of Sellia there is a large group of *Sternbergia lutea sicula* and some plants have such large flowers that they might well be classified as the sub-species *lutea*.

20.9 **South of Rethymnon:** A road runs from the centre of Rethymnon towards Spili. The first part is not especially interesting but after about 14km a turning to the west takes one to Agios Vasillios and after Agios Ioannis it bends southwards towards the coast. Near here it passes typical maquis vegetation and then through a small gorge. On the east side of the gorge one may see a fair quantity of sternbergia flowering in October. It was hastily identified as *Sternbergia sicula* but it may well be the recently described species *S. greuteriana* which is similar but has brighter green leaves without a glaucous central stripe. It is rather special to Crete, particularly in the east, but also grows on the neighbouring islands of Kassos and Karpathos. The usual chasmophytes growing on the rocks with the sternbergia here include *Alyssoides cretica, Ptilostemon chamaepeuce, Rosularia serrata* and *Verbascum arcturus (Celsia arcturus)*.

Eventually the road runs down to the seaside resort of Plakias. At the time of writing a holiday company runs a well informed botanical tour from this area. The many interesting species to be seen from here include a number of Cretan endemics and other interesting plants such as:

Aristolochia cretica	*Silene sieberi*
Biarum davisii	*Stachys spinosa*
Campanula tubulosa	*Symphytum creticum*
Dianthus fruticosus creticus	*Ranunculus creticus*
Muscari spritzenhoferi	*Tulipa cretica*
Petrorhagia candida	*T. saxatilis*
Scorzonera cretica	

All of these are mentioned in Polunin (1980) and Sfikas (1987) though one will have to look for the symphytum under *Precopiana cretica* - a plant that seems more like a borage than a symphytum with narrow, bent-back, blue or white perianth segments. It is usually found in the mountains whereas the similar *S. insularis*, which is also found on the island, is a lowland plant.

20.10 **Mount Ida:** Although this is the highest mountain group of the island, it is not the most interesting botanically. This is largely because of excessive grazing which has turned much of it into a barren waste and there is a considerable risk that a similar fate awaits the Levka mountains. Nevertheless it is worth a visit for some interesting plants still can be found, especially near the summits. The easiest approach is from Heraklion taking the road out westwards and turning off southwards to the village of Tillissos and from there to Anoyia. A rough track from here leads to the Ideon Andron - a cave claimed as the birthplace of Zeus. It lies on the edge of a high plateau, similar but considerably smaller than the Omalos. Here one is not far from the highest point of the island, Mount Psiloritis 2456m and the associated peaks of Angathias 2424m and Mavri 1981m and others. Near the summits, around the edge of the melting snow, in spring one may see *Alyssum idaeum, Corydalis rutifolia, Crocus sieberi, Draba cretica* and *Scilla cretica* (= *Chionodoxa cretica*). The alyssum is endemic to Crete and although it is named after Mount Ida it does occur on some other high places in the island, but it is rare - a typical dwarf alpine plant with greyish leaves and small yellow flowers. It is possible to confuse it with *A. siculum* which was recorded from here in 1971 though its main site is the Madonie mountains of Sicily and on the Greek mainland. Other species occurring later in the summer include:

Anthemis abrotanifolia	*Ononis spinosa*
Cichorium spinosum	*Ranunculus brevifolius*
Asperula idea	*Scabiosa sphaciotica*
Astragalus angustifolius	*Silene saxifraga*
A. creticus	*S. variegata*
Hypericum trichocaulon	

The anthemis is endemic to the higher mountains of Crete. It forms compact clumps that produce pinkish capitulae, often without ray florets, in June and July. The ononis has numerous thin, and often sparsely branched, stems with needle like thorns and bears small pink flowers; it is widespread throughout the Mediterranean and sometimes grows at low levels. Another plant to be searched for here is *Horstrissea dolinicola* - a new species of a new genus which has recently been discovered (Egli et al. 1990). It is a member of Daucaceae (Umbelliferae) that has a carrot-like root buried rather deeply and only the extremities of the branches appear above ground with pink flowers - an interesting adaption to cope with the excessive grazing of this region.

It is possible to get along the western side of the Ida massif by taking the road to Prasies that leads south from the main road just east of Rethimnon. Polunin (1980) states that the orange-red flowered *Tulipa orphanidea* may be found near the monastery of Asomaton (Assomaton). It is a species which is fairly widespread in Greece but not very often seen in Crete. To the south west of here lies Mount Kedros 1777m which seems to be the only site for the recently-described *Dianthus pulviniformis*. It grows on the limestone cliffs and is a woody-based species with heads of small pink flowers in summer similar to another Cretan endemic *D. juniperinus*. Both species are illustrated by line drawings in Sfikas (1987). In this region I have also seen *Bellis longifolia* which is another Cretan endemic. It grows in damp, grassy areas and closely resembles the common daisy but has narrower and longer leaves. It is similar to *B. margaritaefolia* of Sicily and Calabria and thrives in similar habitats (13.9). Another plant seen here is *Salvia pomifera* which is recognised from *S. fruticosa* (= *S.triloba*) by its narrower and greyer leaves. It is of special interest for the frequent oak apple-like galls caused by an insect parasite which give it the name 'pomifera' - apple-like. These are considered a delicacy and sometimes candied as a sweetmeat in Greece - presumably 'grubs and all'!.

20.11 **Around Heraklion:** Heraklion, which is called Iraklio on some maps, is the capital city of the island and like others of its kind has become very busy during the last twenty years with much traffic and parking difficulties. All visitors, however, should go there to see the extraordinary innovative relics of the Minoan civilisation in the Museum of Antiquities. As a bonus to plant enthusiasts there is a fine blue-flowered specimen of the Brazilian jacaranda tree (*Jacaranda ovalifolia*) near the entrance door.

Two main roads lead south out of Heraklion; the westerly one passes through Agia Varvara to the cultivated plain of Messara at Agii Deka and the archeological site of Gortina and from there westwards to the evocative site of Festos. The easterly one goes to

Knossos and reaches the Messara plain at Pirgos. Both of these pass through hilly country and are excellent routes for botanising. Amongst the numerous species to be seen in this area one may well encounter the following:

Anchusa azurea *Lathyrus aphaca*
Anemone coronaria *Leopoldia comosa*
Arum concinnatum *Malva cretica*
Asphodelus aestivus *Pallenis spinosa*
Bellardia trixago *Prunus prostrata*
Convolvulus althaeoides *Ranunculus asiaticus*
Echium plantagineum *Tetragonolobus purpureus*
Gagea graeca *Thymelaea hirsuta*
Gladiolus italicus *Tragopogon porrifolius*
Glaucium flavum *Urginea maritima*
Gynandriris sisyrinchium *Verbascum macrurum*

Most of these are widespread Mediterranean species. The prunus is especially plentiful in parts of Crete where it forms low bushes covered with beautiful pink flowers in spring followed by small red 'plums'. The verbascum is a white-felted species which produces compact spikes of small yellow flowers in May and June, some 1-2m in length and more or less unbranched. It can be an interesting and curious sight (16.5). The form of *Ranunculus asiaticus* usually seen here, and in most of Crete, has white flowers which are sometimes flushed with pink.

Orchids are plentiful by the roadside in this region and one can expect to see the following in spring:

Anacamptis pyramidalis *Orchis collina*
Ophrys cretica *O. italica*
O. fusca *O. papilionacea*
O. lutea *Serapias lingua*
O. tenthredinifera

Ophrys cretica is, of course, special to Crete though it is also found on some of the surrounding islands. It is a rather distinct species with a well marked lip that is pointed and has pronounced side lobes. It resembles *O. kotschyi* from Cyprus. Another 'bee orchid' to search for here is the newly described *Ophrys gortyna* (Albertis and Albertis 1989) which is endemic to Crete and presumably found near the archeological site of Gortina near Agii Deka. It belongs to the *O. sphegodes* group and has a rather broad dark lip with parallel markings. *Orchis collina* which used to be called *O. saccata* is an uncommon species with a wide distribution throughout the Mediterranean. It has a short, blunt spur and the lip is usually a dull mauve colour but the form here often has a green-margined, or occasionally, a completely green lip.

The more easterly road leading south out of Heraklion goes to Knossos which every visitor to the region will want to see. Some orchids grow amongst the ruins and also *Arum creticum*. A few kilometers south of Knossos a road to the west takes one to Arhanes (Achanes) and the 811m hill of Iouhtas (Youktas). Along the roadside one may see *Ballota pseudodictamnus, Cyclamen creticum, Phlomis lanata, Silene gigantea* and *Trifolium stellatum*. In November there is a chance of finding the rare and curious little *Biarum davisii*. Further up the hillside there are limestone rocks near the small Church of the Transfiguration where several chasmophytes grow such as *Alyssoides cretica, Dianthus juniperifolius, Ebenus cretica* and *Ptilostemon chamaepeuce*. Orchids found here include *Ophrys cretica, Ophrys bombyliflora, Orchis papilionacea* and *Serapias vomeracea*.

A short trip out of Heraklion well worth making, is to take the old road westwards to Marathos and Fodele. In this region there is extensive garigue with many of the common shrubby species such as:

Cistus incanus creticus	*Phlomis fruticosa*
C. salvifolius	*P. lanata*
Erica arborea	*Quercus coccifera*
Lavandula stoechas	*Spartium junceum*

A feature of the region in spring is that every large bush seems to have a nightingale singing in it. Amongst the shrubs in places grows the common bracken *Pteridium aquilinum* - a wide-ranging species that one surprisingly meets up with in many parts of the Mediterranean. Some relatively common herbaceous and smaller-growing plants here are:

Anemone coronaria	*Gladiolus italicus*
A. hortensis	*Lupinus angustifolius*
Cerinthe major	*Orobanche ramosa*
Gagea graeca	*Ranunculus asiaticus*

Some less common orchids here include *Orchis anatolicus, O. boryi* and *O. prisca*. The last of these is similar to, or a sub-species of, *O. patens*. It is a rather tall species, growing to nearly 50cm, with unspotted leaves and mauve flowers that have large sepals blotched with green in the centre.

20.12 Lassiti plateau and Dikti mountains: The Lassiti (Lasithi) plateau is a high area of level land used for growing potatoes and other crops and surrounded by impressive mountains. It has a grid of irrigation channels built during the Venetian occupation and fed by wells from which the water was originally pumped by hundreds of windmills with canvas sails. Unfortunately, these have been superseded by electric pumps but the area is still important for the crops it produces. It can be reached either from Heraklion or Agios Nikolaos but usually by taking the road southwards that leaves the New Road just west of Malia. The route climbs fairly steeply through Mohos and Gonies and on the way one may see many of the plants mentioned in 20.11 and some fine specimens of *Dracunculus vulgaris*. These are imposing plants one can sometimes locate before seeing them on account of their disgusting smell. The form in Crete is generally classed as the variety *creticus* with more distinct white markings on the leaves and purple-blotched stems but this is not recognised in Flora Europaea.

Approaching the plateau and around it there is rather dense

garigue often comprised mainly of the very prickly *Genista acanthoclada*. Another plant which is plentiful here *Daphne sericea* with flowers that are white or pink but fade to a dirty brown colour though they have a pleasant scent. The main orchids here are *Orchis pauciflora, O. tridentata, O. quadripunctata* and occasionally the giant orchid *Barlia robertiana*. The *O. tridentata* is a varied and often particularly fine form being short-growing and with flowers that are rather large and especially well-coloured for the species. In the autumn one may well encounter *Sternbergia greuteriana* but one should check it carefully for *S. sicula* also grows here. There is a road that circumvents the plain and several villages on it where one may find accommodation if one wishes to start off early to climb the mountains. One can also engage a guide in one of the villages to take one up the mountains though it is fairly straight forward to find ones own way.

To get to the top of Dikti 2148m and the other higher peaks such as Lazaros 2085m and Afendis Hristos 2141m there is a track southwards near the village of Avrakondes. It is rough but no mountaineering skill is needed. After a short distance one passes the church of Agio Pneuma where climbers sometimes spend the night rather than staying in a village. Around here in March and April the air is scented with flowers of the wild sweet violet *Viola odorata* which grows plentifully amongst the stones. From here on there are extensive screes with an interesting flora. *Iris cretica* (=*I. cretensis*) is common forming compact clumps with its small dark mauve, marked flowers in spring. It is usually classed as a form of *I unguicularis* but is distinctly smaller in all its parts. The typical, large-flowered form can also be found on the island but usually at lower levels. Another interesting plant here is the recently described *Arum idaeum* - like a dwarf form of *Arum creticum* with a milky-coloured spathe and a short dark-purple spadix. Some other species to be found include:

Anemone heldreichii	*Myosotis refracta*
Cichorium spinosum	*Romulea ramiflora*
Cyclamen creticum	*Verbascum spinosum*

Towards the summit, around the edges of the melting snow in spring, grows a very beautiful form of *Crocus sieberi* with especially large white flowers that have a yellow throat and varying bands of purple on the outside and sometimes on the inside of the corolla. Some pure white flowered specimens with yellow centres look as fine as the garden variety known as 'Bowle's White'. Other plants in flower with the crocus are: *Aethionema saxatile creticus, Centaurea raphanina raphanina, Draba cretica* and *Scilla nana* (= *Chionadoxa nana*). The aethionema has white or pale pink flowers and the draba is a small species with light yellow flowers that was first described from Mount Lazaros in this range. There are other interesting and rare species to be seen later in June and July when all the snow has melted including:

Alyssum lassiticum	*Sedum tristriatum*
Berberis creticus	*Scorzonera mollis idaea*
Crocus oreocreticus	*Scutellaria hirta*
Daphne oleoides	*Silene thessalonica dictaea*
Lysimachia serpyllifolia	*Vincetoxicum creticum*

The alyssum is a grey-leaved species only found here. The crocus flowers in autumn with rather small lilac-purple flowers. The lysimachia has yellow flowers and is endemic to eastern Crete. The scorzonera is a dwarf species which has only been recorded from Mount Lazaro. The silene has a woody stock and greenish-yellow flowers and is endemic to the Dikti range. The vincetoxicum, also an endemic to the area, produces dull yellow flowers that are bearded within.

There are good views on a fine day from the summit of all of these mountains and one has a good chance of seeing the larger raptors, especially golden eagles, griffon vultures and the occasional lammergaier.

20.13 The extreme Eastern part of Crete: Many tourists spend a holiday at Agios Nikolaos and this is a good centre from which to explore the Dikti mountains. From here one also has access to the Sitia mountains (Orini Thripi) which are not as high as the three main groups but rise to 1476m at Afendis Kavoussi. The region is briefly discussed by Polunin (1980) and one should look for such species as *Viola scorpiuroides*, a shrubby plant with greyish leaves and yellow flowers that has one of its only sites outside North Africa here.

Just east of Sitia on the north coast and towards the fortified Monastery of Toplou there are stands of the Cretan date palm *Phoenix theophrastii*. It is somewhat shorter growing than the true date palm and tends to produce a number of trunks from a single rootstock. The fruit is blackish when ripe, fibrous and inedible. It is sometimes planted in other parts of the island and occurs occasionally in some eastern Aegean islands and south-west Turkey.

Cyprus 1. Brassica hilaronis 2. Acer obtusifolium 3. Centaurea aegialophila 4. Bellevalia trifoliata

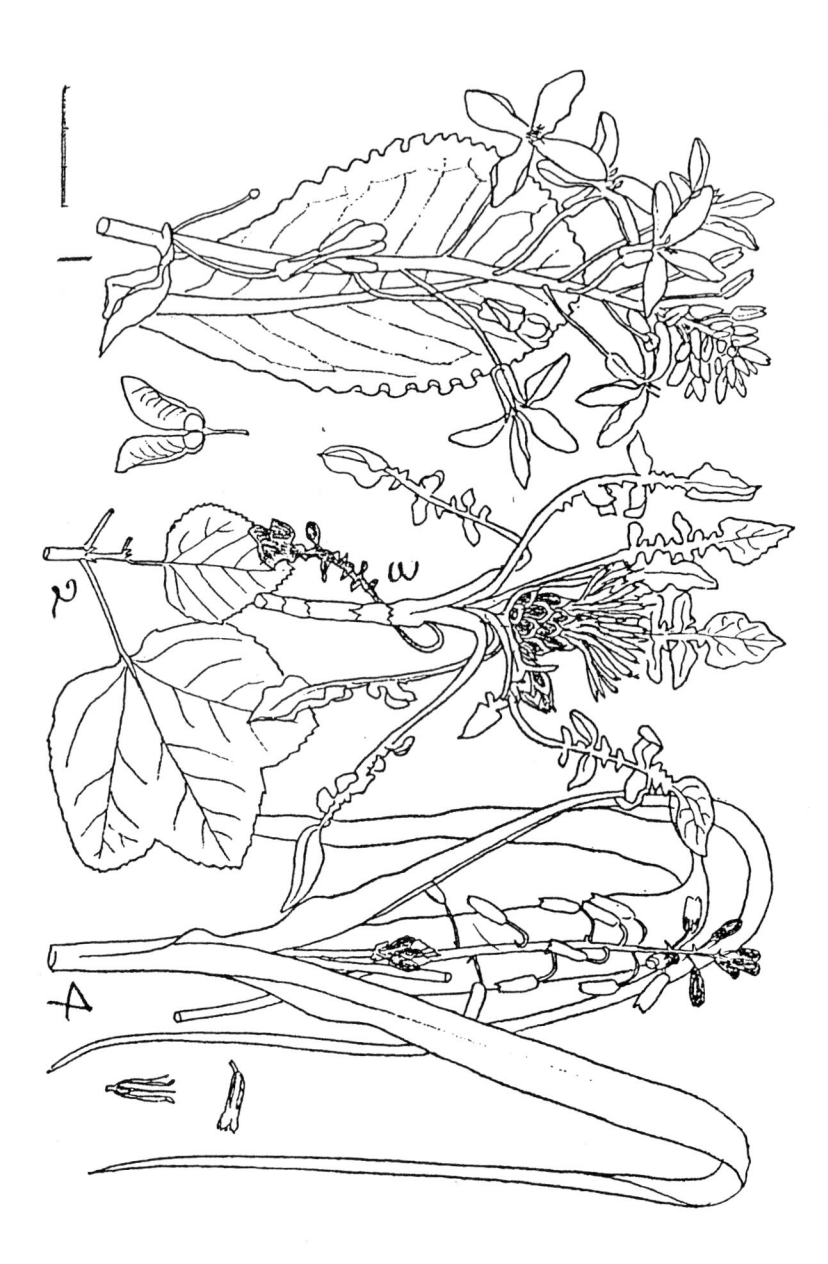

Cyprus 1. Bellevalia nivalis 2. Ranunculus millefoliatus leptaleus 3. Scilla cilicica 4. Cistus parviflorus

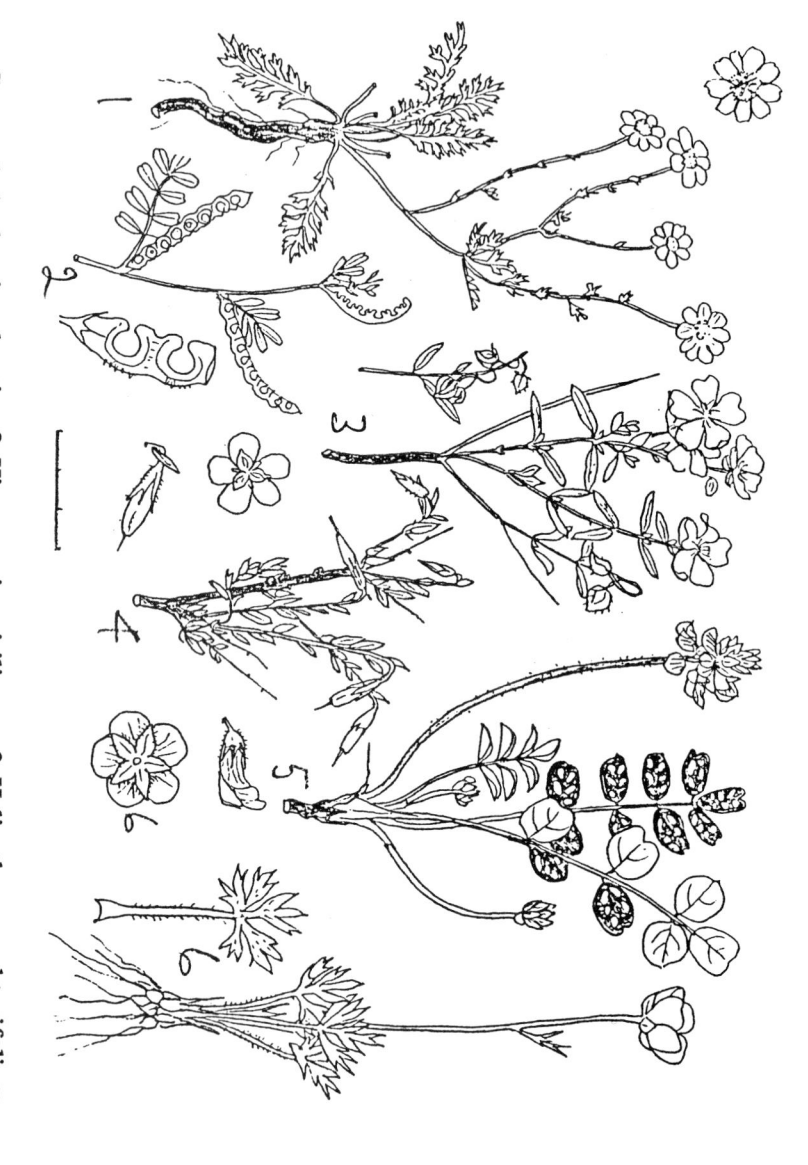

Cyprus 1. Anthemis palaestina 2. Hippocrepis unisiliquosa 3. Helianthemum obtusifolium 4. Onosma fruticosa 5. Onobrychis venosa 6. Ranunculus paludosus

21. RHODES

21.1 Spring comes early to Rhodes which is nearly as far south as Malta and has no high mountains to destabilise the weather. It is a popular holiday centre for sunseekers of all nationalities wishing to escape from the northern winter. One may find that the hotel staff there speak Swedish as well as they do English.

It is not a large island, roughly lenticular in shape and 80km long by 30km wide; about the same size as Minorca. Most visitors stay at Rhodes town (Rhodos in Greek) in the extreme north of the island or one of the seaside resorts along the east coast. Any one of these venues is a convenient base from which to study the flora. Bus services to many parts are fairly frequent from Rhodes town but the use of hired transport is, as always, a very considerable advantage for plant spotters. If one feels up to the challenge then a motor scooter or motorcycle is inexpensive and convenient for the distances to be covered are not large.

Getting to Rhodes is simple for it has an international airport so one can fly there direct from Britain. Late March or April is the best time to see most plants in flower but a few specialities bloom later and some in the autumn and winter. In one's spare time the Crusader's city of Rhodes, which was greatly restored under Mussolini's rule, the Greek and Roman remains at Lindos and the Doric ruins of Ancient Kamiros are all well worth visiting.

21.2 **The terrain:** Rhodes has a backbone of hills, roughly down its long axis but somewhat nearer the north than the south coast. It is hilly rather than mountainous and rarely sees any snow but there are three substantial highlands which are, from north to south, Mt. Profitis Ilias 798m, Mt. Attavyros 1215m and Mt. Akramitis 825m. These are all more or less clothed with trees and the first of them, although it is not the highest, is easily accessible and by far the more interesting for botanising. The highlands are mainly composed of limestone but some of the lower areas are of later sedimentary deposits.

21.3 **Literature:** The small Wisley Handbook No 9 by Huxley (1972) and 'Flowers of Greece and the Aegean by Huxley and Taylor (1977) are both helpful to the first time visitor. One may also find the book on flowers of Crete by Sfikas (1987) and Polunin's (1980) treatise on the flowers of Greece and the Balkans useful though neither make special reference to Rhodes. By far the most complete work for the serious enthusiast is 'A survey of the flora and phytogeography of Rodhos, Simi, Tilos and the Marmaris Peninsula' by Annette Carlström (1987). It is from Lund university, written in English and available from some specialist bookshops. It lists all species found on the island and gives distribution maps but has no descriptions.

The Island of Rhodes is not included in the area covered by Flora Europaea, instead it is dealt with in Davis's (1965-85) ten volume flora of Turkey and the east Aegean Islands.

21.4 **Rhodes town, and the North-East:** Although most tourists gather at Rhodes town, it is not the best place to seek wild plants but along the shore there one will probably meet up with *Carpobrotus edulis* and the yellow horned poppy *Glaucium flavum*. Between rocks at the end of the beach where the road leads to the airport grows the autumn-flowering *Cyclamen graecum* which is rare on the rest of the island. In the Turkish graveyard by Mandraki harbour one can find *Cyclamen persicum*, a spring flowering species which also is not widespread here.

A good road leads out of Rhodes town along the east coast southwards to the resort of Faliraki Beach and here grow shore plants typical of most parts of the Mediterranean, including:

Anthemis rigida   *Glaucium flavum*
Cakile maritima   *Medicago marina*
Cichorium spinosum   *Pancratium maritimum*
Eryngium maritimum   *Verbascum sinuatum*

Two less common shore species which one has a good chance of seeing here are *Erysimum crassipes* and *Matthiola longipetala*.

The first is a rather typical treacle mustard with yellow flowers and the second is similar to *M. bicornis*.

From Faliraki southwards through Afantou to the charming village of Archangelos there are areas of garigue which should perhaps here be called by its Greek name 'phrygana'. This originally meant simply common pasture land. No doubt, even from the earliest days the grazing areas in Greece comprised many small shrubs. The main ones here are:

Cistus creticus	*Lithodora hispidula*
C. parviflorus	*Salvia fruticosa*
C. salvifolius	*Sarcopoterium spinosum*
Erica manipuliflora	*Satureja thymbra*
Genista acanthoclada	*Thymus capitatus*

Cistus parviflorus is the most abundant rock rose here and its neat habit and small pink flowers are most charming. The island's name Rhodos means a rose and it would be interesting to know whether it refers to this cistus rather than the only true rose species to be found on the island - the rare white-flowered *Rosa phoenicia*. The genista provides the prickles when one walks through the garigue. The lithodora (O = *Lithospermum hispidulum*) is also common in Cyprus and Crete. The thymus is sometimes called *Coridothymus capitatus* and is common throughout the Mediterranean, from east to west. Amongst these one may expect to find the following:

Anchusa aegyptica	*Mandragora autumnalis*
A. hybrida	*Onobrychis caput-galli*
Asphodelus aestivus	*Ranunculus asiaticus*
Biscutella didyma	*Salvia verbenaca*
Convolvulus althaeoides	*S. viridis*
Leopoldia comosa	

Anchusa aegyptica is an uncommon species, rather like *A. arvensis* but with pale-yellow flowers. The ranunculus here is nearly always the brilliant scarlet flowered form. Orchids are also fairly widespread in the garigue and one should look for some of the specialities of the region. These include *Orchis sancta* which is similar to the bug orchid *O. coriophora fragrans*, which is also found here, but with flowers of a dull pink or mauve that have an unspotted lip with strongly toothed side lobes. Another is *Ophrys regis-ferdinandii* (= *O. speculum var regis-fernandii*), a form of the mirror orchid in which the lip is narrow and with curled back margins so it appears almost completely blue. Other less common species here include *Ophrys argolica, O. reinholdii, O. scolopax heldrechii, O. scolopax orientalis*.

Many of these orchids also grow by the roadside with other interesting species such as *Arum dioscoridis, Acanthus spinosus, Dracunculus vulgaris* and *Petrorhagia velutina*. The arum is an impresive species with a velvety-black base to the spathe; it is essentially a plant of the eastern Mediterranean and especially common in parts of southern Turkey. The petrorhagia resembles a dianthus with small pink flowers that occur in a head but with bracts that give the appearance of an inflated calyx bearing a single flower, since only one is open at a time.

At Kolibia, about 4km north of Archangelos, there is a turning west to Epta Pigai (Seven Springs). This leads to a wooded area with a running stream and is very popular as a picnic area for the people of Rhodes during the heat of the summer. In the cool shade of the trees one can see the maidenhair fern *Adiantum capillis-veneris* and, in places, carpets of *Cyclamen repandum rhodense* flowering in spring. This is the white-flowered form only found on Rhodes. It is possible to continue along this side road through Elousa to get to the mountain Profitis Ilias, which will be discussed in 21.6.

21.5 Lindos and the South: Thousands of tourists visit the Greek and Roman ruins at Lindos and the route may become rather congested. A quiet walk off the main track into the town itself to see the neat buildings is well worth while. The countryside here and, if one feels energetic, the walk to Cape Milianos and Cape Mirtias provide a good opportunity for botanising. On rock faces grow a number of interesting chasmophytes, such as:

Campanula rupestris anchusaeflora   *Ptilostemon chamaepeuce*
Helichrysum orientale   *Rosularia serrata*
Inula heterolepis

The campanula seems to be endemic to Rhodes but its correct naming poses a problem. It is illustrated in Huxley and Taylor (1977) under the above name but Carlström (1987) calls it *C. hagiella* which grows also in Turkey. It resembles the *C. rupestris* found in southern Greece on Lykavitos (17.4) and *C. topaliana* on Parnassos (16.6) but the flowers have a longer corolla. However, even in Rhodes it is variable and forms with more bowl-shaped flowers can be found. Whatever its name it is a beautiful plant that tends to be short-lived and is unfortunately not very amenable to cultivation. The ptilostemon has clustered, narrow, basal leaves and rounded capitulae resembling a centaurea with mauve florets. The helichrysum has silvery leaves and relatively large lemon-yellow flowers. The rosularia is a sempervivum-like plant with a flat rosette - a genus which is more widespread in Turkey and found also in Cyprus.

Other plants to be seen on the Lindos peninsula include *Campanula rhodensis* which is endemic to Rhodes, the small island of Khalki and resembles the more widespread *C. drabifolia* (17.5). One may also see *Convolvulus dorycnium* - a shrubby species with hairy leaves and relatively small pink flowers. The rather similar but more attractive *C. oleifolius* is also found here and scattered throughout the island; it is prostrate, has more silvery leaves and somewhat larger pink flowers.

From Lindos the road continues to the southernmost end of the island which has more open country and is worth investigating though there are fewer plant species to be seen there than in the north. The region is, however, of special interest to ornithologists. One may encounter garigue with much the same composition as described above but with some additional shrubs less common elsewhere on the island:

Calicotome villosa
Fumana arabica
Juniperus phoenicia

Lavandula stoechas
Myrtus communis
Pistacia lentiscus

The juniper is especially common in the south, but rare in the north. In coastal regions here one may see also *Juniperus oxycedrus macrocarpa* with its large spherical cones - up to 2cm diameter. The lavandula is the only member of the genus to be found on Rhodes. Other plants that grow amongst the shrubs include the yellow-flowered *Fritillaria rhodia* (21.6) endemic to the island and the only member of the genus found here. Other monocotyledenous species occasionally seen include *Gynandriris sisyrinchium*, *Gladiolus italicus* (O = *G. segetum*) and *G. anatolicus*.

21.6 **The highlands of the North-West:** Leaving Rhodes town along the western shore one soon comes to a turning south to Mount Filerimos 267m, which has a monastery near the top and is a fine viewpoint. One may see *Cyclamen persicum* in pine woods here. Continuing along the main costal road past the international airport there is a turning in a south-east direction to Petaloudes known as 'The Valley of the Butterflies'. In summer this is one of the wonders of nature with congregations of millions of colourful Jersey tiger moths (*Euplagia quadripunctata*). The best time for this show is July and August, but everybody else on the island goes there at this time and few wild plants are to be seen in flower. Continuing further along the main coastal road past the airport at the village of Kalavarda there is a road south to Salakos which leads over Mount Profitis Ilias and this is the most interesting area in the whole island for seeing plants. As one rises up to Salakos there are areas of Maquis which include both *Arbutus unedo* and *A. andrachne*. With them grow:

Calicotome villosa
Colutea insularis
Laurus nobilis
Phillyrea latifolia

Quercus coccifera
Rubus sanctus
Spartium junceum
Styrax officinalis

The colutea is an interesting endemic bladder-senna similar to the more widespread *C. arborescens* and may in fact be simply a form of this species (see Carlström 1987). The blackberry *Rubus sanctus* and *Smilax aspera* cause thickets in places and underneath one may find the orchid *Limodorum abortivum*. By the roadside and in olive orchards grow many other orchids and in spring there is a good chance of seeing *Anemone coronaria*.

A little further along, the road skirts the top of the hill and the scrub becomes more like woodland with much *Pinus brutia* and scattered specimens of other trees such as *Cupressus sempervirens, Liquidambar orientalis, Platanus orientalis, Quercus aegilops* (= *Q. macrolepis*). The liquidambar is not common here; it is primarily a tree of southern Turkey which lies no more than 50km to the north (23.13). On the floor of these woods there are special plants to be found, in particular carpets of *Cyclamen repandum rhodense* and isolated specimens of the endemic *Paeonia clusii rhodia* with glistening white flowers in April and May. One may also see *Colchicum macrophyllum* with large, pleated leaves (up to 35 x 14cm) resembling those of a veratrum and somewhat disappointing pale mauve flowers in autumn. It is in fact a rather smaller-leaved version of the plants of this species found in Turkey. Another autumn crocus that grows in Rhodes is the more attractive *Colchicum variegatum* which also flowers in autumn and has wavy-edged leaves and flowers with distinct chequered markings. It may also be found in the drier north-eastern parts of the island. A great speciality to be searched for here is Comper's orchid *Comperia comperiana* (= *C. taurica*) which was first recorded in the region by two independant observers in 1981 - one of the records being given as near Salakos. It flowers rather later than most orchids, being at its best in May, and can hardly be confused with any other species. In addition one should look here for the giant orchid *Barlia longibracteata* that flowers in February and March and *Orchis morio picta* that comes on later. Other orchids on Rhodes include *Op. candica* and *Op. reinholdii*.

Taking the road south through Ebonas gives good views but is not more interesting floristically. There are substantial *Pinus brutia* woods and in clearings in spring one may find areas that were once cultivated covered with the common but beautiful *Anchusa azurea* (= *A. italica*) and making a fine picture when visited in the sunshine by hundreds of yellow swallowtail and southern festoon butterflies. It is difficult to get to the summit of the highest peak Mount Attaviros and the effort is hardly worth while botanically, for it is hard-grazed and carries little but thistles such as *Carlina corymbosa* and *Onopordon bracteatum*. Occasional specimens of *Anemone blanda* and *Arabis alpina* (= *A. caucasica*) may be seen there. However, the lower slopes are one of the best areas to see the endemic *Fritillaria rhodia* already mentioned. It is a small species with yellow, single, bluebell-shaped flowers and somewhat similar to *F. forbesii* (23.10). Continuing along the road in a south-west direction one passes Mount Akramitis and finally to the interesting ruins of Monalithos castle before going to the southernmost end of the island.

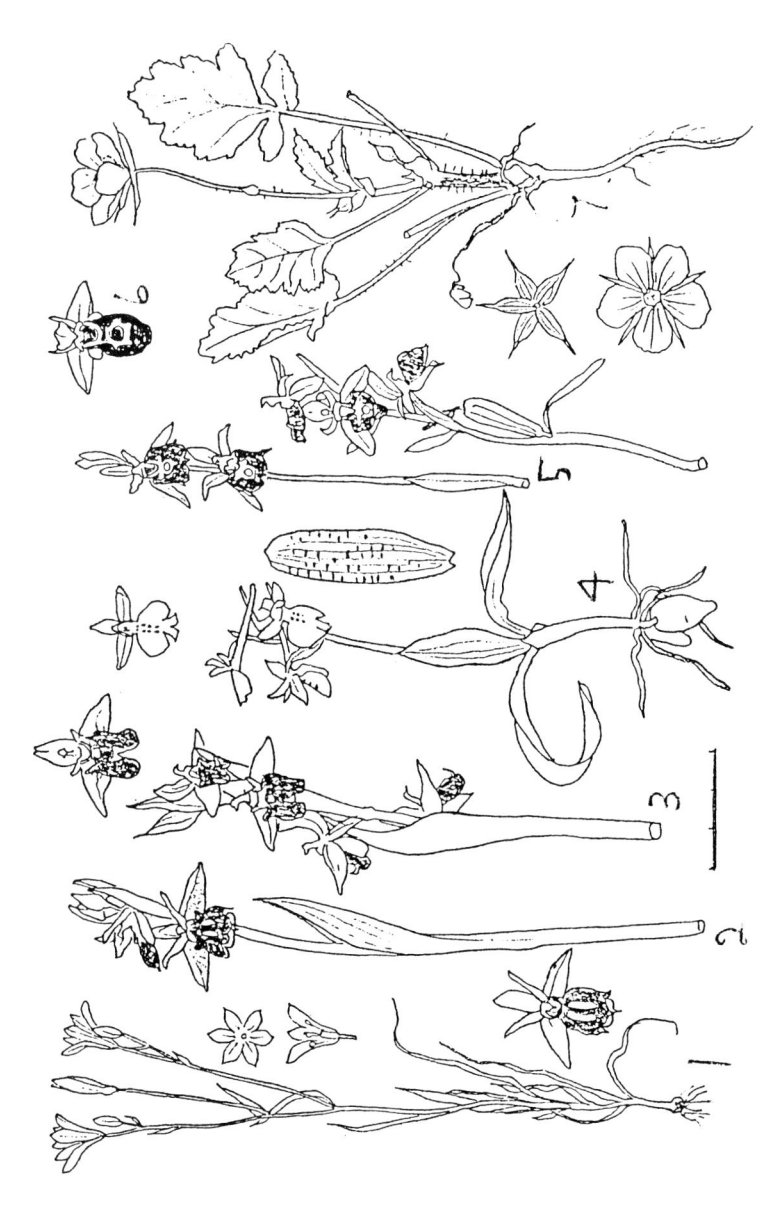

Cyprus 1. Gagea graeca 2. Ophrys sintenisii 3. Ophrys bornmuelleri grandiflora 4. Orchis anatolica 5. Ophrys argolica elegans 6. Ophrys kotschii 7. Erodium gruinum

Cyprus 1. Scabiosa prolifera 2. Nigella damascena 3. Ranunculus cadmicus var. cyprius

22. CYPRUS

22.1 This is the third largest and the most easterly of the Mediterranean islands. It is also the most southerly of the larger ones with its south coast at a lower latitude than Crete, Malta and Tunis and about the same as that of the northern limit of the Sahara desert. Although its flora is basically typical of Mediterranean Europe, Cyprus is home also to a number of species found in the Middle East. A typical example of this Levantine connection is *Scabiosa prolifera* - a distinct stocky annual with pale yellow flowers that is also common in Syria, Israel (24.5) and Lebanon. Cyprus has some 120 endemics and, probably because the grazing is not excessive, it is especially rich in orchids with over 50 different species and subspecies. Like most other Mediterranean islands Cyprus has had a very chequered political history. It is large enough to have been of interest as a strategic base since classical times, but has never been sufficiently populated to be able to defend itself against more powerful covetous neighbours. As a result many different peoples have occupied the land in the past, including the ancient Egyptians who may have been responsible for introducing the fruit-eating bat (*Rousettus aegyptiacus*) and the locust, both of which can be seen there today. Fortunately the locust never reaches plague proportions in Cyprus.

The island obtained its independence in 1960 but Turkey declared the northern half to be Turkish territory and moved its army there in 1983 thus dividing the island politically into two different halves. Although it is possible to visit both parts of the island, the present border regulations make it difficult to cross from one to the other. Visitors can fly direct from Britain to the southern part but they have to travel via Turkey to get to the northern region. Holiday makers can be assured of a friendly welcome in both the north and the south of the island. Many of the inhabitants speak English, the traffic drives on the left and British visitors are made especially welcome.

Spring comes early to Cyprus and the best time to see lowland plants is from early March to mid-April. Patches of snow

Cyprus 1. Arabis cypria 2. Tulipa cypria 3. Lithodora hispidula 4. Anthemis cretica 5. Hyocyamus aureus

persist on the Troodos mountains until the end of May and most mountain plants are not at their best until early May or later. A number of interesting species flower in autumn and early winter including the endemics *Cyclamen cyprium* and *Crocus veneris*.

22.2 The terrain: The most obvious feature is the Troodos mountains in the southern part of the island. These are composed of rounded peaks of igneous rocks that rise to 1950m at Chionistra. Snow lies on the higher parts here for four months of the year and this area is especially interesting for its endemic plant species. The lower parts of the Troodos range and much of the coastal areas are composed of limestone and provide a good base to see many orchid species. Within this region, and indeed scattered throughout the island, there are outcrops of serpentine.

By contrast with the Troodos mountains, the 80km long Kyrenia range, lying parallel to the north coast and mostly no more than 5-8km wide, is composed of hard limestone. The peaks here reach to only 900m and are rarely covered with snow but they are rugged and spectacular like the Dolomites. This part of the island has been less developed than in the south and abounds in interesting plants, especially on the northern side of the range for the southern faces are very much drier and have sparser vegetation.

Between the Troodos and Kyrenia ranges lies the Mesaoria plain, a vast, flat, alluvial region which is largely treeless and of importance for the production of cereal crops.

22.3 Literature: The relatively inexpensive booklet 'Nature of Cyprus' (Georgiades 1989) is a very helpful introduction to the flora and fauna of the island and can be bought there and from some specialist booksellers in Britain. It is especially useful for its list of endemic plant species and the coloured illustrations of some of them. The author has also written a two volume work entitled 'Flowers of Cyprus - Plants of Medicine' (1985, 1987). A recent book entitled 'Wild Flowers of Cyprus' (Sfikas 1990) describes and illustrates more species. It is pocket size but tends to fall to pieces on frequent usage. Any reader who owns a copy of 'Flowers of

Greece' (Huxley & Taylor 1977) will find it helpful to identify the general run of species but it does not, of course, deal with those that are special to the island.

For any serious student of the flora, the recent two volume 'Flora of Cyprus' (Meikle 1977,1985) is essential. This work is moderately bulky but portable and the present price is very reasonable for a treatise of this calibre. Another book of special interest is 'Studies on the Vegetation of Cyprus' by the Norwegian Holmboe (1917). Although it was published in Bergen it is written in English though, unfortunately, it is not easy to obtain nowadays. Sites of special botanical interest are described in it in some detail but sadly many of them now have been destroyed by 'development'. Also, the naming of plants is somewhat out of date.

22.4 The South - Limassol to Larnaca: Many visitors coming to Cyprus will be based at Limassol or one of the nearby coastal resorts east of the town. This is a good area from which to explore most of the southern region governed by the Greek Cypriots. Near the shore, about 10km east of Limassol, lie the evocative ruins of ancient Amanthus. Scattered bushes here include:

Crataegus azarolus   *Phagnalon rupestre*
Helichrysum conglobatum   *Pistacia lentiscus*
Lycium ferocissimum   *Sarcopoterium spinosum*
Olea europea sylvestris

The crataegus is a common 'hawthorn' of the island. In spring it has large, showy white flowers with a somewhat 'fishy' scent. The fruit is larger than the haws of our native species, yellow in colour with a touch of red and edible when ripe though hardly a delicacy. Trees of this species are sometimes grown as an orchard crop in parts of the Levant and perhaps also still in Sicily. The Lycium has spiny branches bearing creamy white, tubular flowers, generally in winter. Amongst these scattered bushes grow a number of herbaceous plants and annuals common throughout the island including:

Ajuga chamaepitys *Oxalis pes-caprae*
Bellardia trixago *Poterium verrucosum*
Chrysanthemum coronarium *Pallenis spinosa*
Convolvulus althaeoides *Plantago cretica*
Carlina involucrata cypria *Papaver rhoeas*
Crupina crupinastrum *Reseda alba*
Clematis cirrhosa *Ranunculus asiaticus*
Calendula arvensis *Silene vulgaris*
Lagoecia cuminoides *Silybum marianum*
Mercurialis annua *Trifolium stellatum*
Notobasis syriaca *Tragopogon sinuatus*
Orlaya daucoides *Tamus communis*
Orobanche ramosa *Vicia hybrida*

The ajuga is the eastern sub-species *palaestina*. The carlina is a Cyprus endemic sub-species which differs very slightly from the type but the rest of these plants are common to most parts of the Mediterranean. Amongst them one can find a number of bulbous and tuberous-rooted species, such as:

Anacamptis pyramidalis *Cyclamen persicum*
Allium neapolitanum *Ornithogalum narbonense*
Asphodelus aestivus *Urginea maritima*
Arisarum vulgare

The cyclamen grows in vast quantities in the north of the island but is less common here where it seems to seek shelter from stone walls that give it some protection from foraging animals eating the tubers - pigs relish them hence the old name, 'sow bread'.

An interesting outing from Limassol is to drive eastwards towards Larnaca. The road passes through orchards and vinyards and is lined in places with the tall flowering stems of the giant fennel in April. Some 20km from Limassol is the 7,000 year old stone age village of Khirokita lying a short distance north of the main road. It is well worth a visit though one perhaps may only see such plants as the mandrake *Mandragora officinalis* there. The road continues to Larnaca and a short distance to the south west of this town, near the international airport, there is a salt lake which is popular with bird watchers. By its western shore lies the fascinating Hala Sultan Tekke - an uninhabited Mohammedan monastery. In the vicinity the woods of pines and eucalyptus near here grow a number of species of orchid to be seen in flower during March and April:

> *Ophrys carmeli* *Ophrys scolopax orientalis*
> *Op. fusca* *Orchis morio picta*
> *Op. fusca fleischmannii* *O. saccata*
> *Op. lutea*

Other plants include:

> *Bellardia trixago* *Helichrysum conglobatum*
> *Centaurium pulchellum* *Lagoecia cuminoides*
> *Ferula communis* *Phagnalon rupestre*
> *Helianthemum obtusifolium* *Pallenis spinosa*

The helianthemum, an endemic rock rose with pale yellow flowers, is commonly seen throughout the island. In the short turf near here, and in woodland clearings there are thousands of plants of *Orchis coriophora fragrans* in a range of colour forms from white to pink and dark red. They flower somewhat later than many orchids and are at their best in late April and early May. Other plants which grow here include: *Asparagus stipularis, Asteriscus aquaticus, Ornithogalum narbonense* and *Scabiosa prolifera*. The last of these is a very distinct annual with flat heads of straw-yellow flowers. It grows mainly in waste places and by the roadside, often in extensive drifts, and is somewhat of a feature of the island. It is one of those species shared between Cyprus and the Levant. In late January and February, the poppy anemone *Anemone coronaria* makes a fine show here.

From Limassol there are several other interesting short 'outings'. One is to the Yermasoyia valley along a road a few kilometers east of the town that leads northwards from the main road. The vegetation is fairly lush here and not excessively grazed and is well worth careful examination. During a hasty visit there in April 1983 the following were noted:

Allium cupani	*Onosma suffruticosa*
A. subhirsutum	*Ornithogalum narbonense*
Anacamptis pyramidalis	*Ophrys scolopax*
Arum dioscoridis	*Orchis coriophora*
Hippocrepis unisiliqua	*O. sancta*
Lithodora hispidula	*Ranunculus asiaticus*

Several of these will be dealt with in more detail later in this chapter. The ranunculus occurs in very large numbers, mostly the white-flowered form but some have flowers tinged with pink. Only one specimen of the holy orchid *O. sancta* was seen; this uncommon species resembles *O. coriophora* with a slightly larger, and more 'ragged' flower that has an unspotted lip. It grows sparsely in other parts of the island, including the area around Larnaca but is fairly common in southern Turkey and Rhodes (21.4).

To the south of Limassol lies the Acrotiri peninsular, culminating in the south-east at Cape Gata. When Holmboe (1914) wrote his botanical treatise of Cyprus this was an area of special interest. Now, the northern part of the peninsula has been taken up by fruit plantations and the southernmost end is inaccessible. However, taking the road signposted as "Lady's Mile", one can still get to the village of Acrotiri itself which is near the western border of a fairly large salt lake. Through the tall reeds and thickets of the local blackberry species *Rubus sanctus* one can sometimes see flamingos. In a eucalyptus wood near here there used to be many plants of *Ophrys apifera var. bicolor* - an unusual form with a greenish lip and no dark markings.

22.5 The South-West: After passing through the base of the Akrotiri peninsular, the main road westwards out of Limassol keeps close to the coastline on its way to the popular holiday resort of Paphos. Some 10km from Limassol and shortly before reaching the restricted Episkopi garrison zone one comes to the archeological sites of Curium and the Temple of Apollo. In this area there are interesting patches of stony garigue on either sides of the road that are well worth detailed examination. Here one may also see griffon vultures circling overhead, for they roost unmolested in the ruins nearby. Typical shrubby and herbaceous species of the garrigue include:

Calicotome villosa   *Micromeria nervosa*
Cistus parviflorus   *Onosma fruticosum*
C. salvifolius   *Origanum majorana*
C. creticus   *Salvia fruticosa*
Fumana arabica   *Sarcopoterium spinosum*
Helianthemum obtusifolium *Thymelaea hirsuta*
Juniperus phoenicea   *Thymus capitatus*
Lithodora hispidula

The helianthemum is endemic and widespread on the island. The lithodora (= *Lithospermum hispidulum*) is also a common component of garigue here; it is a much-branched shrublet with mauve flowers. It occurs also in Crete, Rhodes and parts of the Levant but is not generally seen throughout the Mediterranean area. The onosma is another Cyprus endemic that is widespread on the island - an untidy looking shrublet with small narrow yellow flowers from which the style protrudes.

Unfortunately many of the areas of garigue are rapidly being eroded by building projects and by land 'improvement' though luckily they have escaped excessive grazing in most parts of Cyprus. In addition to the above shrubs one may also come across:

Helianthemum stipulatum   *Teucrium cyprium*
H. syriacum   *T. divaricatum*
Lavandula stoechas   *T. micropodioides*
Micromeria myrtifolia   *Thymus integer*
Onosma caespitosum

For further information on some of these one should consult Meikle (1977). Amongst the shrubs there are a number of bulbous and tuberous rooted plants to be seen, especially:

Allium cupani cyprium　　*Muscari parviflorum*
Allium trifoliatum　　　　*Narcissus serotinus*
Colchicum pusillum　　　*Romulea tempskyana*
Gagea sp.　　　　　　　　*Scilla autumnalis*
Gladiolus triphyllus

The *Allium cupani* form is a small plant with tight heads of dull, purple-red flowers that do not open wide; it is endemic to Cyprus. *Allium trifoliatum* is another small species which rather resembles the widespread *A. subhirsutum* but has charming, pink flowers. The colchicum is fairly widespread in the eastern Mediterranean and produces its small flowers in autumn. Yellow-flowered gageas are difficult to identify and several grow on the island including the endemic *Gagea juliae* which is illustrated in Georgiades (1989). Most gladiolus species are also difficult to name with certainty, but not this one. It is endemic to Cyprus and a distinct and attractive species. The plants are rather short-growing, with spreading foliage and a few large flowers that superficially resemble a freesia and have a similar scent. It is easily distinguished from the only other species which grows on the island, the much taller - *G. italicus*. The tiny *Muscari parviflorum*, sometimes referred to as baby's breath, the colchicum, narcissus and scilla all flower in autumn.

The classical site of Curium lies close to the shore which is worth exploring for its plant life. A special feature here is *Centaurea aegiolophila* that forms flat rosettes of leaves in the sand and has stemless heads of mauve flowers. It also grows in Crete and is similar to *C. raphanina* - a mountain plant from there. After the Episkopi garrison site the road along the south coast passes through a dry, rocky landscape and out to sea near a rock called Petra tou Romiou is where Aphrodite is said to have arrived in Cyprus. In fine weather it is a pleasant site, now much frequented by tourists.

A plant that is noticeable on the earth banks and rocks by the sea here is *Lycium schweinfurthii*, a shrub with arching spiny branches and smallish mauve tubular flowers in late autumn and continuing until summer. Although the cliffs and slopes behind Petra tou Romiou are very dry on the eastern side, to the west they carry more fertile garigue which is worth investigating.

22.6 **Around Paphos:** Continuing along the road from Limassol one eventually comes to Paphos (now frequently transcribed as Pafos on the island) which is the country's main tourist region and has been greatly expanded in recent years. It has some fine archaeological remains but is otherwise not a particularly interesting town. The flora in the immediate vicinity is not special though Georgiades (1989) says that *Tulipa agenensis* can be seen here in cultivated fields and in vineyards. This species which is sometimes called *T. oculis-solis* (eye of the sun), has red flowers with a yellow and black centre. It is distributed throughout the Mediterranean area, even as far west as France, where it was perhaps introduced from the Levant by the crusaders. The native *T. cypria* is similar but mainly found in the north of Cyprus.

The 'Tombs of the Kings' site just to the west of Kato Paphos is well worth a visit and of interest botanically for it is rather extensive and only lightly grazed. An number of typical garrigue plants occur here, including *Cyclamen persicum*. A plant especially noted on the site is *Anchusa aegyptica* - a fairly typical alkanet with sprawling stems covered with bristly hairs but, unlike most other species of the genus, it has pale yellow, instead of the usual blue, flowers.

22.7 **Paphos to Polis:** One can take the most direct road from Paphos northwards via Stroumbi to Polis but a more interesting route is to drive westwards along the coast, past banana plantations to Coral Bay and then up the hill to Peyia. As the road climbs out of this village it passes through a small area of typical local woodland which has been left amongst the extensive vinyards of the region. The main trees are *Pinus brutia* with an undergrowth of *Juniperus phoenicea, Calicotome villosa, Pistacia lentiscus, Cistus salvifolius* and *Cistus monspeliensis*. The last of these is not found throughout the island but it is plentiful here. One also comes across a few trees of *Cupressus sempervirens var. horizontalis* which is a wild form of the typical, upright-growing, Mediterranean cyprus. Many orchids grow in the limestone here. During the autumn, and even at Christmas, the ground is covered with thousands of flowers of the tiny buttercup *Ranunculus bullatus cytheraeus* which is considerably smaller than the form of this species found in the western Mediterranean

The road continues to Kathikas and from there until one reaches Polis most of the ground is taken up by vineyards. The roadsides are often lined with fennel and the very tall, and very prickly *Echinops spinosissimus* whose stems attain 3m when they reach their climax. Its typical 'globe thistle' flower heads vary from near white to a metallic blue when they flower in late summer and autumn.

Polis is one of the main starting places to examine the Akamas peninsula (22.8) but one may also take the coast road that leads north-eastwards from here towards Pyrgos. Some of the side roads in a southerly direction take one into the less cultivated Paphos forest area. About 5km from Polis there is such a surfaced road to Makounta and on to Lyso. Here are scattered patches of interesting garigue with much *Thymus capitatus, Sarcopoterium spinosum*, many orchids and *Anemone coronaria*. One of the next valleys along the coast road is to Giala which has good crops of citrus fruits and even ripening dates. From here a rough unsurfaced road leads into the woods on the southern side.

22.8 The Akamas: This peninsular has remained more or less uncultivated and is one of the most celebrated natural sites of the whole island. A backbone of hills of volcanic rock rising to some 300m fall fairly steeply towards the sea on the north east coast but the limestone rock slopes much more gently from the coast in the south-west and towards the northern tip. The volcanic rocks are clothed with maquis which is dense in places and consists mainly of

the following shrubs:
    Arbutus andrachne   *Juniperus phoenicea*
    Bosea cypria   *Styrax officinalis*
    Ceratonia siliqua

The bosea is a particularly interesting and distinct endemic species belonging to the family Amarantaceae which also includes 'love lies bleeding'. It grows to 1-1.5m tall, has lanceolate leaves and insignificant tiny yellow flowers followed by masses of attractive red berries in the autumn. Smaller under-shrubs of the maquis include:

    Calicotome villosa   *Lithodora hispidula*
    Cistus creticus   *Phagnalon rupestre*
    C. ladanifer   *Phlomis cypria*
    C. parviflorus   *Salvia fruticosa*
    C. salvifolius   *Thymus capitata*

Most of the cistus species are common on the island but the large-flowered *C. ladanifer* is only found in a few localities. The phlomis is another Cyprus endemic and resembles a rather small version of the familiar *P. fruticosa* which also ocurs on the island but is rare here. Interesting bulbous and herbaceous species to be found are:

    Bellis sylvestris   *Ranunculus bullatus*
    Cyclamen persicum   *Scilla cilicica*
    Narcissus serotinus   *Tulipa cypria*
    N. tazetta   *Urginea maritima*

The scilla and tulipa are rare species which will be discussed later (22.12, 22.13). Orchids also occur here, often in large quantities, but are even more profuse in the garigue. However one should look for the rare *O. punctulata* in the maquis area especially when it is on limestone. This is a tall-growing species which blooms early in February with yellowish flowers resembling the shape of those of

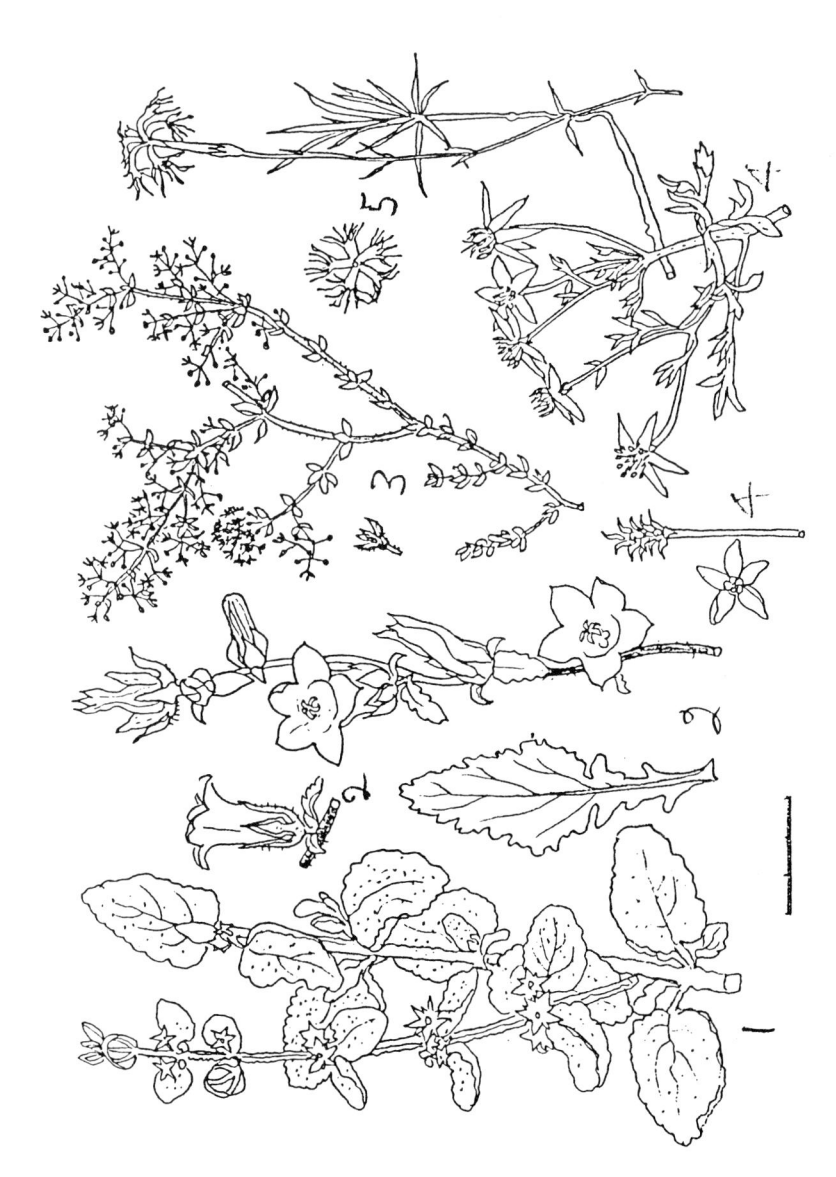

Turkey 1. Ballota pseudodictamnus 2. Campanula lyrata 3. Galium canum ovatum 4. Ranunculus sericeus 5. Dianthus crinitus

the monkey orchid and having a scent like lily of the valley. It is rare everywhere but found also in parts of Greece, Turkey and the Levant.

There is an area of woodland in the south west of the Akamas composed chiefly of *Pinus brutia* and with the shrubby species listed above. Most of the rest of the region consists of low garigue on limestone stretching from near sea level to the higher land. This garigue is primarily composed of *Juniperus phoenicea, Pistacia lentiscus* and *Sarcopoterium spinosum* with many orchids and *Arisarum vulgare* sheltering round the edges. The area has been declared as a Nature Reserve but this is presumably for the benifit of the plants rather than the birds since yellow, blue and black expended shot-gun cartridges litter the ground in places and are sometimes more numerous than the orchids which include: *Ophrys fusca, O. scolopax, Orchis sancta, Serapias orientalis* and, in autumn, *Spiranthes spiralis.*

Access to the Akamas is not straightforward and for the sake of the wild plants it is a good thing that the intention seems to be to keep it that way. There are no surfaced roads into the undeveloped part and one either has to walk or drive on rough, or very rough, roads which are reasonably accessible in an average car if the surface is dry and one takes the road slowly. A four-wheeled drive vehicle is a considerable advantage and can easily be hired locally. There are two main ways into the Akamas; through the south from Paphos area and through the north from the small resort of Latsi west of Polis.

The road from Paphos to the tourist site at Coral Bay and then on to Agios Georgios is surfaced and takes one past orange and banana plantations. The area around Cape Drepanon at Ag. Georgios is worth examining for plants. In the turf by the sea there are thousands of flowers of the small *Scilla autumnalis* during September and November together with the small form of *Ranunculus bullatus* and *Colchicum pusillum*. Here and there one may see plants of the delicate *Narcissus serotinus*. Nearer the shore grows *Asparagus stipularis* and *Lycium schweinfurthii*. Even nearer the sea, often in the sand by the shore, the rather neat small dandelion *Taraxacum aphrogenes* is in flower at the turn of the year. The unsurfaced road past Ag. Georgios takes one some 6km to Lara Bay, famous as a nesting site for the green turtle, and then another 10km further into wilder country.

From the north side at Latsi a surfaced road leads westwards to the 'Baths of Afrodite' which is a small cave with dripping water where one can appropriately see the maidenhair fern *Adiantum capillus-veneris*. Near here is a Cyprus Tourist Organisation notice of the nature trail and a large notice warning of periodic military exercises. When the red flag shows it is inadvisable to proceed further. The track along the coast is very well defined and takes one about 6km, to Fontana Amorosa near the end of the peninsula at Cape Arnauti. A few of the roadside plants have been labelled and one of these is especially interesting - *Pistacia x saportae*, a rare hybrid between the two common species *P. lentiscus* and *P. terebinthus*. It is a large shrub or small tree to 7m high, bears red fruits in winter, and is evergreen like the first of these parent species. This route takes one mainly through maquis. Another, smaller trail leads one higher up through the bushes.

One may also branch off the road from Latsi before the Baths of Afrodite where a surfaced road goes to the village of Neokhorio (Neon Chorion). From here one can drive along a rough, unsurfaced road to a picnic area in the pine woods referred to as Smyies site. If one has the right kind of vehicle and the right weather it is possible to drive in a south westerly direction from here to join up with the road past Lara Beach on the other side of the peninsula.

22.9 Chrysorroyiatissa Monastery: This interesting small monastery, sometimes spelt Khrysorroyatissa, lies on a hill 1142m high and some 25km north-east of Paphos. There are several possible roads to it and one should aim for the village of Pano Panayia. Shortly before entering here there is a group of poplar trees of a kind seen planted also in other parts of the island. These closely resemble the habit of the Lombardy poplar but the bark is grey or nearly white in some cases. Like it, it is indeed a variety of the black poplar *P. nigra* usually referred to as var. *afghanica* (= *thevestina*) but is female as compared with the male Lombardy poplar. It is commonly planted in North Africa, Turkey and parts of the Balkans. It does not seem to develop the white colouring of the trunk and branches when grown in Britain.

The monastery stands in an extensive region of vineyards and the few monks still there produce a vintage for sale. Just above the monastery is an area of natural woodland of some considerable botanical interest. In addition to the pines one may see the oak species *Quercus calliprinos* and *Q. infectoria veneris*. The first resembles a tall-growing kermes oak *Q. coccifera* which may grow to 10m; it is often considered simply as a form of that species. The second is a rugged, robust tree which is semi-evergreen and has large acorns up to 5cm long; it used to be an important component of extensive woods on the island but is now most usually seen as a roadside specimen tree. In the wood are also small trees or bushes of *Acer obtusifolium* and the deciduous *Pistacia terebinthus*. Other shrubs here include *Berberis cretica, Rhus coriaria* and various cistus species. The interesting yellow-flowered endemic *Odonites cypria* is widespread here and one can also find another endemic species *Jurinea cypria* - both flower in late summer and are well illustrated by Georgiades (1989)

In the autumn one may be fortunate to find the small *Crocus veneris*, another Cyprus endemic which is distinguished from the other two species of the island by its white flowers in autumn, not in spring. An even more unusual plant is *Scilla morrisii* only known from this small corner of the world. It has rather long sprawling leaves and milky white flowers in January to March and is best sought for on moist shady banks.

22.10 The Troodos Mountains: A good way to get to the highest part of the Troodos mountains from Limassol is to take the road towards Episkopi and after about 4km turn northwards at Ypsonas along the Kouris valley. It can also be approached from Nicosia by following the road to Morphou and branching off to the south-west a few kilometers after passing the village of Peristerona.

From Limassol one passes first through limestone country

and in mid- February to April it is worth making a diversion to the west towards Omodhos and Kilani to see the profusion of orchid species. The countryside here is of rolling hills with stone walls and somewhat resembles the Cotswold hills of England if it were not for the scattered vineyards. The orchids to be seen include:

Barlia robertiana	*Ophrys scolopax*
Ophrys carmeli	*O. scolopax orientalis*
O. fusca	*O. sphegodes mammosa*
O. fusca fleishmanii	*O. sph. transhyrcana*
O. fus. iricolor	*Orchis italica*
O. lutea	*O. laxiflora*

Ophrys carmeli is something of a speciality here, but it so closely resembles *O. scolopax orientalis* that it is difficult to distinguish between the two. The difference between *O. sphegodes ssp mammosa* and *ssp. transhyrcana* is also somewhat blurred. The last is often a particularly tall plant with rather large, well-spaced flowers and tends to flower later than *mammosa*. Both *Ophrys carmeli* and *O. sphegodes transhyrcana* are uncommon species with a limited distribution but both also occur in fair quantity in Israel (the first is abundant on Mount Carmel (24.5)) and indicate the affinity that the flora of Cyprus has with the Levant.

Other species of interest in the region include *Geranium tuberosum, Helianthemum salicifolium* and *Onobrychis venosa*, the last of these is endemic to Cyprus. It has a rosette of attractively marked leaves and heads of pale, straw-yellow flowers and occurs throughout the island, mainly in dry places.

Another way to approach this area is from Paphos. There are several possible routes but the one that leads off in a north-east direction some 15km east of Paphos along the main road to Limassol is worth taking. For most of the way it follows the west bank of the River Dhiarizos in the gravelly bed of which grows tamarix and much *Vitex agnus-castus* with plants of *Arum italicum* thriving below. In other places by the river one may see the only wild willow species of the island *Salix alba* and the alder *Alnus orientalis* that has ovate, pointed leaves and impressive bunches of long elegant male catkins at the turn of the year. Eventually the road rises to Malia or Ayios Nikolaos to join the Troodos road at Omodhos. It is, however, worthwhile making a short diversion to cross the river between the villages of Kefhares and Ayios Ioannis. The area by the bridge here especially is worth examining. On the steep banks by the road grow hundreds of bushes of *Ptilostemon chamaepeuce var. cyprius* - an endemic variety that has especially prickly bracts around the heads of purple hardhead-like flowers.

Back on the road from Limassol to Troodos around Trimiklini one begins to see the woodland or tall maquis that covers much of the lower slopes of the mountains and it becomes thicker as one climbs towards the village of Platres that has hotel accommodation and is a useful centre from which to explore the region. By the roadside grow clumps of the attractive endemic *Acinos troodi* (O = *Satureia troodi*) covered with its mauve flowers in April and plants of the rather tall, hairy bugle *Ajuga orientalis*. These are often accompanied by *Astragalus lusitanicus* which is a tall-growing member of the genus with white flowers followed by large inflated pods. It has an interesting and limited distribution and another form of it grows in the proximity of Cape St. Vincent in Portugal (4.4) - hence the specific name *lusitanicus*. Another plant seen here is the small shrub *Lithodora hispidula* that has already been mentioned (22.5). The woodland is largely comprised of *Pinus brutia* but in places it is more like a tall maquis of *Arbutus andrachne, Quercus alnifolia, Q. infectoria, Styrax officinalis*. The Golden Oak (*Q. alnifolia*) is an attractive endemic species which grows to a large bush or small tree. It is evergreen and the obovate leaves have a coating of golden indumentum on the underside. A number of orchids grow in these woods and flower from late April until June including the rare *Epipactis troodi* and *Platanthera chlorantha ssp. holmboei*. The first is a fairly ordinary looking helleborine with a reddish lip and once thought to be endemic to Cyprus but has now been found in Turkey. The platanthera is a typical greater butterfly orchid with green flowers. It also has been found now in Turkey and Lebanon and closely resembles the North

Dorystoechas hastata Termessos, Turkey. (p.407)

Tulipa armena var. *lycia* Baba Dag, Turkey. (p.423)

Iris haynei Mt. Gilboa, Israel. (p.440)

Centaurea pullata Nefza, Tunisia. (p.463)

African species *P. algeriensis*. Other orchids here are *Neotinea maculata, Orchis anatolica, O. quadripunctata*. Occasionally one may see *Dactylorhiza sulphurea pseudosambucina* (= *D. romana*) but it is more common in woods near to the Makheras Monastery on the eastern edge of the Troodos - only the yellow-flowered form seems to grow in Cyprus.

A good road leads to Chionistra 1951m (Khionistra, Olympus), the highest part of the Troodos range. It does not have the atmosphere of a high mountain for there are no jagged rocks and light woodland and scattered trees grow right to the top. The trees are mainly *Pinus nigra pallasiana* and *Juniperus foetidissima*. The second of these is a rugged tree that has a substantial trunk in old specimens and superficially resembles a cupressus rather than a typical juniper. It is common in parts of Turkey (23.9). Snowdrifts usually persist here until the end of April when round their edges a number of attractive plants can be seen in flower, including:

Arabis purpurea   *Ornithogalum chionophyllum*
Chionodoxa lochiae   *Ranunculus cadmicus*
Corydalis rutifolia   *Telphium imperati*
Crocus cyprius

Chionodoxa, as it was once called (the genus is now included in *Scilla*), is primarily a genus centered in Asia minor, the only species that grow wild in Europe are two from Crete. The corydalis is a charming small, few-flowered species also found in Turkey (23.9). The crocus is not the only sping flowering one on Cyprus, the other *C. hartmannianus* has longer flowers and anthers tinged black and usually grows in woods of *Pinus brutia* that are found somewhat lower on the mountain. Both are Cyprus endemics. An autumn flowering species *C. veneris* also occurs on the island and resembles a small form of the Cretan *C. boryi* (22.9). The ranunculus, a neat buttercup with yellow flowers, is the variety *cyprius* and another endemic plant. It differs from the type species that grows in Turkey by having a pronounced reddish colouration to the underside of the leaves.

Later in the year an extraordinary number of interesting endemics come into flower on the high slopes of the Troodos, including:

Allium troodi   *Dianthus strictus v. troodi*
Alyssum troodi   *Onosma troodi*
Colchicum troodi   *Salvia willeana*
Cyclamen cyprium   *Saponaria cypria*

For a detailed description of these it is necessary to refer to Meikle (1977,1985). The alyssum is of interest in that it is capable of growing on outcrops of serpentine rock which occur in places here. This is a silicate of magnesium which frequently carries traces of copper and other elements that make it toxic to most plants. Many of such species are members of the Cruciferae (or, to be up to date the family Brassicaceae) and include *Arabis purpurea* and *Alyssum akamasicum* already mentioned. The cyclamen, which has white flowers before the leaves unfold in December, is not so common here as in the Kyrenia range (see 22.12).

In brutia pine woods below the summit of Chionistra the ground is so covered with needles that it is very sparsely colonised. One plant that thrives here in large stands is *Paeonia mascula* and its red flowers make a beautiful show in May and June. Other plants that accompany it include the very prickly *Berberis cretica*, *Crocus hartmannianus*, *Ranunculus cadmicus* and the butterfly orchid *Platanthera chlorantha*.

Another area worth visiting in the Troodos is the Caledonian Falls. Here one can see the mauve rock cress *Arabis purpurea* growing in abundance on the rocks and in damp areas the celandine *Ranunculus ficaria ficariformis* with the local violet *Viola sieheana*. Holmboe (1914) noted several interesting plants here including the tiny, lobelia-like *Solenopsis minuta* (= *Laurentia minuta*, *L. tenella*, *L. bivonae*) and the scarce marsh helleborine *Epipactis veratrifolia*, a giant that grows to over 1m tall and flowers in late May and August.

22.11 The Cedar Forest: At one time Cyprus, like most other parts of the Mediterranean, was covered with trees but most of these have been cut for fuel or building. In Cyprus the main tree on the higher ground was probably the endemic cedar *Cedrus libani brevifolia*. It closely resembles the Cedar of Lebanon but is generally a smaller tree and has shorter needles, somewhat like the form in Turkey (23.10). In the north west of the Troodos some specimens still remain and have been supplemented with artificially raised and planted specimens, through the encouragement of Winston Churchill.

The easiest access to this unique forest is via the Chionistra area to the Kykko Monastery from where a winding road takes one about 5km to the reserve. One route passes through a dry area near Pedhoulas and its cherry orchards. The road from the monastery is unsurfaced and it may take 40-60 minutes to travel this short distance. Along here one passes through woods of *Pinus brutia* intermixed with typical garigue or thin maquis composed mainly of *Arbutus andrachne* and *Quercus alnifolia*. A very special plant to be seen here is *Orchis anatolica var. troodii*. It is more robust than the type species, with the spur curved upwards and lateral sepals suffused with green. It tends to flower earlier, being at its peak generally in early April. The type species also grows on the island and there are intermediate forms but here the plants are usually typical of the variety *troodii*.

Inside the forest is quiet and impressive with pure stands of the cedar though in some places it is mixed with *Pinus brutia*. In spring the ground between the trees is studded with gageas including the endemic *G. juliae*, though plants of this genus are notoriously difficult to identify with certainty.

22.12 The Kyrenia Range: This rugged narrow range of hard limestone, about 90km long, runs roughly parallel to the north coast of the island and is sometimes referred to as the Pentadactylos mountains. The highest steep rock cliffs are mainly treeless but lower regions carry substantial woods of *Cupressus sempervirens* with maquis that includes much *Arbutus andrachne*. The north face and the summits are more interesting botanically than the south side, which has a much lower rainfall and fewer plant species. To visit the more interesting parts one needs to be based at, or near, the coastal town of Kyrenia or Girne as it is called by the Turks. Many of the place names here differ from those on Greek Cypriot maps and the Turkish alternatives are given in parentheses.

Package holidays are available to Kyrenia and places 15-20km further west along the coast as at Lapithos (Lapta). It is easy to walk to the cliffs of the range from here. Along the coast and on roadsides leading to villages one may see *Bosea cypria* an interesting and distinct endemic shrub belonging to the family Amarantaceae and is sometimes cultivated on the island as a hedge. In spring it is an unprepossessing plant with spreading branches bearing privet-like leaves and insignificant small greenish flowers, but in autumn it bears a wealth of beautiful red berries. Other roadside plants here include: *Althea setosa, Hypericum triquetrifolium* and *Lamium moschatum*. The last of these closely resembles the common white dead-nettle. Its flowers are usually tinged pale pink and the leaves are variegated with pink or white centres so rival those of a coleus for their decorative effect. It is a species of the eastern Mediterranean and is common in parts of the Peloponnese (17.5).

Higher up are grassy terraces with some carobs and olives and a number of shrubs growing amongst the rocks, including: *Cistus salvifolius, Lithodora hispidula* and *Sarcopoterium spinosum*. Other interesting species here are:

Anemone coronaria	*Lathyrus aphaca*
Bellis sylvestris	*Ranunculus asiaticus*
Clematis cirrhosa	*R. millefoliatus leptaleus*
Lagoecia cuminoides	*R. paludosus*

The anemone is mainly the form with flowers of varying shades of mauve. The clematis is an interesting small evergreen climber that has saucer-shaped, cream-coloured flowers during the winter. It is found more or less throughout the Mediterranean though it is especially common in the Balearic Islands (8.6) and was once referred to as *C. balearicus*. The *Ranunculus asiaticus* makes a special feature here, covering the ground in places with its beautiful flowers in April. Not merely white as in the southern part of the island but also yellow, orange, pink and red. The *Ranunculus millefoliatus* is a dwarf sub-species endemic to the region around Lapithos, hence 'leptaleus'.

A number of bulbous and tuberous-rooted species grow with the above:

Allium neapolitanum   *Cyclamen persicum*
Arisarum vulgare   *Gagea sp.*
Arum dioscoridis   *Gynandriris sisyrinchium*
Bellevalia nivalis   *Muscari inconstrictum*
B. trifoliata   *Ornithogalum pedicellare*

The arum is a very fine plant with a spathe that is dark, velvety purple on the inner surface - it is essentially an Asiatic species which is fairly common in southern Turkey but relatively rare in Cyprus. *Bellevalia nivalis* is an uncommon species with pale blue flowers, somewhat like a small version of the more widespread *Bellevalia trifoliata*. Only scattered plants of the cyclamen grow here, they are more often seen in the eastern parts of the Kyrenia range. The muscari is a rare, miniscule grape hyacinth, usually no more than 8cm tall and with dark-blue flowers that turn black as they wither. A number of orchids may also be seen here, some of which are special to the region:

Ophrys bornmuelleri   *Op. lutea murbeckii*
Op. fusca fleishmanii   *Orchis italica*
Op. fusca iricolor   *O. morio libani*

The first of these is an impressive plant, often referred to as a sub-species of *O. fuciflora*. It has relatively large flowers with velvety, chocolate brown lip practically without markings, especially in the variety *grandiflora* that is common here. It is also found in

Israel, Lebanon, and south-east Turkey. *Op. fusca fleishmanii* has a similar distribution but is a small-flowered and rather insignificant form. *Orchis libani* is fairly distinct variety or sub-species of *O. morio picta* with a near white lip. As its name suggests, it is essentially a plant from Lebanon though it is fairly common in Cyprus.

Higher up the mountain near the rock faces one may see the endemic umbellifer *Ferulago cypria* and, growing in rock crevices, another endemic, *Rosularia pallidiflora*. The last of these somewhat resembles a houseleek with small glands on the leaves and yellowish-white flowers. *R. cypria*, a similar species but with white flowers, found here is also confined to Cyprus. The genus is particularly widespread in southern Turkey. There are interesting butterflies to be seen here, including the small Gruner's orange tip (*Anthocharis gruneri*), Cyprus meadow brown (*Maniola cypricola*) and the fine festoon butterfly (*Allancastria cerisyi cypria*).

Another way to approach the Kyrenia range from the north coast is to take the main road southwards out of Kyrenia (Girne) that leads to Nicosia (Lefkosa). Some 5km from the town a track to the west leads to the crusader castle of St. Hilarion, perched in the hills. Two special plants to be searched for amongst the stonework are *Brassica hilaronis* and *Scilla cilicia*. The first is a typical cabbage-like plant with white flowers and is said to be the wild parent of the cauliflower though it is not the only one to have this distinction - *Brassica insularis* from Corsica and Sardinia shares the claim (11.9). The scilla resembles *S. sibirica* of our gardens but the foliage precedes the flowers and may be 20cm long at flowering time in spring. It is found in a few parts of northern Cyprus and around Mersin in the south of Turkey. The rocks also provide a home for other interesting species including:

Acer obtusifolium *Hyoscyamus aureus*
Antirrhinum majus *Rosularia pallidiflora*
Arabis cypria

The arabis closely resembles the other endemic species *A.*

purpurea but is invariably found on limestone and never on serpentine rocks.

Several other roads from Kyrenia go high into the range, such as the one leaving eastwards to Arapköy that crosses the mountains to Kythrea (Degrmenlik). Many of the orchids already mentioned grow in the lower parts and in April, high up near the Buffavento Castle, amongst the stunted cupressus trees, there are thousands of plants of a tiny form of *Orchis anatolica*, some with white flowers, and also *Ophrys argolica elegans* which is rather special to Cyprus and differs from the type by its smaller size and three-lobed lip. Here it is still in flower in April but down by the coast it starts to bloom late in February. *Cistus salvifolius* and *C.creticus* grow along with *Salvia fruticosa* and the endemic *Sideritis cypria*. Looking into the sky, one may see the magnificent imperial eagle and identify it with certainty, for the golden eagle is not reported from Cyprus, though it might just be confused with a griffon vulture or the smaller lesser spotted eagle.

Travelling eastwards along the coast road out of Kyrenia for some 40km one comes to Dhavios (Mersinlik) and from here one can drive up the mountain to the Kantara Castle. Along this road in spring, and indeed along the coast near here as well, one may see thousands of *Cyclamen persicum* by the roadside, growing in rocky areas and even coming up through the asphalted road surface - it is one of the special sights for the plant enthusiast visiting Cyprus. The flowers are more delicate than the cultivated form and generally have a pronounced perfume. *Scilla cilicica* also grows on the castle walls here. Along the coast one may see the fruits of *Citrullus colocynthus* littering the roadside and looking like a cartload of oranges that had been overturned. This species is related to the watermelon and probably a native of India from where it was imported originally for its purgative properties. It is now established in many parts of the southern Mediterranean from Spain to Israel.

22.13 Morphou (Güzelyurt) and the North-West: The road westwards along the coast from Kyrenia turns to the south past the western extreme of the Kyrenia range and a right fork at

Dhiorios (Tepebasi) takes one towards the town of Morphu (Güzelyurt). Around Dhiorios and for some 4km southwards there are pine woods which shelter a number of species of orchids including *Neotinea maculata* and other plants such as: *Lithodora hispidula, Onosma fruticosa, Ophrys argolica elegans*. West of this area is a region noted for *Tulipa cypria* but in 1986 it was 'out of bounds' for military reasons, though children were selling the tulip in bunches by the roadside. It occurs in arable fields and scattered throughout Cyprus, especially in the north but is urgently in need of protection. It resembles the widespread *T. agenensis* (= *T. oculus-solis*), also found on the island, but has purple-crimson, rather than bright red flowers. Around the town of Morphou there are extensive citrus orchards.

Cape Kormakiti (Koruçam), in the extreme north-west of the island, is fairly remote and worth a visit. The coast road from Kyrenia takes one to Liveras (Sadrazamköy). Along this road orchids are fairly abundant and include the charming Cyprus bee orchid *Ophrys kotschii* with a large, well-marked, almost black and white lip. This is by no means the only place where it occurs for it grows throughout the Kyrenia range and in other parts of Cyprus though it is decidedly uncommon in the south.

22.14 The Mesaoria: This central plain is a wedge-shaped region between the Kyrenia and Troodos mountains stretching roughly from Morphou to the east coast. As a relatively flat, fertile and treeless part of the island it is important for cereal crop production. At one time it was known to botanists as an area of special interest for its weed flora but since the regular use of herbicides most of the more unusual species have disappeared. It is bordered in the north by the southern slopes of the Kyrenia mountains which are very much drier than the rest of the range and at the western end form strange hummocks of grey marly soil that carry chalk and gypsum. Plant life is rather sparse in this inhospitable environment but there are several species of interest to those who like unusual plants:

Daucus dureana   *Pteranthus dichotomus*
Iberis oderata   *Salvia crassifolia*

The pteranthus is an especially strange member of the Caryophyllaceae found only in the southern extremes of the Mediterranean region, particularly the dry areas of North Africa (25.6).

Most visitors will traverse the Mesaoria to visit the enchanting ruins of Salamis and the Monastery of St. Barnabus nearby and to note any plants on the way such as the several forms of *Adonis annua* and in places the unusual corn-marigold *Coleostephus myconis* - sometimes the white-flowered form. Neither Blamey & Grey-Wilson (1993) or Meikle (1985) include it in Cyprus.

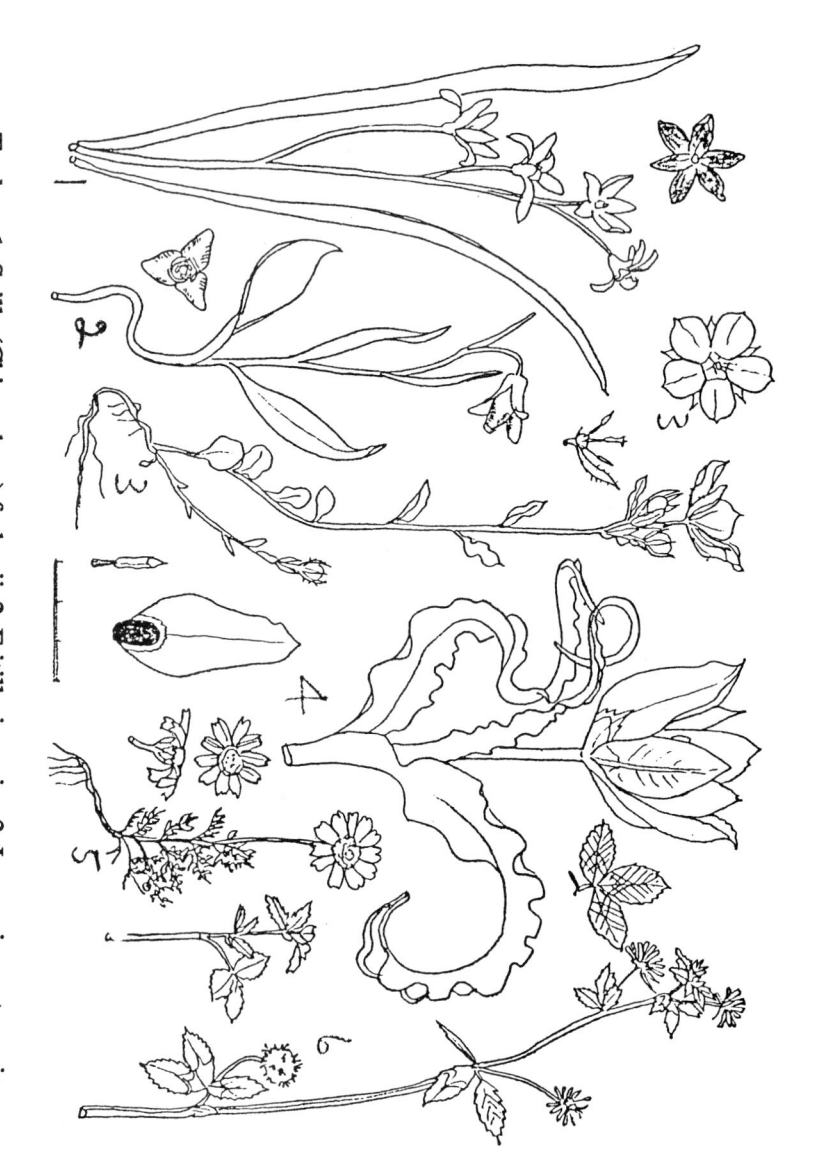

Turkey 1. Scilla (Chionadoxa) forbesii 2. Fritillaria carica 3. Legousia pentagonia 4. Tulipa armena var. lycia 5. Anthemis rosea var. carnea 6. Trifolium resupinatum

23. SOUTHERN TURKEY

23.1 From a botanical standpoint Turkey is extremely exciting for it has a large number of common Mediterranean species mingled with others from Asia. The geology and topography of the country are especially complex leading to many different and varied environments. Plants tend then to become localised in comparatively small areas and isolated from neighbouring similar species with which they might hybridise. This situation encourages the development of numerous endemic forms making Turkey the most likely Mediterranean country, within the limits of this book, where one may find previously unrecorded, or even new, species. However, grazing by sheep and goats is often excessive and newcomers to the region may be disappointed to find large areas which seem to be steppe-land devoid of vegetation, but on closer examination it will often be found that they are certainly not lacking interesting plants. Expeditions to the higher and less accessible regions are likely to be especially rewarding

The best time to see the coastal flora is late March and April. However, the weather at this time of the year can be somewhat unpredictable and stormy in the south-west resorts due to the effects of nearby mountains so it is advisable to come well prepared for rainy conditions. The mountain flora is at its best from May to early July. Later in the year many lowland places are parched and desert-like.

Turkey has encouraged tourism in recent years and it is now easy to travel in the western part of Anatolia, which is of great interest to Mediterranean plant enthusiasts. Package holidays are available at coastal sites in the south west around such towns as Marmaris, Fethiye and Antalya and these are excellent areas from which to study the flora. Also, one may readily make ones own travel arrangements between hotels or 'pansiyons' using the excellent bus system or a hired car. However, car hire is relatively expensive at present and not universally available in all towns so it is advisable to check on this matter before leaving home. Driving in Istanbul may seem frenetic but there is little traffic to contend with on most country roads and in the smaller towns. The free Turkish Ministry of Tourism map (1:1850,000) shows the main roads and other maps of larger-scale (1:1600,000) are readily available. However detailed local maps are difficult to find and are best sought from a map specialist before leaving Britain. Turkish place names are standardised and the traveller does not have the same difficulty locating places there as in Greece. Most Turks are particularly hospitable and helpful to visitors, and many speak German.

23.2 The terrain: As in many other parts of the Mediterranean the coastal areas of Turkey are backed by substantial hills and mountains. The impressive Taurus mountains (Toros Daglari) which run parallel to the south coast, give way, west of Antalya to a series of ranges running north-east to south-west. These are, from east to west, the Bey Daglari (Kizlarsivrisi 3070m), Ak Daglari (Uyluk 3015m), Boncuk Daglari (Boncuk 2185m) and Gölgeli Daglari (Çiçekbaba 2294m). 'Dag' (plural 'dagi') means a mountain and 'daglari' a mountain range - the 'g' is not pronounced and ç said like our 'ch'. Many of these peaks, though not readily accessible, are of great botanical interest and would make very good subjects for enterprising excursions. Even in the height of summer, after most of the snow has melted in June and July, species of *Acantholimon* and other genera of interest may be seen in flower. Baba Dag 1969m, a smaller mountain to the south-west of Fethiye is fairly accessible and well worth a visit in late April or May.

The high valleys between the main ranges are largely treeless, overgrazed and desolate but they have a special remote attraction and close inspection will show that they are by no means completely devoid of vegetation. Forests occur in places and extend to 1,600m on some mountains. They mainly comprise pines but large areas of impressive indigenous cedars occur. On the lower slopes one finds deciduous species, including the oriental plane and the sweet gum *Liquidambar orientalis*. The lower woods, in particular, provide a habitat for many interesting small-growing species.

An area north and east of Antalya is important for the

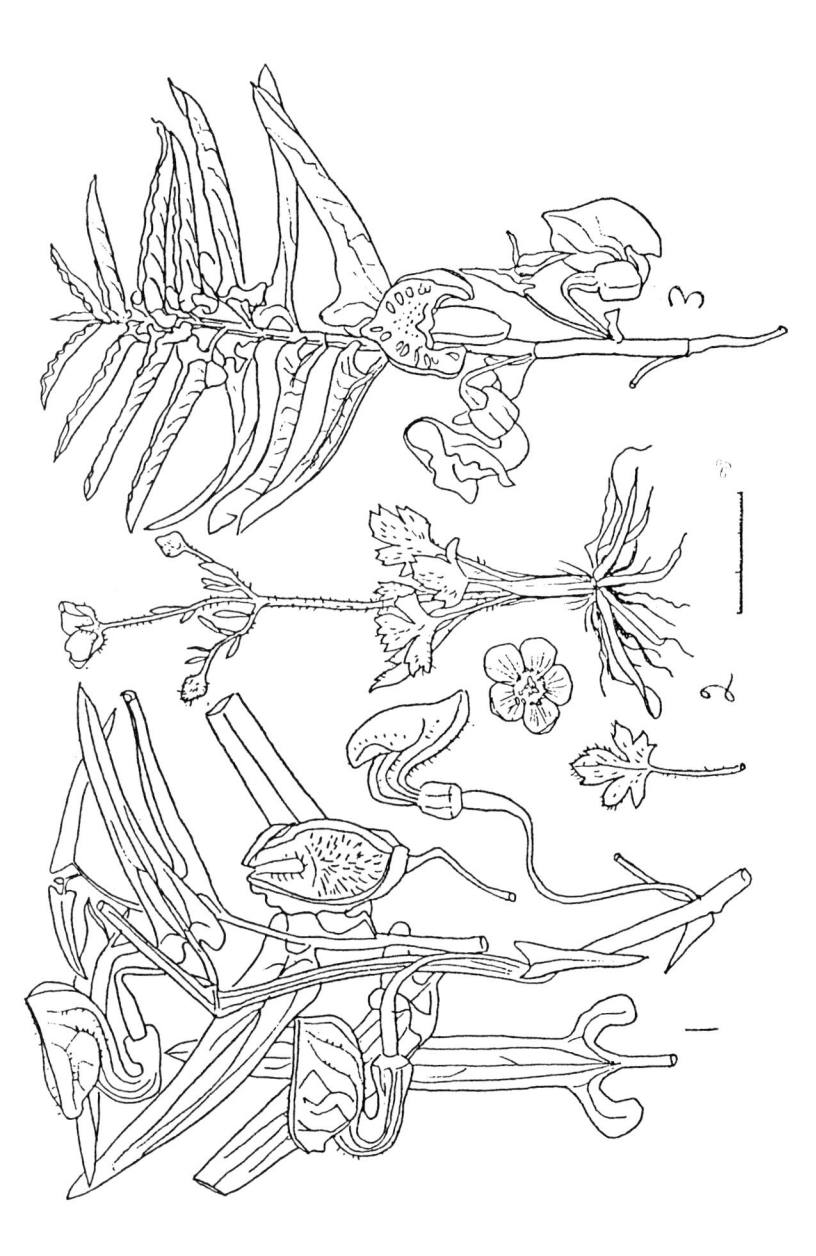

Turkey 1. Aristolochia poluninii 2. Ranunculus agyreus 3. Aristolochia maurorum

cultivation of cotton and vegetables; similar but smaller regions may be found in a number of other sites near the coast.

23.3 Literature: There is very little popular literature on the flora of Turkey. The only reasonably priced and portable book on the subject in English deals with the bulbous plants of the region (Mathew and Baytop 1984). This has some fine photographs and gives information on species distribution but it is concerned with only a small group of the many thousands of interesting species to be found in the country. The recently completed 'Flora of Turkey and the east Aegean Islands' (Davis, P.H. et al 1965-1988) is a magnificent work - amongst the very best of all scientific floras. However, it consists of 10 volumes, is bulky and expensive and only available for study in the specialist libraries of some botanic gardens and universities. It has very limited illustrations but some useful distribution maps especially in the later volumes.

Three articles on the flora of the region by the late Professor Davis, who was the editor of and main driving force in the production of the above flora, are well worth reading. They are entertaining as well as informative. All are published in the Journal of the Royal Horticultural Society and it should not be difficult to locate copies for perusal before embarking on a visit. The first two are entitled 'A Journey in South-West Anatolia' (Davis, P.H. 1949) and it would be interesting to organise an expedition to follow in his footsteps. The third is called 'The Taurus revisited' (Davis, P.H. 1951).

If one possesses 'Flowers of Greece and the Balkans' (Polunin 1980) this is well worth taking on a visit to the area, as it describes a number of the less common plants to be found there. However, some species differences are small and subtle and one must be on one's guard against making hasty decisions on the nomenclature of Turkish plants.

23.4 Antalya: This pleasant small coastal town is blessed with a magnificent view over the sea to the Bey Mountains. It has its own airport, some good hotels and is a convenient centre from which to look for plants. In one part, near the harbour, several small waterfalls cascade down cliffs to the sea. On these rocks grow *Ptilostemon chamaepeuce* and *Dianthus actinopetalus* - like a form of the better known *D. rupicola*.

Some 25km to the north-west of Antalya, off the road to Korkuteli, are the ruins of Termessos. These lie in woods on a mountainside and the area is a national park, noted for its flora and fauna, called Düzler Çami. At the entrance to the park there is a museum largely dedicated to the fauna of the district. An 8km track leads from here through the woods to the highest point which has maquis type vegetation. The deciduous trees include the Turkey oak *Q. cerris*. Amongst the smaller growing woody species one may find *Arbutus andrachne, Fontanesia phillyreoides, Ilex aquifolium* and *Juniperus foetidissima*. The fontanesia is an uncommon privet-like shrub of the Oleaceae family growing to 2-3m with numerous, small, greenish-white flowers with protruding stamens in June. Amongst these shrubs grow:

Arabis ionocarpa	*Fibigia eriocarpa*
A. vernalis	*Lilium candidum*
Bellevalia sp.	*Muscari sp.*
Cardamine graeca	*Ornithogalum sp.*
Clematis cirrhosa	*Orchis anatolica*
Colchicum macrophyllum	*Phlomis samia*
Cruciata levipes	*Symphytum orientale*
Doronicum orientale	*Vicia narbonensis*
Dorystoechas hastata	*Viola kitaibeliana*

Of the crucifers (or members of the Brassicaceae) *Arabis ionocarpa* is a small plant with white flowers and the cardamine also has white flowers. The fibigia grows to a metre or taller with felted leaves and small yellow flowers followed by bat-shaped fruits. The most interesting plant amongst those cited above is the dorystoechas - the only species in this genus and only found within this very limited area of the world. It smells like and resembles garden sage. The leaves are, however, 'hastate' (like the head of a halberd - with two pointed lobes projecting at the base) and the narrow inflorescences of many small white flowers stand up from the bush like christmas tree candles - quite unlike those of the garden sage. The symphytum has small white flowers and, like many others of its group, is usually badly eaten by insects. *Vicia narbonensis* is not uncommon throughout the Mediterranean and closely resembles a small, mauve-flowered broad bean. In fact it may be an ancestor of that vegetable whose origin has never been satisfactorily established. The viola is a minute annual heartsease pansy, often with yellow flowers but here they are pale mauve with a yellow centre.

23.5 **East of Antalya - Alanya:** The small town of Alanya is a popular seaside resort with fine sandy beaches to the east, sometimes described as the 'Turquoise Coast'. It has a castle associated with Alâeddin Keykubad and a 'magic cave' where sufferers of respiratory diseases may have a cure providing, according to the local inhabitants, they do not die from a heart attack through the heat and humidity - the Turks like a joke! A visit to the castle is quite rewarding botanically. Climbing the rocky path there one passes fine bushes of *Styrax officinalis* and the evil-smelling but showy leguminous shrub *Anagyris foetida*. Smaller plants include:

Ajuga bombycina *Rosularia globulariifolia*
Asphodelus microcarpus *Ruta chalepensis*
Hyocyamus aureus *Salvia viridis*
Leopoldia comosa *Vicia hybrida*
Phagnalon rupestre

The ajuga is a charming and special plant. It resembles the common 'ground pine' *A. chamaepitys* with similar yellow flowers in early summer but is more compact and the leaves are covered with long white hairs so that the shoots look like woolly balls. All the herbarium specimens in official collections seem to have been obtained from near the castle here but the species does grow in stony places in other parts around Alanya. Rosularia is a genus that is rather special to Turkey, though species also grow in Cyprus (22.12) and Crete. They are houseleek-like plants with somewhat bell-shaped white flowers and a flat rosettes of leaves with sticky hairs. The salvia is short-growing, usually with a tuft of purple bracts at the top of the inflorescence. A form, generally referred to as *S horminum* is a well-known 'annual' in British gardens and sometimes has pink or white, instead of purple, bracts.

Walking eastwards along the shore from the castle towards the tourist hotels one can see *Medicago marina* growing in the sand at the upper reaches of the beach. Less common species in the same area include *Campanula stellaris, Trifolium pamphylicum* and *Valantia hispida*. The campanula is a charming small annual with relatively large mauve flowers and the trifolium is a clover with conical heads of crimson flowers; it is in its homeland here - the old kingdom of Pamphylia. There is a marshy area behind the beach surrounded with *Iris pseudacorus* and the moisture-loving *Veronica anagalloides* (= *V. anagallis-aquatica*). In grassy areas nearby grow plants of a particularly attractive form of *Arum dioscoridis* with dark, velvety spathes in April that are green only towards the tip. Other plants here include the chaste tree *Vitex agnus-castus, Melilotus sprunerianus* and the bright pink *Silene papillosa* together with more *Campanula stellaris*.

Just to the north of the main coastal road are rocky areas colonised by a number of shrubs such as:

Calicotome villosa	*Euphorbia characias*
Cistus incanus creticus	*Myrtus communis*
C. salvifolius	*Olea europea*
Ephedera fragilis	*Phlomis leucophracta*
Erica manipuliflora	*Quercus coccifera*

The calicotome is the common species in this area though in the western Mediterranean it is often replaced by the similar *C. spinosa*. The phlomis is an interesting and uncommon member of the genus which is very well represented in southern Turkey and produces a number of natural hybrids. This one resembles the common Jerusalem sage but has leaves with an attractive wavy margin and covered with silky silver hairs. Another similar phlomis here with leaves having yellowish hairs and brownish flowers was identified as a hybrid *Phlomis lunariifolia x leucophracta*. The genus would merit further investigation. A number of smaller species, typical of the region, grow amongst the shrubs:

Agrostemma githago   *Orobanche crenata*
Ajuga chamaepitys   *Plantago cretica*
Anagallis arvensis   *Scrophularia pinardii*
Centaurium maritimum   *Serapias vomeracea*
C. pulchellum   *Sherardia arvensis*
Gladiolus anatolicus

The agrostemma is the corn cockle (O = *Lychnis githago*), a showy cornfield weed that often used to be seen in Britain but is now rare as a result of the use of herbicides. It is sometimes offered in seed catalogues and grown in our gardens as an 'annual'. The two species of centaurium are similar, small, more or less unbranched, annuals with grassy leaves and yellow and pink flowers respectively. The gladiolus is endemic to southern Anatolia. It closely resembles the widely distributed *G. communis* and is sometimes classed as a sub-species of *G. illyricus* - this is a difficult genus taxonomically. The orobanche is a robust species, usually with white or straw-coloured, mauve-veined flowers that have a strong scent like carnations. It is parasitic on leguminous plants and frequently causes serious damage to broad bean crops, which are popular here. The scrophularia is an unprepossessing figwort with a limited distribution in the region.

In the hotel lawns at Alanya one may see *Phyla filiformis* (= *Lippia nodiflora*) a creeping plant and member of the Verbenaceae originating from South America and sometimes used as a substitute for grass as we plant 'chamomile lawns' in Britain. Its main advantage is its ability to stand very dry conditions though when it escapes cultivation it is thoroughly at home in wet places. Accompanying the phyla in grassland one may also see the widespread *Trifolium resupinatum*, an annual clover with flat heads of pink flowers. *Geranium purpureum* grows as a weed in the flower borders here and in many parts of Turkey as well as in waste places. It closely resembles our herb Robert but has smaller and darker flowers.

Some 100km along the coast eastwards from Alanya lies the country town of Anamur. The road there passes several small banana plantations before reaching Gazipasa and from there on the rocky, pine clad, slopes of the Taurus mountains come down rather steeply to the sea. Anamur is well known for its storks; nearly every house seems to have a nest and in spring one can often see large numbers of these magnificent birds soaring in the warm upcurrents. At the town, and indeed in most parts of Turkey, one can obtain a drink of 'ada çay' (island tea), which is an infusion of locally collected dried herbs. Various sideritis species are generally used and produce a pleasant drink but certain salvias are occasionally substituted and produce a concoction which some people may find unpleasant.

At a short distance from Anamur lie the ruins of the old Roman settlement of Anamurium that was later adopted by the Byzantines. There is very little grazing on this site and the vegetation is grassy and lush with scattered bushes of *Styrax oficinalis* and *Laurus nobilis*. Other species include *Campanula stellaris, Coronilla parviflora, Mandragora autumnalis, Onosma alboroseum* and *Trifolium pamphylicum*. The onosma, which is fairly common in Turkey is a robust, attractive species with grey foliage and large white flowers that turn crimson purple as they age. It may be the same as the plant called *Onosma albo-pilosum* sometimes grown in our gardens.

23.6 West of Antalya - The Bey Mountains: The area immediately to the west of Antalya has the highest peaks in the region under discussion but they are not easily accessible to the average tourist. However it is not difficult to traverse them by travelling to Korkuteli and then down the valley between the Bey and Ak Daglari to Elmali and thence to the coast at Finike. The area *en* *route* around the village of Çobanisa, may at first seem desolate and devoid of vegetation except for a few small pines and junipers but a closer examination will reveal many interesting small plants. Around the rocks and cultivated areas in spring here one may see:

> Aristolochia maurorum   Moltkia caerulea
> Aubretia pinardii   Onosma frutescens
> Ceratocephalus falcatus   Ornithogalum nutans
> Geranium tuberosum   Prunus prostrata
> Lathrys aphaca   Ranunculus argyreus
> Leontice leontopetalum   Wiedemannia orientalis
> Moltkia aurea

The aristolochia is a low-growing species with large, sinister Dutchman's pipes also found in Israel (24.10). The ceratocephalus is a tiny annual ranunculus-like plant with relatively large heads of hooked fruits; it is also found in cultivated areas throughout most of the Mediterranean. The lathrys is our yellow vetchling but here the flowers are nearly white; the group splits into a number of sub-species in Turkey. The ranunculus is a showy buttercup with some elongated tubers amongst the roots and rather similar to the more widespread *R. flabellatus* (O = *R. paludosus*). Both the moltkias are beautiful Asiatic members of the Boraginaceae with heads of yellow and mauve-blue flowers respectively. They grow as roadside weeds in parts of Turkey and would be delightful garden plants if they were not almost impossible to grow in cultivation. The onosma is a common species throughout the country and found also in Greece and parts of the Levant. It has yellow flowers. The wiedemannia resembles a sophisticated red dead-nettle - it does not seem to be found in Europe.

The road from Elmali descends to near the sea at Finike where it joins the coastal road through Kumluca back to Antalya. Along the route by the coast there are wooded areas and a number of interesting plant sites where one may find:

> Carduus acicularis   Lagoecia cuminoides

Coringia grandiflora　　*Smilax aspera*
Cymbalaria longipes　　*Viola kitaibeliana*
Dorystoechas hastata

The cymbalaria (O = *Linaria longipes*) resembles a robust form of the ivy-leaved toadflax; it is rather special to this area but is occasionally found also in Greece, including Crete. The viola is a diminutive pansy with mauve flowers. The smilax is commonly found throughout the Mediterranean, usually as a component of garigue and maquis but here it climbs the trees like ivy and can reach a height of 10m.

23.7 **Tahtali Dag:** This isolated mountain rises to 2366m behind the village and coastal resort of Kemer some 40km south of Antalya. It is of special botanical interest but some planning and effort is needed to climb to the top. I have never reached the summit myself but writings by the late Professor Davis (Davis 1949, Davis 1951) give an enthusiastic account of some of the plants to be found there. In 1947, when Davis made his first ascent, there was no suitable road to Kemer and he was obliged to travel by boat from Antalya. There is now a good road and an additional rough secondary road that turns off westwards some 5km from Antalya, passes to the west of Tahtali Dag and joins the main coast road at Kumluca.

A small gorge called Kesme Bogaz leads towards the mountain and under the plane trees here grow *Dorystoechas hastata, Echinops ritro, Phlomis bourgeaui* and *Phlomis chimerae*. The last of these is a rare species distinguished by its purple flowers. A number of interesting chasmophytes may be seen on the rock faces, particularly a very local globularia now called *Globularia davisiana*. It is a shrub 1m tall with mauve-blue flowers in long inflorescences instead of the usual near globular flower heads of this genus. It is probably not yet in cultivation and seed collection would be well worth while. With it grows an interesting poppy that has not yet been named and *Erica sicula* (O = *Pentapera sicula*) found also in western Sicily and Malta.

Climbing to the summit is somewhat of an expedition but many special plants, flowering in June, await the enthusiast especially after traversing the woods of pine and cedar. Here, above the tree line at about 1,800m one may see:

Acantholimon acerosum *Ricotia davisiana*
A. ulicinum *Rosa glutinosa*
Asyneuma lycium *Salvia caespitosa*
Omphalodes lucillae *Sedum sempervivoides*
Lamium cymbalariifolium *Viola crassifolia*
Pterocephalus pinardii

The pterocephalus is a small scabious-like plant. The ricotia is a very special species of the Brassicaceae family only known from this small corner of the world. It is low-growing with fleshy leaves covered with dense indumentum and lilac-pink flowers in July.

23.8 **Fethiye and Ölü Deniz:** Fethiye is a small town and port some 30km east of Dalaman. It has a marina but is not often noted as a holiday site for foreign visitors, whereas Ölü Deniz that lies on a bay some 10km to the south is a relatively new custom-built package holiday resort. It has a beautiful site surrounded by mountains dominated by Baba Dag 1969m to the east. There is a sandy beach which extends westwards to a lagoon that gives the site its name Ölü Deniz meaning 'dead sea', presumably because the water there is always calm.

The area around the lagoon is classed as a reserve and one pays a small charge to enter and to park a car there if necessary. The sandy soil surrounding the lagoon is furnished with pines, lentisk and a few other lower-growing shrubs including myrtle. The area between the shrubs and the surrounding beach is a very profitable for botanising and one may see many species including:

Anagallis arvensis *Inula crithmoides*
Cachrys cristata *Leopoldia weissii*
Campanula drabifolia *Misopates orontium*

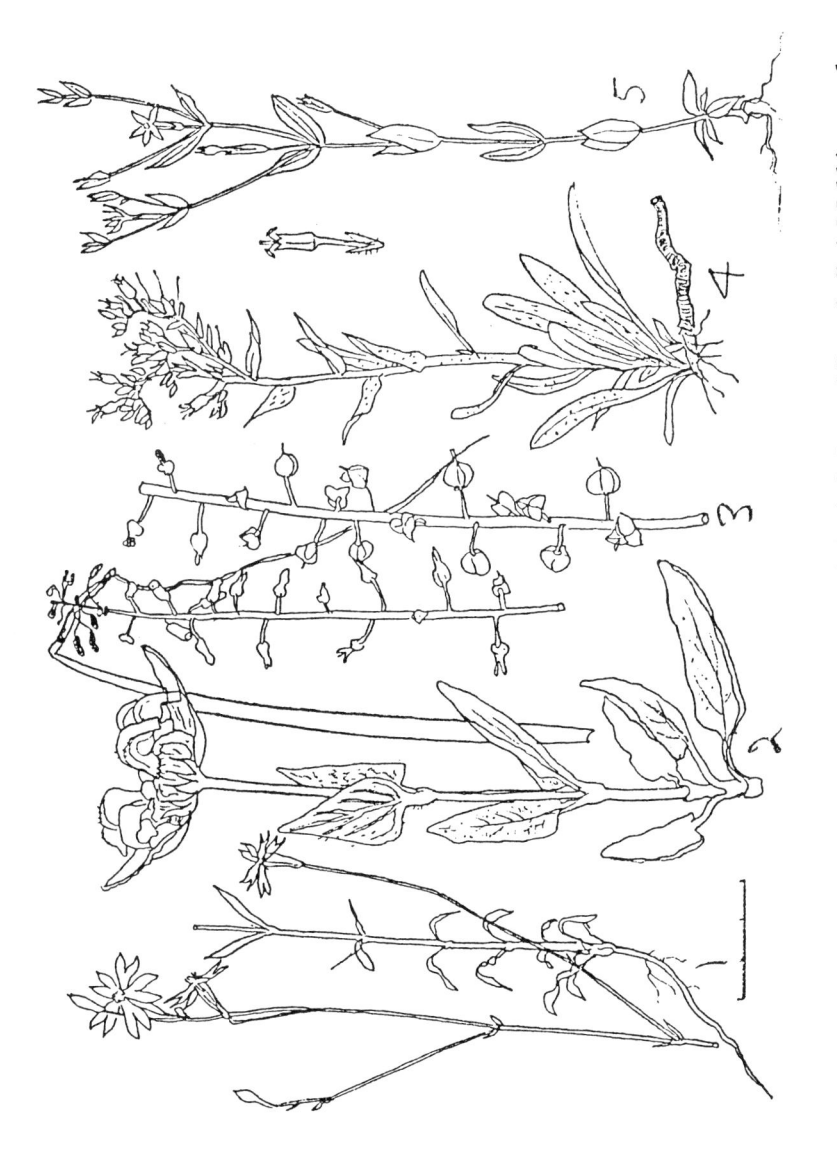

Turkey 1. Silene echinospermoides 2. Phlomis lycia 3. Leopoldia weissii 4. Moltkia coerulea 5. Centaurium pulchellum

Cynanchum acutum
Daphne gnidioides
Dianthus crinitus
Dittrichia viscosa
Echium angustifolium
Euphorbia parialis
Glaucium flavum
Hedypnois cretica

Ononis natrix
Ruscus aculeatus
Scrophularia lucida
Senecio vernalis
Silene echinospermoides
Smilax aspera
Valanta hispida
Vitex agnus-castus

With the exception of the dianthus and cachrys, all these species are dealt with in Polunin 1980. The dianthus is a typical 'pink' with white or very pale pink flowers that have narrow petals frayed into numerous strands at the ends. The cachrys is a robust and attractive umbellifer with finely cut leaves. Some of the other species also deserve special mention. *Daphne gnidioides* is very similar to the more familiar *D. gnidium* and replaces it here; it has narrower leaves which are usually spine tipped. The cynanchum, often called 'stranglewort' is a climber with heart-shaped leaves, a milky juice and horn-like fruits typical of the family Asclepiadaceae. Hedypnois rather resembles a small specimen of calendula but with the stems conspicuously swollen below the flower heads. The leopoldia (= *Muscari weissii*) is a typical tassel hyacinth with rather long inflorescences of dingy brown flowers and a head of purple sterile florets; it often takes the place of *L. comosum* in Turkey and northern Greece. The ruscus here is the butcher's broom with especially small and narrow leaf-like cladodes sometimes referred to as the variety *angustifolius*. The silene has white, deeply-cleft petals.

There are reeds and rushes around the lagoon including *Phragmites australis*. Wild celery *Apium graveolens* grows in wet areas as it does at home. In one, dry area there are clumps of the impressive grass *Saccharum ravennae* (= *Erianthus ravennae*) which grows to 2-3m tall. It is related to sugar cane and is found in other parts of the Mediterranean, especially Italy and parts of southern France, where it is referred to as 'roseau de Ravenne' - the Ravenna reed.

On the seaward side of the beach a few other species may be seen in the sand and grassy area behind, especially *Cionura erecta*, a sub-shrubby member of the Asclepiadaceae with heart-shaped leaves and white flowers. At a distance it reminds one of a suckering lilac bush. With it grow :

Anchusa aggregata   *Medicago marina*
Echium angustifolium   *Ononis natrix*
Euphorbia parialis   *Papaver dubium*
Knautia integrifolia   *Vaccaria pyramidata*
Lotus halophilus

The knautia is an annual 'scabious' common in the hedgerows here and vaccaria (O = *Saponaria vaccaria*) is another showy annual with heads of pink gypsophila-like flowers that have a distinctive inflated white and green calyx. It is sometimes called cow basil (vacca means a cow) and is said to be good fodder. The papaver is the widespread 'long-headed poppy' with relatively small light red flowers.

The sand at the east end of the beach gives way to cliffs that fall down to the sea and here one may see many of the plants associated with the numerous rock faces in the region, including:

Campanula lyrata   *Origanum onites*
Euphorbia dendroides   *Phlomis lycia*
Genista acanthoclada   *Ptilostemon chamaepeuce*
Inula heterolepis   *Satureia thymbra*
Onosma frutescens

The campanula is a very handsome rock-hugging species with relatively large blue flowers. It is endemic to Turkey but belongs to a complicated group of species and resembles *C. rupestris* that grows in Greece on the way up to the Lykavittos in Athens (17.4). The inula and origanum are both grey-leaved small shrubs that are common in the region and sufficiently distinct that they may be recognised whilst passing in a bus. Both flower in the height of summer with yellow and white flowers respectively. The phlomis is the main one found in this particular area but is one of many confusing species in southern Turkey. It has relatively small heads of yellow flowers.

Growing by the base of the rocks to the east end of the beach one may see *Ballota pseudodictamnus lycia*, a form of horehound which is endemic to south-west Turkey. A new road leads from the extreme south east corner of Ölü Deniz, southwards along the coastal cliffs to Kidirak - another purpose-built holiday centre. By the roadside here one may encounter *Chenopodium vulvaria* (= *C. foetidum*), a species which is a widespread weed in the Mediterranean and most of northern Europe including parts of Britain. It closely resembles the common fat hen but smells strongly of rotting fish. Some other plants to look out for here are *Convolvulus siculus, Linum strictum* and *Malva cretica*. The first of these is an annual species with small blue flowers. The linum is a diminutive flax with small yellow flowers. One may also encounter the attractive tall *Delphinium staphisagria* which produces its spikes of purple-blue flowers in May. On the cliffs above the road grow plants of *Galium canum ovatum* and *G. graecum*; the first having flowers that are very dark reddish brown or nearly black and those of the latter are much paler. Here also grows the very spiny annual thistle *Picnomon acarna* (= *Cirsium acarna*).

Past the building site one comes to a wooded path where the face of the mountain has stands of *Cupressus sempervirens*. Amongst the rocks here one may find:

Campanula lyrata	*Phillyrea latifolia*
Cyclamen trochopteranthum	*Ranunculus creticus*
Helichrysum orientale	*Veronica lycica*
Matthiola sinuata	

The cyclamen species is endemic to this region of Turkey. It has a rounded, patterned leaf which is dark red on the undersurface and, in autumn, dark pink flowers with petals so twisted that they remind one of a ship's propeller. It seems to be allied to *C. coum* and, until recently, was named *C. alpinum*. The ranunculus is an attractive and rather large-flowered buttercup with leaves and stem covered with soft hairs. As its name implies it is a Cretan speciality, but is also found in Rhodes. The veronica is endemic to the area but closely resembles the widespread, white-flowered *V. cymbalaria*.

23.9 Around Hisarönü: The village of Hisarönü lies about 3km north of Ölü Deniz and some tourists stay here and regularly make use of the dolmus to visit the shore. A walk westwards from the village here takes one to the ruined Greek village of Kaya. On the way there one may encounter a number of roadside species such as *Lamium moschatum, Legousia speculum-veneris, Notobasis syriaca* and *Orobanche ramosa*. A special plant to be searched for in April and May is *Aristolochia poluninii*, a rare endemic species confined to this small area and only discovered ten years ago. It has large, brownish Dutchman's pipes with the usual objectionable smell.

Somewhat further along the road under pine trees grow the orchids *Ophrys fusca, O. fuciflora, Orchis sancta* - the last of these is an uncommon species in most parts of the Mediterranean but quite frequently met with in this area. One may also expect to see *Gladiolus anatolicus, Gagea graeca* (O = *Lloydia graeca*) and *Petrorhagia velutina* . Further along, to the south of a straight stretch of road there are pine trees with a substantial undergrowth of shrubs including:

Anagyris foetida   *Salvia fruticosa*
Cistus incanus creticus   *Sarcopoterium spinosum*
Genista acanthoclada   *Styrax officinalis*
Osyris alba

Amongst these grow several smaller plants including the orchids mentioned above and also *Orchis anatolica, O. coriophora* and *Ophrys holoserica*. Other interesting species here are *Asparagus aphyllus, Asphodelus aestivus, Cyclamen graecum* which flowers in autumn, and *Smilax aspera*.

23.10 **Baba Dag:** Although this is not one of the highest mountains in the region it rises to nearly 2,000m, has a particularly rich and interesting flora and is fairly accessible to the average tourist. A short distance from the village of Ovacik, about a kilometer north of Hisarönü, an unsurfaced road leads to the summit of Baba Dag. Under the right conditions it is possible to drive to the top in any well-serviced car, but a four-wheeled drive vehicle is a considerable advantage and it is advisable to carry a shovel to remove any fallen stones, earth or snow from the road. If one does not fancy the drive oneself then it is possible to hire a taxi or dolmus to take one part of the way and to walk from there. To make the most of the botanising it is advisable to arrange at least two visits, one to the lower half and another to the higher reaches.

The lower slopes are largely covered with *Pinus brutia*. These gradually give way to the cedar *Cedrus libani* - a short-needled form sometimes referred to the sub-species *stenocoma* and similar to, though not identical with the one found on Cyprus (22.11). The upper limit of the cedars is around 1,600m and above here there are scattered bushes or small trees including *Acer sempervirens, Amelanchier parviflora, Fraxinus angustifolia* and *Juniperus foetidissima*. The acer is the Cretan maple, somewhat similar to our field maple *A. campestre*. The amelanchier, or snowy mespilis, is rather upright-growing, smothered with white flowers in April and has small black fruits later in the summer. It is endemic to Anatolia and close to *A. vulgaris* (= *A. ovalis*) (10.4) found in Europe. The juniper is an interesting species which inhabits high places also in Cyprus and the Epireus. Although its specific name means 'stinking' the fruits have a very strong and pleasant smell like gin. There are few tall bushes of any kind above 1,800m.

At the start of the climb, The forest has an undergrowth of styrax in places and patches of *Colutea melanocalyx* a bladder senna with yellow flowers. In clearings and by the roadside one may see:

Anemone blanda *Geranium lucidum*
Anthemis rosea *G. purpureum*

Arabis vernalis   *Salvia tomentosa*
Colchicum macrophyllum   *Saponaria calabrica*
Cyclamen trochopteranthum *Verbascum bellum*

The colchicum has very large pleated leaves and pale mauve mottled flowers in autumn. It is also found in Crete and a rather smaller-leaved form of it in Rhodes (21.6). The verbascum is one of several species that are endemic to Anatolia.

The mixed forest of pines and cedars from about 1,000m hold many interesting species:

Alkanna orientalis   *Leontice leontopetalum*
Astragalus depressus   *Paeonia mascula*
Centaurea cyanus   *Phlomis grandiflora*
Fritillaria acmopetala   *Ranunculus agyreus*
F. forbesii   *R. ficariiformis*
Geranium tuberosum   *Sternbergia candida*
Lamium garganicum

The alkanna has relatively large, yellow flowers. *Fritillaria forbesii* is endemic to south west Anatolia with small, yellow, campanulate flowers resembling the shape of a bluebell and 4-6 very narrow leaves. The paeonia is the fine red-flowered species (= *P. corralina*) found scattered sparingly throughout the hilly regions of the Mediterranean. The phlomis is also endemic to the area and has yellow flowers that are considerably larger than those of *P. lycia* which is common lower down. It also flowers later in June. The sternbergia is a distinct and very special plant limited to this corner of the world and found for the first time as recently as 1976 by the late Oleg Polunin. It has scented white flowers in January and February and the leaves are narrow. After flowering it could well be mistaken for a narcissus. A few orchids may also be found in this zone, especially *Limodorum abortivum, Ophrys reinholdii, O. fusca, Orchis anatolica* and the exciting *Cephalanthera epipactoides*.

Many of the old cedar trees are impressive and their canopy

is often dense so that there is very little undergrowth. Coming out of the forest one has fine views eastwards to the Ak Daglari, which carry substantial snow coverings well into May. From here onwards the track winds upwards through rocky terrain with scattered shrubs and interesting small-growing and alpine species

Acantholimon acerosum	Geum heterocarpum
A. ulicinum	Lithospermum incrassatum
Anemone blanda	Prunus prostrata
Arabis caucasica	Ptilotrichum cyclocarpum
Aubretia deltoidea	Ranunculus cadmicus
Chionodoxa forbesii	Rosularia libanotica
Corydalis densiflora	Satureja spinosa
C. rutaefolia	Scorzonera mollis
Crocus species	Tanacetum praeteritum
Euphorbia herniariifolia	Tulipa armena var.lycia
Fritillaria carica	Veronica lycica
Geranium tuberosum	Viola heldrechiana

The arabis resembles the more common *A. alpina* but the stem leaves have arrow-shaped bases clasping the stem. The aubretia is the species which has given rise to our garden forms. The chionodoxa (N = *Scilla forbesii*) is an attractive species endemic to south west Anatolia with rather large dark blue flowers that often have a white centre. It is allied to the more familiar *C. luciliae* (O = *C. gigantea*), common in home gardens and may simply be a form of it. *Corydalis rutaefolia* is a relatively few flowered species with leaves reminiscent of rue as its name implies. It is also found in Crete, Rhodes and a sub-species in Cyprus. There are probably several different crocus species here but I have never seen any of them in flower; the most probable early spring-flowering species are the purple *C. antalyensis* and the yellow-flowered species *C. chrysanthus, C. flavus*. The fritillaria, which is probably the sub-species *F. carica serpenticola*, is somewhat similar to *F. forbesii* found lower on the mountain but is shorter growing and has few broader twisted leaves and the flowers are often brownish on the outer segments. The geum *G. micropetalum* (O = *Othurus heterocarpus*) is a short-growing species with small, pale yellow flowers in May and June. It is also found in Greece and the Apennines of Italy. The ptilotrichum (O = *Aurinia rupestris* or *Alyssum rupestre*) resembles a white-flowered draba and is also found in Greece and the Balkans. The satureja is a rare, small spiny shrub with white flowers in late summer. The ranunculus is rather special to Turkey but a form that has leaves that are reddish coloured occurs on the Troodos mountains of Cyprus (22.10). The tanacetum is endemic to south west Anatolia. The tulip is a delightful short-growing plant with fairly large red flowers and attractive leaves that have very wavy margins. The species is rather widespread in Turkey and further east in Iran but is variable. The viola is another miniature annual species similar to *V. kitaibeliana* and with minute, white, tinged mauve flowers.

Davis on his earlier journeys (23.3) recorded *Atraphaxis billardieri* from this area. It is a rare but rather insignificant small spiny shrub belonging to Polygonaceae, also found sparingly in Crete and mainland Greece; the genus is mainly represented in the Levant and North Africa. He also eulogised about a species of echinops found here of which he said "the heads can be nearly 6 inches across - incredible spheres of jade green". This is *Echinops emilae* and obviously well worth looking for. Another plant to be seen here is *Eremurus spectabilis* first noted in this area in 1991 by Mr Nicholas Turland. It grows on a south-west facing slope near the summit and overlooking Ölü Deniz in a large group of hundreds of plants that produce their tall spikes of pinkish-white flowers in late May and June.

23.11 **Fethiye to Finike:** The main road 400 leaves Fethiye eastwards then passes the fascinating archeological sites of the ancient kingdom of Lycia including those of Xanthos and Letoon. Whilst these are certainly worth a visit they are of no special botanical interest though one may see groups of the impressive *Acanthus spinosus* by the wayside.

The main reason for taking this route would probably be to

look for *Ophrys lycia* a newly defined species described in Buttler (1991) and recorded only from grassy areas between Kas and Finike. It closely resembles species of the *sphegodes* group and some enthusiasts may have difficulty in agreeing that it is not, in fact, a mere form of that species. The flower has a rounded brownish-purple lip with an H-shaped pattern and back petals and sepals that are pale pink or purplish. It flowers in March and April.

23.12 Fethiye North-Eastwards towards Korkuteli: Driving eastwards along the main road 400 out of Fethiye after some 20km one comes to a turning north-eastwards that eventually joins the new 350 road leading to Korkuteli and on to Antalya. This route passes through a picturesque high valley which has the 'remote' feeling of many such roads in Turkey. Much of the land is very heavily grazed and one would expect it to be devoid of plants but *Iris unguicularis* can be seen by the roadside. In places there are pine woods where, in April and May, one can find fine specimens of the spurred helleborine *Cephalanthera epipactoides*. It resembles *C. longifolia* but grows to 1m tall carrying up to twenty-five conspicuous white flowers.

Some 40km north of Kermer there is an old stone bridge crossing the Seki river and a small gorge here is worth exploring. In April one may see *Aubretia deltoidea*, *Ornithogalum nutans*, *Osyris alba*, and many plants of the starry-flowered buttercup *Ranunculus sericeus*. Near the bridge a road leads off eastwards to the typical unspoiled village of Seki where, around the buildings grow large stands of woad *Isatis sp.* and fine groups of *Ornithogalum nutans* in the local cemetery. The exact species of the isatis is difficult to determine - there are twenty-six in Turkey! In the stony, heavily-grazed land adjoining the village are areas with kermes oak, juniper and the very prickly *Berberis cretica*. Smaller shrubs include *Astragalus angustifolius* and *Jasminum fruticans* which is a widely distributed species found as far away as the Algarve (4.4). Herbaceous plants to be seen here include the attractive, yellow-flowered skullcap *Scuttelaria orientalis*.

A rough or very rough track rises from the village over the

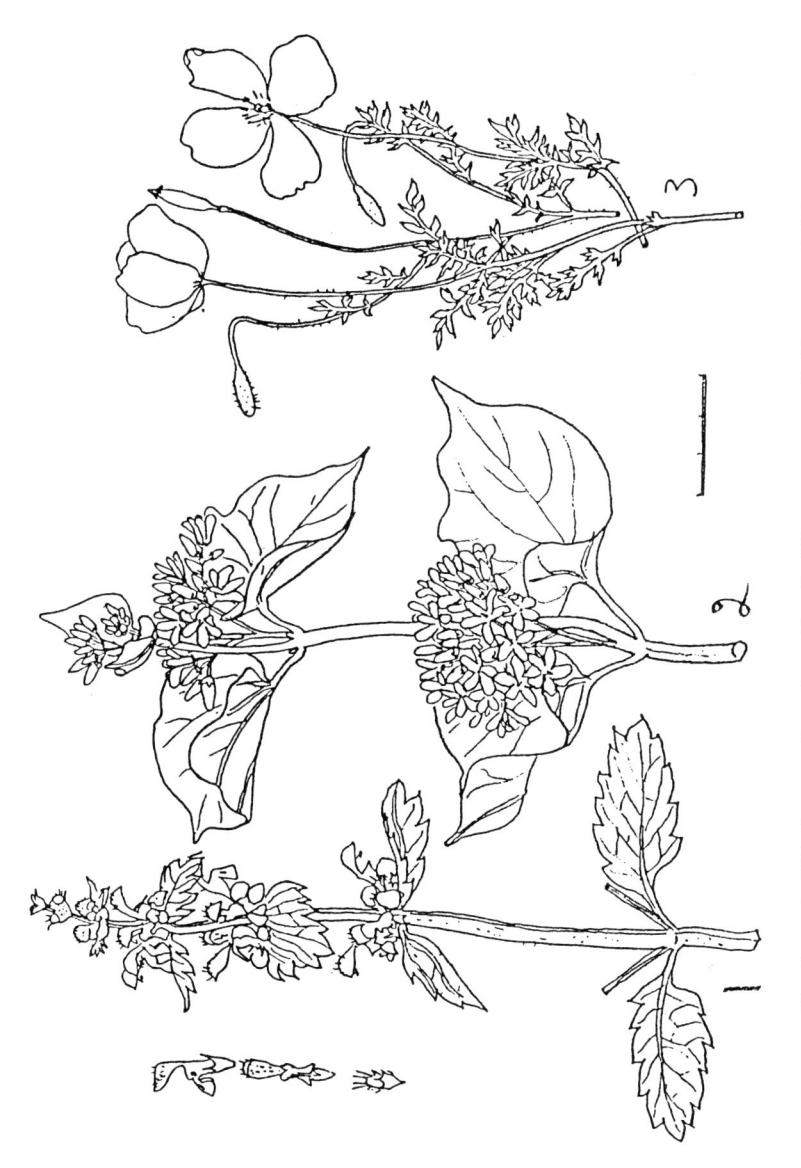

Turkey 1. Wiedemannia orientalis 2. Cionura erecta 3. Papaver dubium

Ak Daglari and down to Elmali. It is not easily negotiated without a four-wheel drive vehicle, especially in wet weather but well worth attempting. There are still patches of snow at the higher parts of the route in May and around them one may see:

> *Bulbocodium versicolor*   *Geum heterocarpum*
> *Corydalis rutaefolia*   *Primula vulgaris*
> *Crocus sp.*   *Scilla bifolia*
> *Euphorbia kotschyana*   *Solenanthus stamineus*

The bulbocodium is a small, spring-flowering, colchicum-like plant with pale mauve flowers which open before the leaves. The species here is an eastern type similar to the more familiar *B. vernum* sometimes grown in home gardens. It is easily possible to confuse the bulbocodium with *Colchicum triphyllum* that also grows in this area. The foliage is similar but the perianth segments in bulbocodium are split nearly to the base, whereas in colchicum there is a distinct corolla tube. The crocus is probably *C. chrysanthus* though the similar *C. flavus* is also found in this area. The primula is, of course, the primrose; it frequently grows as a mountain plant in the Mediterranean. The solenanthus (O = *Cynoglossum stamineum*) is an uncommon species with cymes of small, pale mauve, tubular flowers in May and June and also grows in southern Greece on Mount Taiyetos.

If, instead of turning off the road to Seki, one continues northwards a road leads in a north-westerly direction to Altinyayla and Gölhisar. Along here one passes first through a small gorge with pines, *Juniperus foetidissima* and *Platanus orientalis*. In places there is a thick undergrowth of *Berberis creticus* and some *Anthemis rosea*, *Saponaria calabrica* and *Smyrnium connatum*. After the gorge one comes to a high valley which is almost treeless except for scattered specimens of *Juniperus foetidissima* and *Pyrus amygdaliformis*. By the stream here grow thousands of *Dactylorhiza* orchids with unspotted leaves which may be *D. osmanica* (= *D. cilicica*) but one will need to see them in flower in June to be sure of the identification. The road then rises over the

Bonkuk Daglari to a pass Dermil Geçidi 1595m before dropping down to Antinayla. At the higest point there are scattered pines and a number of interesting species of small plants grow in the rocky ground between them, Including:

>Acantholimon ulicinum　　Erysimum kotschyanum
>Anemone blanda　　　　　Fritillaria carica
>Chionodoxa forbesii　　　　Myosotis sp.
>Colchicum triphyllum　　　Prunus prostrata
>Crocus baytopiorum　　　　Ranunculus sp.
>C. chrysanthus　　　　　　Viola kitaibeliana.

The colchicum is a small species that flowers as the snow melts here in March and April; it is widespread in mountainous regions of the Mediterranean and was first described from Spain. It may, however, be confused with bulbocodium. The crocus also flower as the snow melts and their precise identification needs checking. The first has beautiful ice-blue flowers with a paler centre. The form of *Viola kitaibeliana* here has small, bright yellow flowers with a darker centre. Many of these tiny viola species have specialised in a short growing period and produce their seed in a few weeks growth by being minute plants and having some cleistogamic flowers that produce seed without pollination. In this way they escape crop cultivations and excessive grazing, even under adverse climatic conditions.

23.13 Dalaman to Fethiye: The road from the international airport at Dalaman winds, for the most part, through attractive woodland scenery with impressive rock outcrops. Pines form the largest proportion of the trees but there are also deciduous species, especially in the valley, including *Liquidambar orientalis*. The last of these superficially resembles the unrelated oriental plane *Platanus orientalis* but is more upright in growth and has smaller and finer leaves. It may also be confused with an acer but has alternate, instead of opposite, leaves. It is more or less confined to the southwest of Asia Minor and does not occur wild in Europe though it very closely resembles the American species *L styraciflua* grown in home gardens for its beautiful crimson autumn colouring. Plants of this genus are referred to as sweet gums on account of the fragrant resin obtained from the bark and used as one of the main constituents of 'friar's balsam'.

Parts of this region have acidic soils, indeed the liquidambar shuns calcareous regions. Here are areas of fairly tall maquis type vegetation between the pines including:

Cistus salvifolius *Pistacia terebinthus*
Erica arborea *Sarcopoterium spinosum*
E. manipuliflora *Spartium junceum*
Phillyrea latifolia *Quercus pubescens*

The cistus is an unusually tall-growing and small flowered form of the species. The pistacia is probably the sub-species *palaestina*.

In places one may see the wild vine *Vitis vinifera sylvestris* climbing up trees. It produces small, sour, bluish-black grapes and male and female flowers are borne, unlike the cultivated grape, on separate plants. The chaste tree *Vitex agnus-castus* grows here in moist areas. Smaller plants found in more open spaces include:

Centaurium maritimum *Lavandula stoechas*
Geropogon hybridus *Ornithogalum narbonense*
Hypericum sp. *Polygala venulosa*

The most exciting of these is the hypericum which has shell-pink flowers but at the time of writing I have been unable to find a name for it.

On the rocks one may see the silvery ovate leaves of *Inula heterolepis*, which flowers in June and July, and the bright yellow flowers of an *Alyssum sp.* in April and May. At the base of the rocks in places grow the candytuft *Iberis attica*, usually with white flowers here. There is also an interesting eryngium with toothed kidney-shaped or round, long-stalked basal leaves with blades that stand up vertically and are described in the flora as bizarre. This is *Eryngium thorifolium*, endemic to the area and nearly always found on serpentine rocks. It produces its uninteresting greenish flowers in july.

23.14 Addendum: To the west of Dalaman there are a number of important coastal holiday sites, such as Kusadasi, Altinkum, Bodrum and the Marmaris peninsula and all of these are good areas from which to study the Turkish flora. The plants of the Marmaris peninsula are exhaustively dealt with in the relatively inexpensive work by Carlström (1987) which gives distribution maps but no descriptions or illustrations. There is also a useful list of plants to be seen around Camiçi Gölu (also called Bafa Gölu or Bafa-See in German) by a Swiss botanist (Strasser 1989) with some text in German. It could be useful for enthusiastic visitors to Kusadasi and Bodrum. This is a region where one is most likely to see the rare and extraordinary orchid *Comperia comperiana* though it is not mentioned by Strasser. Its natural terrain is in pine woods but many of the known sites are in old walled burial grounds where it is protected from grazing animals and from humans who in other areas may dig up the tubers to eat or to produce salep - used in making ice cream and as a kind of substitute for coffee. It is sad to think that the continued existence of this rare orchid seems to depend upon the dead!

To the east of Anamur (23.4) lies a new tourist centre at Silifke and this may be a useful site from which to study another region of the Turkish flora, though there is no helpful literature on the region. Even further east along the coast one comes to Mersin and Adana and from near here one can proceed north to Nigde through the celebrated Cilician Gates pass over the Toros Daglari. This is also a very good area for botanising. For the venturesome there are untold treasures to be seen in other parts of Turkey, though they cannot all be described as strictly Mediterranean. Crocus, tulip, fritillaries and iris abound and some insight into what may be seen to whet one's appetite is available in the book by Mathew and Baytop (1984).

Israel 1. Ophrys carmeli 2. Orchis galilea 3. Orchis caspia 4. Adonis aleppica

24. ISRAEL

24.1 Israel is, at present, the most accessible country from which to study plants of the region often referred to as the Levant. Its flora comprises a mixture of typical Mediterranean species with others that are common to western Asia and Africa. Although it is a relatively small country, about the size of Wales, it is host to some 2,500 species of wild plants.

Getting to Israel presents no serious problems. One can fly direct to Tel Aviv from Britain and several package holidays are available throughout the year. It is quite possible to cover the ground adequately by using local transport and staying in kibbutzim, though plenty of free time is required for such an approach. The majority of visitors will wish to travel by hired car and as a word of warning to motorists - it must be said that the traffic in many areas is far denser there than one might expect and driving is competitive. For a visit not exceeding two weeks, during which one can only be expected to see in detail the flora of one type of habitat of this varied country, it is possible to lodge at a single venue and make daily sorties from there. The north part of the country is especially rewarding to Mediterranean plant enthusiasts and one might work from a base in Tel Aviv, Netanya or Haifa.

Bartholomew print a contour map (No.16 Israel with Jordan 1:350,000) which is useful for studying the lie of the land and a touring map of Israel (Survey of Israel 1:500,000) is usually provided with a hired car but it is advisable to look for a more detailed map that gives all road numbers. Signposts in the towns are in both Hebrew and English but Hebrew alone tends to be used in rural areas and it is difficult for those unfamiliar with the language to make much of it. Many or most of the signposts give road numbers.

Potential visitors to Israel may avoid going there on account of the present political situation but many areas of very special interest to Mediterranean plant enthusiasts outside the larger towns and politically sensitive areas are quite safe in this respect for tourists. For those who may feel more comfortable in an organised group there are some botanical tours of Israel from time to time but they are relatively expensive compared with a 'do it yourself' tour (24.3).

Some interesting species flower in winter when the weather can be quite cold and wet, but the time to see the majority of the plants in bloom is from about the middle of March to the end of April. However, on Mount Hermon most species come into flower when the snow melts in May.

Most visitors, whatever their main reason for going to Israel, will want to see the archeological sites of biblical interest but it is not generally realised that the country is exceptionally rich in bird life. Private firearms are severely restricted and although the armed forces are well equipped they do not shoot at birds. As a result the country is an ornothologists paradise. There are bulbuls, Palestine sunbirds, graceful warblers, Tristram's grackles, Smyrna kingfishers and many other species often relatively tame and in considerable numbers.

24.2 **The terrain:** The present State of Israel is bordered on the north by Lebanon, the west by the Mediterranean Sea, the south by Egypt and the east by the valley of the River Jordan, though part of what was Syria in the north-east - the Golan Heights east of the Jordan valley - is at the time of writing claimed by Israel. In spite of its relatively small size, Israel has a very varied climate and topography. For convenience, the country can be considered as being broadly comprised of three regions with Galilee in the north, Samaria and Judea in the centre and the Negev desert in the south.

Galilee is perhaps the most attractive area on account of its charming countryside and it is the best place to look for typical Mediterranean plant species. The rocks are mainly of limestone and chalk forming rounded hills which come to the coast as cliffs in the extreme north-west. To the south of Galilee is the fertile plain of Jezreel and to the east the freshwater Lake of Tiberias - the Biblical Sea of Galilee, now called Lake Kinneret or Yam Kinnaret by the Israelis. The Golan heights to the north-east of this lake stretch northwards to Mount Hermon which, at 2,224m, is high enough to be covered with snow during the winter. The rainfall of Galilee is comparable with that of most parts of the northern Mediterranean region - there are pine and oak forests there and it is green in the spring.

The region of Samaria and Judea also is composed of limestone hills with the capital city of Jerusalem more or less in the centre. The annual rainfall in this part is much less than in Galilee and decreases southwards and eastwards so that pastures give way to the Judean desert and the very saline Dead Sea at 394m below mean sea level. Along the west coast of Samaria and Judea is the fertile Plain of Sharon bordered by extensive sand dunes and compacted sand cliffs - a relic of the quaternary period.

The Negev is a sparsely populated desert area which stretches to a narrow point in the south to the Gulf of Aqaba at Eilat. It is a colourful and interesting area for plants and animals but is not typically Mediterranean in character.

24.3 **Literature:** There is an excellent up-to-date flora of the region called 'Flora Palaestina' by Zohary and Feinbrun (1966-86). It is in English but comprises several bulky, expensive volumes and it is hardly the sort of literature that a casual visitor to the country will wish to carry. 'Flowers of Jerusalem' by Avigard and Danin (1972) has some useful illustrations but is not readily available. By far the most helpful work is the 'Pictoral Flora of Israel' by Plitman et. al. (1982). The text of this publication is in Hebrew with a brief English summary and it has good photographs and distribution maps of 750 species. It is readily available in Israel at most of the larger book shops and not expensive so it is an excellent choice as a souvenir to take home.

The 'Society for the Protection of Nature in Israel' publishes a quarterly journal in English entitled 'Israel - Land and Nature' (Editorial Office 13, Queen Helena St., P.O.Box 930 91008, Jerusalem). There is also an 'Israel Plant Information Centre', usually referred to as ROTEM, at Hargillo Field Study Center, 91076 Jerusalem. In addition to cataloguing wild species, this organisation arranges some study outings.

24.4 **The coastal Plain:** Along the coast south of Haifa there are extensive sandy areas and sand dunes backed to the east by the fertile Plain of Sharon where a variety of fruit and vegetables are grown with the aid of irrigation water from Lake Kinnaret.

A number of familiar shore plants are to be found here including:

Chrysanthemum coronarium *Otanthus maritimus*
Eryngium maritimum *Pancratium maritimum*
Euphorbia parialis *Urginea maritima*

In wet areas there are large stands of the coarse 'bulrush' or reedmace *Typha domingensis*. This grows also in Egypt and is probably the plant amongst which the baby Moses was found - that is, if it really was a bulrush and not papyrus. Here and there one may come across the tiny crucifer *Maresia pulchella* with mauve-lilac flowers about 1cm diameter and south of Tel Aviv *Crucianella maritima*. Both of these grow in the sand dunes. The genus *maresia* is sometimes included in *malcomia* but differs from it by having star-shaped hairs. In similar terrain, just south of Netanya, there are many plants of the attractive, shrubby *Oenothera drummondii* with 8cm diameter yellow flowers. It is one of the several plants that has been given the name Rose of Sharon, but is a poor candidate for the title since it hails from Texas! Another plant from the region which vies for this title is *Tulipa sharonensis*. It has scarlet flowers with a dark centre and wavy leaf margins and is close to *T. agenensis* (= *T. oculus-solis*) found in the south of France and North Spain and may have been introduced there from Palestine by the crusaders. A very special plant to be looked for near where the oenothera grows is *Iris atropurpurea*. This magnificent oncocyclus species produces its dark, mauve-black flowers in March in sandy areas just south of Netanya.

Travelling along the main Tel Aviv (Yafo, Jaffa, Joppa) highway northwards to Haifa one passes the ruins of King Herod's port of Caesarea (Qesari). This is worth a visit though much of the original harbour is now under the sea. Some of the roads leading to it are lined with the endemic *Tamarix aphylla* which can grow to a substantial tree with a considerable trunk. It originates from the Negev and is not to be confused with the smaller species *T. nilotica* that is found wild amongst the sand dunes here. At the ruins one can see the evil-smelling solonaceous shrub *Withania somnifera* and the aptly-named ice plant *Mesembryanthemum cristallinum* that looks as though it is glistening with ice crystals. A few kilometers further north on the main road one passes through a fish-pond area well worth a visit. This is the Ma'agan Mikhael nature reserve noted for its birds. To get there one must turn off east at the signpost to Zikhron Ya'aqov then continue to the traffic lights and turn right. Some two kilometers further along a dirt road takes one under the highway to a kibbutz where one can leave a car and wander on foot between the ponds. It is primarily a place to see birds such as the lesser pied kingfisher and an occasional mongoose but there are a number of shore plants and acres of *Chrysanthemum coronarium* in several colour forms.

24.5 Mount Carmel (Har Carmel): This is a north-west extension of the limestone hills of Samaria which fall to the sea as cliffs at Haifa. It is more of a hill than a mountain and only rises to 528m. The city of Haifa, which is the country's main port, is gradually encroaching on Mount Carmel and there is considerable building in the north reaching to the Druze village of Daliyat el Karmil.

Most visitors to Israel will be staying at Natanya or Tel Aviv rather than Haifa and will approach the area from the south. In this case it is worth turning east along the Zikhron Ya'aqov to Elyaquim road. In spring the rolling limestone hills are covered with a truly magnificent carpet of flowers - perhaps the Biblical 'Flowers of the Field'. These include:

Anagallis arvensis	*Linum pubescens*
Anthemis palaestina	*Plantago afra*
Artedia squamata	*Ranunculus asiaticus*
Chrysanthemum coronarium	*Salvia verbenaca*
Echium angustifolium	*Scabiosa prolifera*
Echium judaicum	*Trifolium clypeatum*
Erodium gruinum	

In many places the scabiosa is the most prominent species. It is a stocky annual plant with heads of pale yellow flowers and a typical Levantine species though it does grow in Cyprus (22.4). The artedia is an umbellifer rather like an orlaya and with heads of white flowers having a group of darker, sterile florets in the centre. The two echiums are similar and difficult to distinguish apart; both have reddish purple flowers. *Erodium gruinum* is a cranesbill with large 2.3cm diameter lavender-coloured flowers. The linum is a particularly fine species that has pink flowers up to 3cm diameter held at the top of the plant in a flattened inflorescence. It also grows in Greece and Cyprus but it never looks so magnificent as in Israel. The *Ranunculus asiaticus* here is the brilliant scarlet form and the salvia has pinkish-lilac, rather than the usual mauve, flowers. The whole effect of this coloured carpet with swallowtail and other butterflies hovering above it is quite stunning. A few weeks later, in May, the above species give way to the pale mauve crucifer *Erucaria hispanica* and the very dark red poppy *Papaver subpiriforme* (N = *P. rhoeas var.oblongatum*).

By the roadside grows the tall, endemic *Cerinthe palaestina* with blackish-purple upper bracts and calyxes and white flowers. It resembles an especially robust and tall form of *C. major*. Here also one may see the relatively short bear's breech *Acanthus syriacus* and the attractive hollyhock *Alcea setosa*.

Turning off north-westwards at Elyaquim the road starts to climb up Mount Carmel. There are cultivated and cattle-grazed fields, patches of garigue and woodland. By the roadside one may see groups of *Lupinus varius* (= *L. pilosus*) a species with blue flowers that have a white upright mark on the standard. It grows in other parts of the Mediterranean, as far west as Portugal, but here it is designated as one of the seventy or so protected species.

The shrubs in patches of garigue include the following, all of which are common in other parts of the Mediterranean:

Calicotome villosa	*Pistacia lentiscus*
Cistus salvifolius	*Salvia fruticosa*
Cistus villosus	*Sarcopoterium spinosum*

In January and February one may find *Iris palaestina* in flower here. It is a member of the juno group, a short growing species with leek-like foliage and white or pale blue flowers having a yellow mark on the falls and borne singly. In shape it resembles a smaller version of *Iris planifolia* which grows in south Spain and Sicily. It is also rather like a pale flowered version of *I. persica*, found in Lebanon, Syria, Iraq and Turkey, and may be a form of that species. In spring *Cyclamen persicum* flowers in profusion in the garigue accompanied by *Gynandriris sisyrinchium, Ornithogalum narbonense*, and *Gladiolus italicus*. A short-growing cornflower *Centaurea cyanoides*, sometimes included under *C.cyanus*, adds a splash of blue. Several orchids can be found in the garigue including a very dark coloured form of *Orchis tridentata* that looks like an anacamptys at a distance, *Ophrys carmeli* and *Serapias vomeracea*. The ophrys is the Mount Carmel orchid and a form of the *Ophrys scolopax* group with the back sepal curved forward. It is fairly common here and also grows in Cyprus (22.10).

In the garigue there are groups of *Quercus calliprinos* which resembles a tall-growing form of the kermes oak *Q. coccifera* though it seldom reaches more than three metres in height. Patches of more substantial woodland also occur here and are particularly good hunting places for plants. The trees are mainly Aleppo pine *Pinus halepensis* but include others such as *Quercus calliprinos, Rhus coriaria, Laurus nobilis* and *Pistacia palaestina*. The last of these grows to some four metres in height and is a form of *P. terebinthus*. Around the edges of these woods and in clearings one can expect to find many species which are typical of the region as a whole:

Anacamptis pyramidalis   *Linum strictum*
Asphodelus microcarpus   *Nigella ciliata*
Campanula rapunculus   *Ophrys carmeli*
Centaurea iberica   *Orobanche ramosa*
Coronilla scorpioides   *Scorpiurus muricatus*
Eryngium creticum   *Smilax aspera*
Geranium robertianum   *Trifolium clypeatum*

Israel 1. Verbena tenuisecta 2. Oenothera drumondii 3. Nigella ciliaris 4. Cerinthe palaestina

Helichrysum sanguineum *Trifolium purpureum*
Lagoecia cuminoides   *Trifolium resupinum*
Lathyrus aphaca   *Trifolium stellatum*
Lathyrus blepharicarpus *Vicia narbonensis*
Linum pubescens   *Xanthium spinosum*

The centaurea has 2.5cm long spines standing out at right-angles from the bracts below the flower heads and white flowers with scattered, purple, fertile florets. *Helichrysum sanguineum* is an eastern species with small crimson flower heads and is called in Hebrew 'Blood of the Maccabees' after the brothers that led the revolt against Greek rule in Jerusalem 165 B.C. The nigella is another eastern species, found in Israel, Turkey and Cyprus. It is an annual with pale, greenish-yellow flowers that are of very interesting stucture when viewed closely.

Some 6km north of Daliyat el Karmil a road leads off westwards to join the main Tel Aviv to Haifa road. This passes through a fine wooded area which is the nature reserve of Bet Oren. The trees here are mainly the native oaks, the evergreen *Quercus calliprinos* and deciduous *Q. ithaburensis*. It is not easy to park along this narrow road but the area would be well worth exploring. Quite recently, *Brassica cretica* has been found growing on rocks in this region around the caves at Nahal Me'arot.

24.6 Mount Gilboa (Hare Gilboa): Travelling south eastwards from Haifa through the Plain of Jezreel (Emeq Yizreel) and past Afula, one comes to the small town of Bet She'an. From here one can drive up Mount Gilboa which rises to 497m near the north edge of Samaria and provides a magnificent view eastwards over the Jordan valley. King Saul was defeated by the Philistines here in 1010 B.C. It is also one of the places where one has a good chance to see the local gazelle. Unfortunately, the summit is not accessible at times because of military restrictions, but there are plenty of areas to explore.

Climbing up the hill the roadsides are lined in places with the Syrian thistle *Notobasis syriaca* and in spring one can see a tall, white-flowered sage *Salvia dominica* and the blue gromwell *Alkanna tinctoria*. On higher ground grows *Ajuga chia* (sometimes considered as a sub-species of *A. chamaepitys*), an interesting pink salsify *Scorzonera papposa* and the dwarf chicory *Cichorium pumilum* together with giant fennel *Ferula communis*. *Fritillaria libanotica* can sometimes be seen here; it has several drooping greenish flowers and is considered by some taxonomists to be a form of *F. persica*. Two treasures to be searched for on Mount Gilboa are *Iris haynei* and *Orchis israelitica*, both of which are in flower from late February to early April. The first is a magnificent, tall oncocyclus species with large, dusky-purple, scented flowers. The orchis which was first discovered as late as 1978, is somewhat like *O. boryi* and has white or pinkish-white flowers that open first from the top of the inflorescence.

After visiting Mount Gilboa it is worth driving northwards out of Bet She'an to the crusader castle of Belvoir. There are good views from here and many plants of woad grow around the castle, this is *Isatis lusitanica* which closely resembles our native *I. tinctoria*.

24.7 Lake Kinnaret region: Following the road northwards out of the Plain of Jezreel from Afula to get to Lake Kinnaret, one can either go through Nazareth and then bear eastwards to Tiberias or past the southern slopes of Mount Tabor through Kefor. Nazareth, with its biblical connections in mind, is a somewhat disappointing small town. However in pine woods nearby, one may see the endemic, comfrey-like plant *Podonosma orientalis* with mauve flowers. The beautiful oncocyclus iris *I. bismarckiana* is reputed to grow here. It flowers in February and March and produces large blossoms with pinkish-white standards and very 'netted' dark falls. It is a shorter growing species than *I. atropurpurea* or *I. haynei*.

Taking the alternative route out of Afula, it is worth driving up the narrow road to the top of Mount Tabor. There are two interesting churches at the top and wooded areas of *Crataegus azarolus*, *Pinus halepensis*, *Pistacia palaestina*, *Quercus calliprinos*, *Styrax officinalis*. On the stone walls one can see *Clematis cirrhosus* and there are plants of *Asphodeline lutea* growing in grassy clearings. The form of the last of these in Israel often has brownish-yellow flowers instead of the clear yellow type seen in western areas of the Mediterranean. The interesting *Lamium moschatum* also grows near here. It is a white dead-nettle in which the leaves are often variegated with lighter patches in the centre of the leaves.

One is likely to approach Lake Kinnaret crossing the high ground on the edge of the Horns of Hittim (Qarne Hittim) en route to Tiberias. This is the site of an ancient, extinct volcanic outcrop and the soil is acid rather than the typical basic limestone of the area, so there is a slight change in the flora. The globe thistle *Echinops adenocaulis* and the salsify-like *Geropogon hybridus* (= *G.glaber*) are common here. Another plant that favours these conditions is the impressive thistle-like *Gundelia tournefortii* with a rosette of large leaves that have orange-yellow midribs and with short flowering stems of small mauve, or yellow, flowers. It is hardy in Britain.

The town of Tiberias is becoming a popular holiday resort and has somewhat of a Swiss-lake atmosphere. One notices the very tall Doum palms *Hyphaene thebiaca* which do grow wild in Israel but here they are planted in avenues. The wood is used in Egypt for making furniture. It is worth examining the banks around the lake and amongst the plants to be found there one can see the crown of thorns bush *Zizipus spina-christi*, not to be confused with the other contender *Paliurus spina-christi*.

24.8 North Galilee, Mount Meron, Hula Reserve:
Northern Galilee is rather less populated and quieter than the country further south and there are some charming Aleppo pine woods. A good way to traverse the area is from 'Akko the old town of Acre. Here on the walls of the crusader castle grow *Matthiola tricuspidata* and *Hyoscyamus aureus*. The last of these is frequently seen on similar fortifications throughout the eastern Mediterranean and was probably planted deliberately as a handy painkiller. A good way to proceed from here is to go north to Nahariyya

and then turn east to Zefat (Safad). As weeds amongst fruit trees and in neglected arable land, one can expect to see in spring the buttercup *Ranunculus millefoliatus*, the pink *Silene palaestina* and the delicate thistle-like *Crupina crupinastrum*. By the roadside one may well encounter the indigenous *Arum palaestinum* which closely resembles *A. dioscoridis* but the spathe is entirely dark purple on the inside and the plant is generally somewhat larger. Furthermore it does not have the repugnant smell of *A. dioscoridis*. It is an impressive plant. In some pastures the tall *Scilla hyacinthoides* grows with an adonis. The first can reach a height of 1m with 50-100 small mauve, star-shaped flowers in a dense inflorescence. The second is a typical annual adonis like *A. annua* but the flowers quite often reach 5cm. diameter and are brilliant red with a dark centre. It is an fine plant which would be an impressive addition to the 'annual' border, but the identification needs verifying - it may be *Adonis aleppica*. Cattle do not graze these two species which are left standing in lush grass.

Approaching Zefat one can find the autumn-flowering *Cyclamen coum* in the woods and on rocky banks and roadsides there are several interesting species including:

Arum hygrophilum *Iris lortetii*
Bellevalia flexuosa *Iris palaestina*
Crocus hyemalis *Lilium candidum*
Crocus olbanus *Muscari neglectum*
Gagea commutata *Narcissus tazetta*
Glaucium grandiflorum *Orchis caspia*
Iris bismarckiana *Sternbergia clusiana*

The arum has long, narrow, arrow-shaped leaves and pale green or white spathes narrowly edged with purple. It is a species mainly found in Syria and flowers in summer. *Crocus hyemalis* has white flowers about Christmas time; it is endemic and quite common in places, especially around Jerusalem. *Crocus olbanus* has narrow, pale lilac segments to the flower and blossoms in autumn and is rather rare in Israel, though plentiful in parts of south-east

Turkey. The gagea is a rather extraordinary-looking species with relatively long, narrow, pointed, yellow petals and is fairly common here. The glaucium has extra large (12cm diam.) scarlet flowers with dark centres and finely divided leaves. Although it is found in north Galilee it is more frequently seen in dry areas of eastern Judea and the north of the Negev. *Iris lortetii*, which is unfortunately rare, is another superb oncocyclus species similar to *I. bismarckiana* (24.7) but with brownish, and less netted, falls. *Orchis caspia* is similar to *O. papilionacea* but has 10-15 flowers on the inflorescence instead of 3-8 in the usual form of the pink butterfly orchid and the lip is smaller and narrower. It is sometimes referred to as *O. papilionacea ssp. bruhnsiana* and is an eastern form that extends to northern Iran. The sternbergia is rather uncommon and found mainly in southern Turkey. Like other species of this genus, the yellow flowers are produced in September to October and in this case on short stems and before the leaves.

Mount Meron (Hare Meron), not far from Zefat, rises to 1208m and is the highest peak in Galilee. It has an orthodox Jewish settlement and the summit is a nature reserve but unfortunately there are military restrictions at times and cameras may not always be allowed. It is an especially good area to find orchids and, because of its height, some early flowering species may be seen there still in bloom in April. Most of the mountain is covered with bushes and small trees, amongst which grow the following species of orchid:

Cephalanthera longifolia   *Orchis caspia*
Epipactis veratrifolia   *Orchis galilea*
Limodorum abortivum   *Orchis italica*
Neotinea maculata   *Orchis tridentata*
Orchis anatolica

The epipactis is a very robust species growing to 1-1.5m tall. It produces relatively large flowers in late May. They are green with a strong reddish overlay. It is an uncommon or rare eastern species found also in Cyprus, Turkey, Syria and Lebanon. *Orchis galilea* is the other special orchid on this list. It usually has greenish flowers somewhat resembling the monkey orchid and, like that species, the flowers open from the top of the inflorescence first. The flowers are produced in January or February at lower altitudes, but may still be in good condition in April on Mount Meron. They have reddish spots near the base of the lip and in some Israeli forms the flowers are entirely dark red. This is a rare species also found occasionally in Lebanon and Syria.

Other species of interest that grow amongst the orchids listed above include *Bellevalia flexuosa, Euphorbia hierosolymitana, Scilla autumnalis* and *Symphytum palaestinum*. The bellevalia is a localised species with pale blue flowers and somewhat resembling the more common *B. trifoliata*. The euphorbia grows as a small shrub with yellowish inflorescence bracts. *Symphytum palaestinum* is an uncommon, localised species of comfrey with white flowers rather similar to *S. ottomanum*.

Turning east at Zefat one comes to a road leading northwards up the valley of the Jordan and some 7km north of the end of Lake Kinneret is a turning eastwards to the Hula reserve. This is a wetland area reminiscent of the Nile delta and has catwalks built through the marshes to observe wildlife. Tall papyrus *Cyperus papyrus* grows to some 4 metres high and there are large groups of the beautiful pale blue waterlily *Nymphaea coerulea*. In shallow water one can see the so-called flowering rush *Butomus umbellatus* and the tall endemic yellow water iris *I. grant-duffi*. There are terrapins, catfish, water buffalos and many interesting birds such as white pelicans, osprey and white-tailed eagle on migration.

24.9 Mount Hermon: This is in the most northerly part of the Golan heights, described on some maps as under Israel military administration. The area is a nature reserve and skiing centre and there are two main peaks, Mount Hermon (Ketep Hermon) 2224m and Mount Harbetarim (Har Habetarim) 1296m. Access is usually allowed and the best time to go there is probably in May. Many of the species seen at lower altitudes also occur here but it is worth looking out for the following:

Arum elongatum
Colchicum hierosolymitanum
Eremurus libanoticus
Iris palaestina
Ixiolirium tataricum

Orchis anatolica
Prunus prostrata
Scilla cilicica
Tulipa polychroma

The plant described as *Arum elongatum* may be a form of *A. conophalloides* and produces the inflorescence on a stem up to 1 metre tall. The spathe is generally brownish on the inside, but yellow in some forms, and the leaves often have begun to wither by the time the plant commences flowering. The colchicum flowers in autumn with star-shaped, pale mauve blossoms and has numerous rather narrow leaves. The rare eremurus has whitish flowers with a reddish brown stripe down the centre of the outside of the petals. The ixiolirium is sometimes called *I. montanum* and has two to five mauve flowers on a stem some 20cm long. It is common in eastern Turkey. *Scilla cilicica* resembles *S. sibirica* of our gardens but has long, untidy leaves at flowering time. It can be more easily seen near the St Hilarion castle in northern Cyprus (22.12). The tulip (= *T. biflora*) has 1-3 pink flowers with a yellow base to the tepals and yellow anthers; it is very similar to *T. humilis* (= *T.pulchella*) which grows in mountains of Turkey and Iran and it may be a form of that species.

24.10 **Judaea and Samaria:** Most of this area comprises what has become known as the 'West Bank' and is one of the most politically sensitive parts of the country. Here there are more limestone hills as in Galilee but it is drier and the rainfall drops away southwards and eastwards. Many of the plants found in Galilee can also be found here but species which are well worth looking for include:

Adonis dentata
Aristolochia maurorum
Caralluma europaea
Colchicum hierosolymitanum

Crocus haemalis
Erminium spiculatum
Ixolirion tartaricum
Onosma gigantea

The adonis closely resembles *A. cupaniana* but with orange-yellow flowers. The aristolochia is a strange herbaceous plant growing to about 30cm tall. It has rather large and lurid-coloured flowers that look somewhat like a minature euphonium and is found also in Turkey (23.5). The caralluma is a strange member of the asclepiadaceae which grows also in Almeria (7.8). The erminium is an aroid with fleshy red spathes on very short stems and much-divided leaves not unlike those of *Helicodiceros muscivorus* (8.8). *Onosma gigantea* is, as its name suggests, a tall-growing species and has yellow flowers.

All visitors will wish to see the beautiful city of Jerusalem built mainly of the local honey-coloured limestone. From here one can go eastwards down into the Jordanean desert to see and to bathe in the Dead Sea and visit the site of Massada and the Dead Sea Scrolls. There are many interesting birds, some residents such as fan-tailed ravens and Tristram's grackles and many migrants. Leopards and gazelles still live here. Amongst the sparse plants, one will see flat-topped bushes of *Acacia raddiana* and the shrubby *Atriplex halimus*, sometimes parasitised by the giant broomrape *Cistanche tubulosa* which is similar to *C. phelepaea* of southern Spain. With luck one may come across *Androcymbium palaestinum*, an ornithogalum-like plant with a rosette of leaves and nearly stemless star-shaped white flowers that are veined with green. It closely resembles the rare *A. gramineum* found in the Cabo de Gata, south Spain (7.8).

24.11 The Negev Desert: This wedge-shaped region stretches from the southern end of the Judean hills to a point at Elat on the Red Sea in the south. It occupies about half of Israel's land area and the rainfall is, of course, very low though substantial storms occur at times. Far from being a boring and lifeless, dry area, it is varied and colourful with interesting birds, mammals and plants. However, it can hardly be classed as Mediterranean in character and will be considered only in a cursory way. A selection of the many special plants that grow here includes:

Israel 1. Podonosma orientalis 2. Linum pubescens 3. Glaucium grandiflorum 4. Salvia dominica

Tunisia 1. Matthiola fruticulosa 2. Pteranthus dichotomus 3. Convolvulus supinus 4. Scorzonera undulata 5. Retama monosperma

Abutilon fruticosum   *Fagonia mollis*
Acacia tortilis   *Forsskaolea tenacissima*
Anvillea garcinii   *Glaucium grandiflorum*
Asphodelus tenuifolius *Iris mariae*
Bassia eriophora   *Leopoldia eburnea*
Bellevalia desertorum   *Leopoldia longipes*
Colchicum ritchii   *Mesembryanthemum nodiflorum*
Dipcadi erythraeum   *Tulipa ampylophilla*

With the exception of *Asphodelus tenuifolius* and *Mesembryanthemum nodiflorum*, these species do not grow in the Mediterranean proper. The abutilon is a shrub with rounded leaves and yellow flowers. The acacia is a small tree similar to *A. raddiana* also found in the Judean desert. The anvillea is a yellow-flowered composite without ray florets and with small grey leaves. *Bassia eriophora* (= *Anisacantha sp.*) belongs to chenopodiaceae and has very woolly stems that lie flat on the ground. The bellevalia has relatively large, hyacinth-like, pale mauve flowers and the colchicum is a small-flowered species with broad, bluish-green leaves that lie flat on the soil at flowering time. The dipcadi is very similar to *D. serotinum*, the 'brown bluebell' of the south of Spain. *Fagonia mollis* is just what it sounds like, a plant closely resembling F. cretica but covered with velvety brownish hairs. The forsskaolea (sometimes spelt *Forskohlea*) belongs to the urticaceae and, like the nettle, has stinging hairs. *Iris mariae* is another beautiful oncocyclus iris, short-growing, with short, downward-curved leaves and large purple flowers. *Leopoldia eburnea* is a tassel hyacinth with a narrow inflorescence of greenish flowers with some purple sterile florets at the apex and in *L. longipes* the flower stalks of the lower flowers are greatly elongated and the flowers a dull, brownish-green colour; it also occurs in Turkey, Russia and Iran. The tulip flowers in March and April with light red flowers and narrow, twisted leaves that are often rolled inwards at the margins.

Tunisia 1. Salvia lanigera 2. Dipcadi serotinum 3. Astragalus caprinus 4. Marrubium alysson

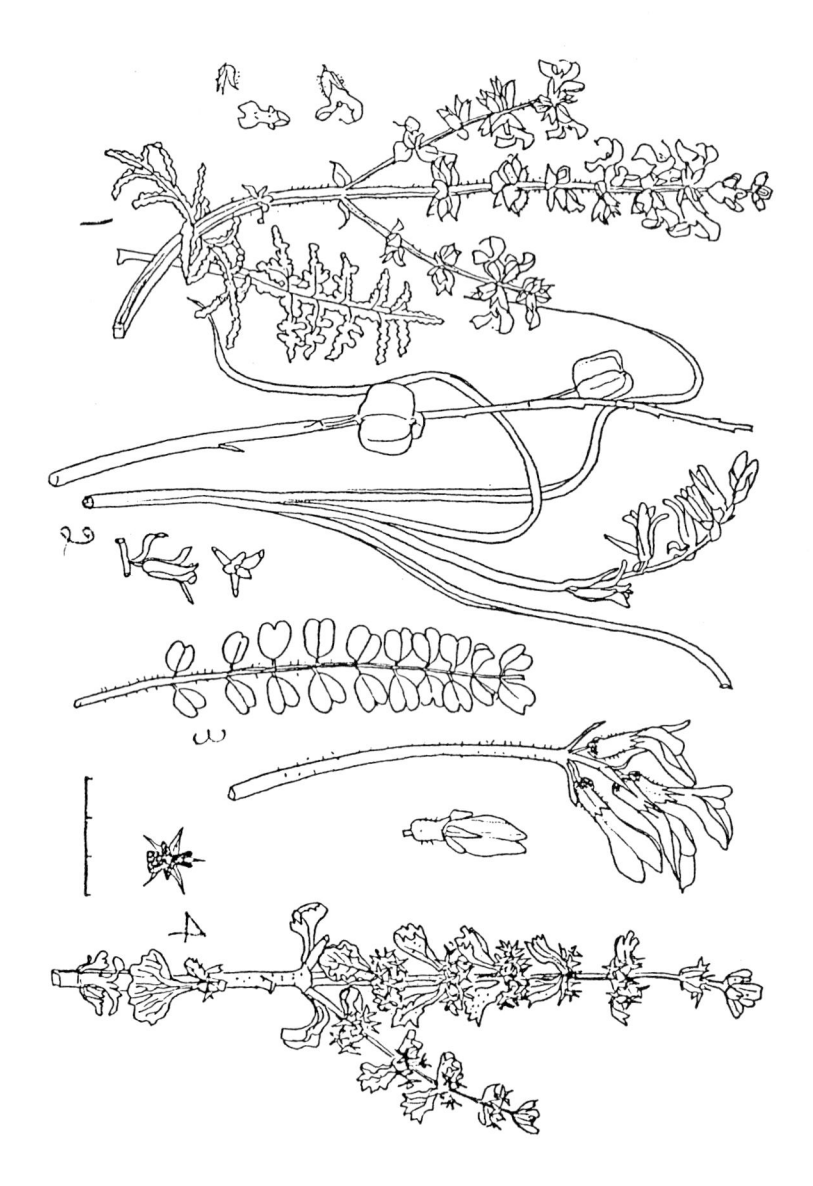

25. TUNISIA

25.1 Tunisia is the most accessible of the North African states bordering the Mediterranean. Package holidays are regularly organised to resorts on the south coast of Cap Bon, such as Nabeul and Hammamet, and further south around Sousse and Sfax. Travelling by car presents few problems, though car hire there is relatively expensive. However, to compensate for this, hotel prices are not high if one is making one's own way by car. French is spoken in the larger towns and by hoteliers, tourist agents, police and other officials but not always in the villages. Tunisia is, of course, an Arab country with a fair proportion of Berbers who speak a dialect of Arabic which is difficult for some Tunisians to understand and whose womenfolk often dress colourfully.

The northern part of the country has a fairly typical Mediterranean climate but south of Sousse the rainfall is much reduced and the south west borders on the Sahara desert. Plant hunters visiting the country will probably recognise some familiar species they have seen in southern Spain or Sicily but often with slight subtle differences and there are, of course, many species which are typically North African, especially in the desert areas. Unfortunately for the plant enthusiast, considerable pressure on land to meet the increasing demand for food resulting from the population increase in recent years has greatly reduced the interesting wild areas of garigue. For this reason the visitor may find the countryside botanically disappointing though one should always be prepared for pleasant surprises.

Several maps of Tunisia are available and Hildebrand's 1:900,000 gives topography, road numbers and details of some of the interesting archeological remains.

25.2 **The terrain:** North of Sousse to Touzeur there are three fairly well-defined ranges of hills lying in a south-west/north-east direction; they are a continuation of the Atlas mountains of Morocco and Algeria. The most northerly of these, following the coast from Bizerta to Aïn Draham, attracts the highest rainfall and the western area bordering on Algeria, known as the Khroumirie, tends to be green and provides a welcome relief for a holiday from the heat and dryness of summer.

The range to the south of this, from Tunis to Le Kef, is much drier and the most southerly of the three. It forms an ill-defined broken line from the most northerly tip of the large peninsular of Cap Bon in the east to the Algerian town of Tébessa and is also rather dry. Most of the hills throughout the country are composed of limestone but there are some volcanic intrusions. In between these ranges a number of crops are cultivated, especially cereals.

South of these three hill ranges, which rarely exceed 1,300m, the climate is extremely dry and to the west of the coastal town of Gabès there is a large salt lake called the Chott el Djerid which varies in size according to the infrequent rainfall but is generally the largest of its kind in North Africa. South of here is mostly desert, either the stony Tunisian desert or the sandy Sahara, and this will only be dealt with in a cursory manner for it cannot be classed as typically Mediterranean.

Near Bizerta, in the north, is the large Lake Ichkeul which has an outlet to the sea and varies in salinity according to the time of the year. It is an important and charming nature reserve and of great interest to bird watchers, especially during the winter months. The extreme north-east tip of Cap Bon is also of interest to ornithologists during the migration period but many of the special plants that were recorded from there in the past have disappeared through excessive grazing

25.3 **Literature:** As one might expect, practically all of the literature on the plants of the area is in French. The most up to date is a flora by Mme. Pottier-Alapetite (1979-81). It consists, so far, of two volumes but is still incomplete and the section on monocots, which may be of special interest to many readers, will be the last published. Another useful but earlier work is that by Murbeck (1897-1900), also in French though it is written by a Swede and published as a collection of papers in the annals of Lund University. For most purposes the list of plants of the area by Bonnet and Barratte (1896) is the most helpful and complete but it has to be stressed that nomenclature has been somewhat changed since it was written and, it seems that some interesting species listed therein have probably been lost owing to increased human population pressures.

Any reader who has an urge to study the plants of the Sahara, with its own interesting species, will find the excellent flora of the region by Ozenda (1977) a great help.

25.4 **Around Hammamet:** The first chance that most visitors to Tunisia have for looking around will probably be in the Hammamet to Nabeul area. The region has been considerably built up to cater for tourists in recent years and most of the wild sites have gone but a search along the thin turf by the seashore may reveal the following:

Calendula aegyptiaca   *Euphorbia parialis*
Carpobrotus acinaciformis *Hypercoum procumbens*
Carpobrotus edulis   *Pancratium maritimum*
Centaurea sphaerocephala *Salvia lanigera*
Diplotaxis erucoides   *Silene colorata*
Eryngium maritimum

The calendula is a particularly attractive form of wild marigold with a spreading habit and flowerheads up to 2.5cm in diameter - often of a luminous orange yellow though the species is somewhat variable. It is more or less confined to North Africa and seems to be widespread in northern Tunisia. The two carpobrotus (O = *Mesembryanthemum sp.*) are common, showy plants found throughout the Mediterranean and referred to as Hottentot figs on account of their edible fruits. They are natives of South Africa and thus nearer their wild haunts here than in Europe. The first has brilliant magenta flowers but those of *C. edulis* are often of a straw-yellow colour. Although the flowers are showy and attractive these are two introduced species which are the most destructive to the natural dune vegetation of the Mediterranean. The centaurea is a rather fine species with mauve florets similar to *C. polyacantha* and

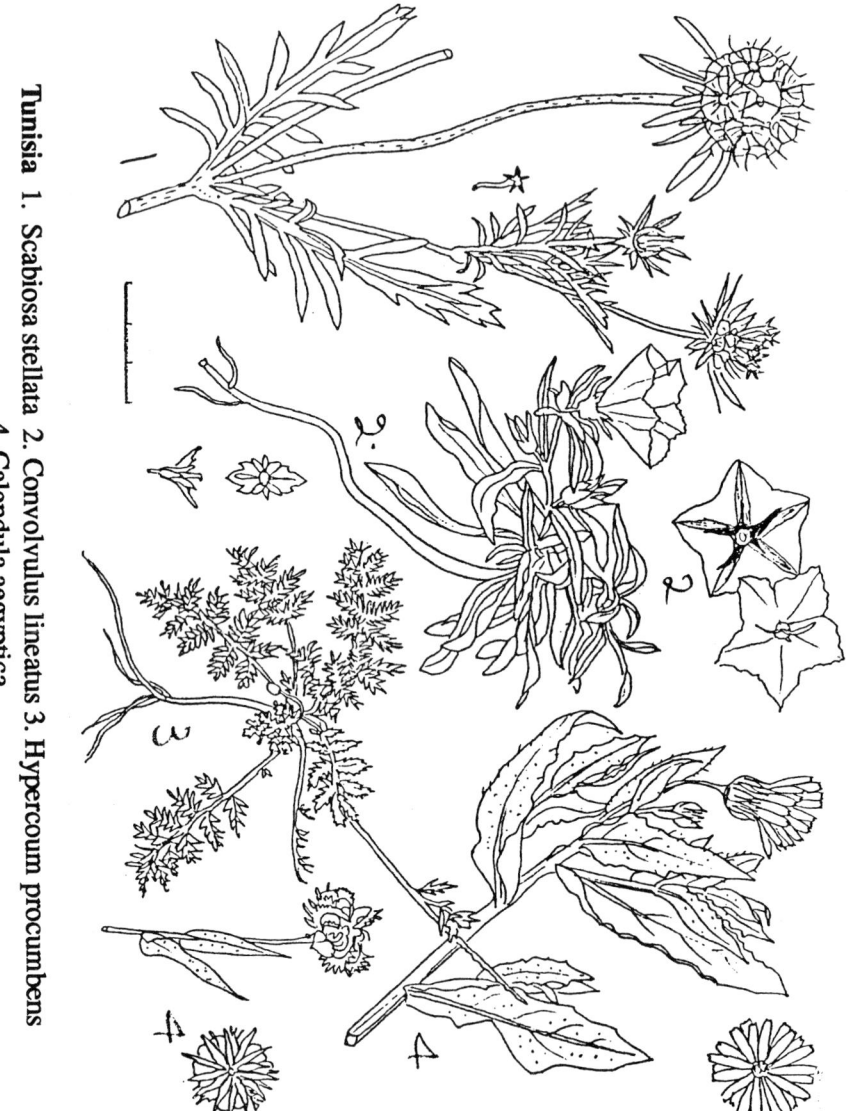

Tunisia 1. Scabiosa stellata 2. Convolvulus lineatus 3. Hypercoum procumbens 4. Calendula aegyptica

limited to the western Mediterranean area. *Diplotaxis erucoides* is a small white-flowered crucifer reminiscent of shepherd's purse but with elongated fruits. The salvia resembles *S. virgata* but has finely-divided leaves.

Further inland one may see the curious little batchelors' buttons *Cotula coronopifolia*, a small composite without ray florets which hails from South Africa but has become established here and in southern Europe (4.10). With it grows the attractive small clover *Trifolium resupinatum* with flattened heads of pink flowers and sometimes *Nonea vesicaria* which is a member of the boraginaceae having tiny purple-black flowers. The so-called tea plant *Lycium barbarum*, grows wild here and is also used for hedges. It is a spiny, solenaceous shrub from China growing to 2m bearing small lilac-coloured flowers and has become naturalised in some European countries, including Britain. .

25.5 Cap Bon: At one time this was a place of special interest for botanists but it is now largely cultivated with cereals, market gardens and vinyards on the south-east coast and much of the rest is over grazed. Every morning colourful landowners and farm workers may be seen leaving Nabeul with their mule and camel carts to travel to Korba and Keliba for the days work. They grow a variety of crops including vines from which an unusual dry muscat wine is produced.

The road 44 leading from Korba in a west north-west direction passes through cultivated fields but there are some rougher areas near Beni Khalled where annual weeds and some more interesting plants may be seen:

Anagallis monelli	*Galactites tomentosus*
Biscutella didyma	*Hedysarum coronarium*
Calicotome spinosa	*Hedysarum spinosissimum*
Chrysanthemum segetum	*Leopoldia comosa*
Convolvulus althaeoides	*Serapias parviflora*
Fedia cornucopiae	*Silene colorata*

After Beni Kahlled one can make towards the northerly coast and the small spa of Aïn Okter. Here are rocky outcrops facing the sea with garigue containing the following shrubs:

Anthyllis barba-jovis   *Juniperus phoenicia*
Chamaerops humilis   *Lygos monosperma*
Cistus crispus   *Phillyrea latifolia*
Erica arborea   *Pistacio terebinthus*

The anthyllis is not very plentiful here. It is a rather uncommon leguminous shrub growing to about 2m high with silvery leaves and heads of yellow flowers. Occasionally it is grown as a wall-shrub in Britain where it is called Jove's beard or silver bush. *Lygos monosperma* (= *Retama monosperma*) is another beautiful shrub, a broom with hanging branches and scented white flowers followed by single-seeded rounded pods. It is an important feature of the coastal flora in the western part of Andalucia (5.6). Between these shrubs grow *Scilla peruviana*, which seems to be fairly common in northern Tunisia. The plants here are relatively small compared with the large, lush specimens found in marshy places of southern Spain and Portugal. Other species here include: *Asphodelus aestivus* and *Urginea maritima*.

From the corniche-like road one has fine views over the bay of Tunis and, after passing through Korbous the road rises inland over a hill where there is interesting garigue. Here, in addition to the shrubs already mentioned, grow:

Cistus clusii,   *Globularia alypum*
Cistus monspeliensis   *Lavandula stoechas*
Erica vagans   *Rosmarinus officinalis*

Cistus clusii is not very widespread in Europe but common in North Africa. It has relatively small white flowers and rosemary-like foliage. It is the most drought-resistant and lime-loving member of the genus.

Scilla peruviana is widespread amongst the shrubs of the

garigue and there are scattered specimens of the brown bluebell *Dipcadi serotinum* - a species which is common in North Africa where it splits up into a number of different forms, including some with dull mauve flowers. Several orchids occur in the garigue, especially *Orchis papilionacea* and *Ophrys lutea* but careful searching may reveal *Ophrys scolopax, O. speculum* and *O. tenthredinifera*. Other plant species in this region are: *Alkanna tinctoria, Asteriscus maritimus, Calendula aegyptiaca, Mercurialis annua, Misoprates orontium, Tuberaria guttata* and *Scorzonera undulata*. The last of these is an especially attractive member of the genus; it is relatively short-growing, and has large (4-5m diameter) flower-heads of pale mauve with a darker centre and leaves with a distinctly wavy margin.

The road joins the main route 26 which leads towards the north-east end of Cap Bon. There are some planted pine woods with large bushes of the white-flowered retama and several interesting species of linaria which are somewhat a feature of Tunisia but difficult to name correctly. Pressing on in a northerly direction one comes to the village of El Haouaria which is noted for its annual falconry festival. To get to Djebel Abiod, which is the highest point on the headland north of here requires a permit during the bird migration period in April and May. This can be obtained from the Direction des Fôrets, 30 rue Alain Savary, Tunis. There is an information centre for wildlife on the road leading northwards out of El Haouaria to Ghar el Kebir, but mainly for birdlife. One might expect to find several very interesting plant species in this area but the land is heavily overgrazed and the chances of seeing most of them are remote. In 1896 Bonnet and Barratte noted the following:

Adonis annua *Lavatera trimestris*
Bellevalia romana *Ornithogalum arabicum*
Campanula dichotoma *Platycapnos spicata*
Dianthus hermaeensis *Scabiosa farinosa*
Iris juncea *Scilla villosa*

The campanula is a bristly annual with intense purple flowers and grows in sandy places and on rocks; it is also found in Sicily and Greece. The dianthus is said to be a Tunisian endemic; it is a form of *D. rupicola* - a woody-based perennial found in southern Italy. *Iris juncea* is a bulbous species with yellow flowers in April/May. It resembles *I. xiphium* and, like many other Tunisian plants such as *Lavatera trimestris* (13.11) is also found in Sicily. The scabiosa and the scilla are species special to Tunisia.

From the road 27 that leaves El Haouaria in a south easterly direction one can get to the coast at Ras el Drak and this is a site worth looking for plants. The road continues southwards to Hammamet through cultivated fields of market gardens, olives and vines. There are some planted woods of *Pinus pinea* and *P. halepensis* on the way. The undergrowth includes *Arbutus unedo*, *Cistus salvifolius* and *C.villosus*. In a few places the ground under the trees is carpeted with an asparagus which is probably *Asparagus tenuifolius* but may be *A. officinalis*. There are several orchids here, including *Ophrys scolopax* and *O. ciliata*.

25.6 Zaghouan and South-West of Hammamet: Leaving Hammamet by the main coast road No.1, which goes from Tunis down to Libya, one soon comes to the village of Bou Fichu. A secondary road leads westwards from here to the town of Zaghouan. It is typical of many others in the area in having a narrow strip of tar macadam down the centre which all traffic tries to occupy so that it is a matter of brinkmanship and nerve to remain on it when meeting an oncoming vehicle. A slightly better route (133), in the same direction, leaves the main road at Enfidaville (sometimes simply called Enfida) a few kilometers further south. Large fields of cereals and other crops flank the roads and to the west one has impressive views of the mountain Djebel Zaghouan 1295m - a jagged peak on the range of limestone hills that extend from Cap Bon to Tébessa. In places the roadside weeds here are especially lush and interesting and include a tall onopordum and two other rather uncommon composites *Silybum eburnum* and *Scolymus maculatus*. The first of these resembles the common milk-thistle but has golden tipped spines and larger flower heads. The scolymus is an attractive, annual, thistle-like plant with milky coloured leaf veins and fairly large yellow flower heads. In a few places there are substancial hedges of *Paliurus spina-christi* which seem to be popular nesting sites for thousands of noisy sparrows.

A few kilometres from the coast the road begins to rise and, especially from the Enfidaville turning, there are large stands of *Convolvolus tricolor* forming patches of blue on the landscape. They are accompanied by plants of *Moricandia arvensis* and *Urospermum dalechampii*. Rocky outcrops which cannot be cultivated contain numerous interesting plants and are especially worth examining. Here one may find:

Atractylis cancellata   *Limonium sp.*
Carrichtera annua   *Mercurialis annua*
Centaurea solstitialis   *Nonea vesicaria*
Convolvulus lineatus   *Pteranthus dichotomus*
Ecballium elaterium   *Scabiosa stellata*
Echinops sp.   *Urospermum dalechampii*

The atractylis is an interesting small, annual, centaurea-like plant with the bracts curled around the flower head in the form of a lantern; it is occasionally found in Spain and the south of France. The carrichtera and pteranthus are essentially North African species - the first an insignificant cruciferous plant with tiny yellow flowers and fruits which turn downwards and the second an extraordinary small annual species of the caryophyllaceae with dichotomously branching and insignificant flowers and, at first sight, may be mistaken for a zygophyllum. The first occurs rarely in Sicily and Sardinia but is found also in Cyprus (22.14) and the second is limited in Europe to the Maltese islands. The convolvulus is one of the finest of the genus with tufted silvery foliage and pale pink flowers. The scabious is an annual sometimes grown in gardens at home for its attractive seed heads sought after by flower-arranging enthusiasts.

In the few places where the ground remains reasonably moist one can see *Arum italicum*, a species of watercress and a large form of scopiurus which was identified as *S. muricatus var. subvillosus*. It has orange-coloured flowers and much larger leaves than the usual type. By far the most interesting plant associations around here are in the remnants of garigue which demonstrate how rich the flora was before cultivations became so widespread. In 1990 there was still a small example by the roadside shortly before entering Zaghouan town. The main shrubs are *Cistus clusii, C.monspeliensis* and some plants which seem to be hybrids between the two species. The common rosemary and heather abound and in the sandy soil a wealth of interesting species:

Anthyllis tetraphylla   *Scorzonera undulata*
Dipcadi serotinum   *Stachys oxymastrum*
Leopoldia gussonei   *Teucrium pseudo-chamaepitys*
Marrubium alysson   *Thymus hirtus*
Ophrys lutea   *Urginea undulata*
Scilla peruviana

The leopoldia (= *Muscari gussonei*) is like a small version of the common tassel hyacinth though the terminal head of sterile florets is very much reduced or absent. It is primarily a North African species with a very limited distribution but found sparingly in Sicily. The marrubium is a labiate with woolly stems and leaves and whorls of small, dark reddish-mauve flowers that have a rigid, prickly calyx. The uginea is smaller than the common sea squill with a bulb about the size of that of a garden hyacinth and with the top at ground level - not half exposed as with *U. maritima*. The most obvious characteristic is the rosette of rather narrow leaves that have a distinctly wavy margin, as the name implies. Its asphodel-like flowers are produced during the winter. Another small, or even smaller, species *U. fugax* is also found in Tunisia. Both of these rare species may occasionally be seen in the south of Spain, Sicily, Sardinia and Corsica (11.8).

It is quite possible to get to the top of Djebel Zaghouan by following the track to the radio station on the summit. One passes through typical dense garigue including the usual *Pistacia lentiscus*

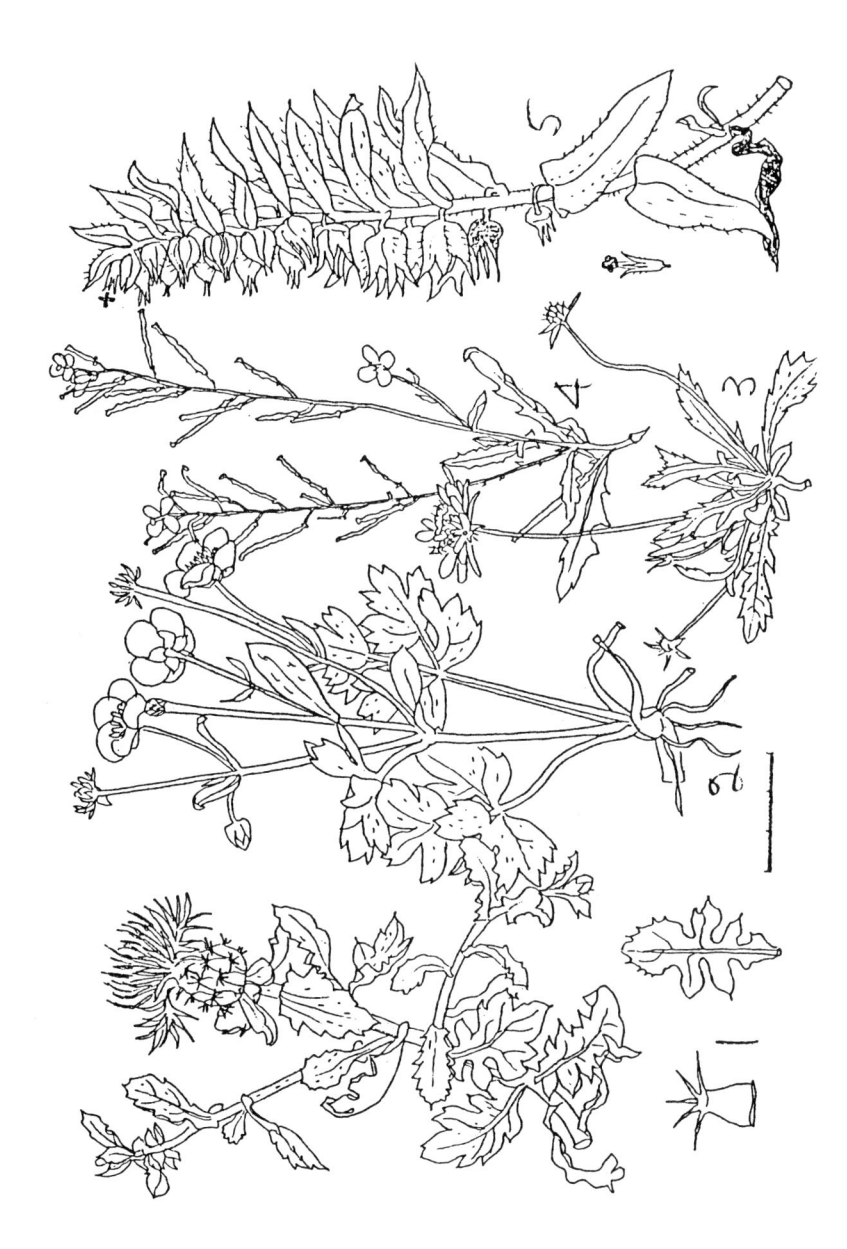

Tunisia 1. Centaurea sphaerocephala 2. Ranunculus rupestris 3. Scabiosa arenaria 4. Diplotaxis erucoides 5. Nonea vesicaria

and *Quercus coccifera* and with *Euphorbia dendroides* in places. Other interesting species include *Asphodeline lutea* and *Tulipa australis*. The area is especially interesting to bird watchers and is said to have nesting sites of several raptors including golden eagle, Bonelli's eagle, Egyptian vulture, black kite and peregrine falcon.

25.7 **Garat Ichkeul:** This is an important nature reserve which mainly comprises the lake (approximately 13x5km), which is partly tidal and varies in salinity according to the season, and its surroundings. To get to the entrance one should make for the village of Menzel Bourgiba and follow the signs for the road that leads in a south-west direction for some 8km to the entrance of the reserve. About 2km from the entrance there is an interesting museum with helpful attendants, if you speak French or Arabic, but the reserve is primarily for the bird life and practically no information can be obtained there on the plants. Nevertheless they do seem to know, and are proud of the fact, that *Euphorbia dendroides* grows by the roadside on the way to the museum. From a plant point of view the Djebel Ichkeul hill near the museum is well worth investigating even though it seems to be frequently grazed by herds of goats. It is mainly tree covered with an undergrowth of wild olive, pistacia and other shrubs. A number of species grow below the bushes and along the verges:

Allium triquetrum	*Narcissus serotinus*
Blackstonia perfoliata	*Narcissus tazetta*
Calendula aegyptica	*Ophrys lutea*
Cyclamen africanum	*Ranunculus paludosus*
Gladiolus byzantinus	*Ranunculus rupestris*
Lobularia maritima	*Selaginella dentata*

The presence of several of these, such as the allium, cyclamen, narcissus and selaginella, indicate that the rainfall is considerably higher than in the Hammamet and Zaghouan areas. The calendula is a beautiful species already mentioned (29.4). The cyclamen is a fairly large autumn-flowering species confined to North Africa; it is not hardy in most British gardens. The more widespread *C. repandum* and the autumn-flowering *C. hederifolium* also occur in Tunisia. *Ranunculus rupestris* is a neat large-flowered buttercup which can also be seen at El Torqual and around the base of the Sierra Nevada in southern Spain.

The hill, Djebel Ichkeul, sometimes referred to as "The Mountain", is a home for jackal, porcupine, mongoose and wild boar but the main attraction for visitors is the huge numbers of wildfowl and wading birds that frequent the lake including some rare species such as the purple gallinule. From the mountain one gets a superb grandstand view of the lake where one can sit with high power binoculars on a tripod and it is also possible to get down to the marshes in places.

25.8 The Khroumirie, Aïn Draham and Tabarca: The Kroumirie is the name given to the mountainous region in the extreme north-west of the country and is part of the Mejerda range which extends from Algeria. Aïn Draham is a small spa high up in the Khroumirie - the Arabic word 'aïn' means a spa. Tabarca is the nearest town of any size to the area and lies just north of Aïn Draham on the coast; it is a seaside resort and noted for the red Mediterranean coral that is collected near here though supplies are now running low.

It is quite possible to visit the Kroumirie from Hammamet or Tunis by car and to return in a day but it is worthwhile giving oneself longer time by staying overnight at Tabarka. Coming from the south one will notice that the vegetation is more lush and although most of the land is cultivated a number of interesting roadside plants may be seen. In places around the village of Nefza the attractive *Centaurea pullata* with pink, mauve or white cornflowers, is plentiful. Another attractive species here is *Verbascum laciniatum* that grows to 1.5m tall or more and has large (3.5-5cm diameter) yellow flowers with a dark blotch on the lower petal.

There are woods on either side of the road 7 from Nefza to Tabarka composed mainly of a mixture of cork oak and *Quercus mirbeckii* which is robust, vigorous, deciduous tree with large leaves that often remain on the branches until January. It is sometimes called *Q. canariensis* but is essentially a North African species known as the Algerian oak, though it also grows wild in Portugal and is perfectly hardy in Britain. The undergrowth here is very thick in places and approaches maquis density. It is comprised mainly of *Cistus salvifolius, Crataegus azarolus, Erica arborea, Phillyrea angustifolia* and *Pistacia lentiscus*. The ground beneath the main shrubs is extensively dug up by wild boar but a number of smaller plants may be seen, including:

Ambrosinia bassii *Polygala microphylla*
Cyclamen africanum *Rosa sempervirens*
Fedia cornucopiae *Ruscus hypophyllum*
Myrtus communis *Smilax aspera*
Osyris alba *Urginea maritima*

The ambrosinia is a tiny aroid of a monotypic genus. It has ovate leaves and small inflorescences bourne more or less at ground level during the winter. The male and female flowers are carried on separate appendages within each spathe. Although it is plentiful under the maquis here, it is rare in Europe and found only in Corsica, Sardinia, Sicily and parts of southern Italy (11.8). The polygala is a tall growing attractive milkwort with blue flowers. The rose is a rather tall climbing species with white flowers.

From Tarbaka the 17 road climbs to Aïn Draham, at first through cultivated and grazed fields and higher up through woods which are similar to those mentioned above. A hazard to drivers along here, and indeed lower down, is the number of boys that rush into the road to try and stop cars to sell their wares which vary according to the season and may be carvings, wild strawberries or coral. The trees are mainly *Quercus mirbeckii* and some plantations of cork oak with bracken and heather as undergrowth. The wild boar diggings are widespread and undergrowth is very limited. However, the following were noted on one visit:

Arbutus unedo *Clematis flammula*

Arum italicum *Cyclamen africanum*
Arisarum vulgare *Myrtus communis*
Bellis annua *Ranunculus ficaria*
Bellis sylvestris *Ranunculus rupestris*
Blackstonia perfoliata *Viola reichenbachiana*

The viola (= *V. sylvestris*) is the typical dog violet and Ranunculus rupestris the large-flowered buttercup already mentioned (25.7).

Several other interesting species have been recorded from around here and it may be worth taking the path to the Barrage de Ben Metir where one may have a chance of finding:

Anemone palmata *Narcissus tazetta*
Fritillaria oranensis *Orchis patens*
Iris juncea *Scilla aristidis*
Laurentia michelii *Scilla numidica*
Lilium candidum *Sternbergia lutea*
Narcissus serotinus

However, one will be lucky to see any of these. The fritillaria is sometimes classed as *F. lusitanica var. algeriensis*. The orchis is a rare species in southern Spain and northern Italy; it is primarily a plant of Algeria. The two scilla species are rare North African endemics. The laurentia is now usually referred to as *Solenopsis minuta*.

It is well worth while continuing on the road southwards from Aïn Draham to Jendouba and to call in at the extensive archeological site of Bulla Regia which was a Numidian (early Berber) settlement taken over by the Romans. It was once in the centre of the wheat growing region in ancient times. The climate has changed since then and the whole area between here and Hammamet is dry and rather unproductive and overgrazed. Two interesting plants to be found here are *Andrachne telphoides* and *Caralluma europaea*. The first is an African member of the Euphorbiaceae with spreading branches having small pointed, ovate leaves and with separate male and female flowers on the same plant. The caralluma is a curious succulent member of the Asclepiadaceae - essentially an African plant but found in a few parts of southern Europe, notably the Cabo de Gata in Southern Spain (7.8).

25.9 South of Hammamet - Sousse, Sfax and the desert: From the main road No.1 south between Sousse and Sfax there are seemingly endless olive orchards and rather few wild plants to be seen. South of Sfax one comes to Gabès and from here route 16 leads westwards to the very large salt lake of Chott el Djerid (25.2). Scattered plants seen on the approach to the Chott will probably include *Zygophyllum album* which is one of a number of North African species of the genus and similar to the Syrian bean-caper *Z. fabago* which is not uncommon in dry areas all around the Mediterranean. There are also several species of *Fagonia*, also members of the Zygophyllaceae, found in the region. Taking the road south from the Chott to Douze brings one to the Grand Erg Oriental of the Sahara desert. At the oasis here, amongst the date palms there are small crops of henna *Lawsonia inermis*, a shrubby member of the loosestrife family Lythraceae with scented red flowers. The dye from the powdered, dried leaves is much used by women in Tunisia to colour their fingernails and hands and was once very popular in Europe to dye hair reddish brown.

South of Gabès, running roughly north to south, are the Monts des Ksour and to the east the stony Tunisian desert. Plants are sparse here and will not be dealt with in detail as most of them can hardly be described as typically Mediterranean. Near to the Berber underground houses at Matmata, which is a popular tourist site, one may encounter:

Astragalus caprinus *Moricandia arvensis*
Carrichtera annua *Muricaria prostrata*
Convolvulus supinus *Perganum harmala*
Erodium glaucophyllum *Scabiosa arenaria*
Matthiola fruticulosa

The astragalus is an attractive species with heads of rather large, yellow flowers held out horizontally. The convolvulus is a prostrate plant with leaves covered with soft hairs and large, pale pink flowers. The matthiola is a local form of the species and so is also the moricandia which seems to grow perennially and to carry sharp, thorny branches. The perganum is a common plant of the hottest and dryest regions of the Mediterranean and has been used to produce the dye 'Turkey red'.

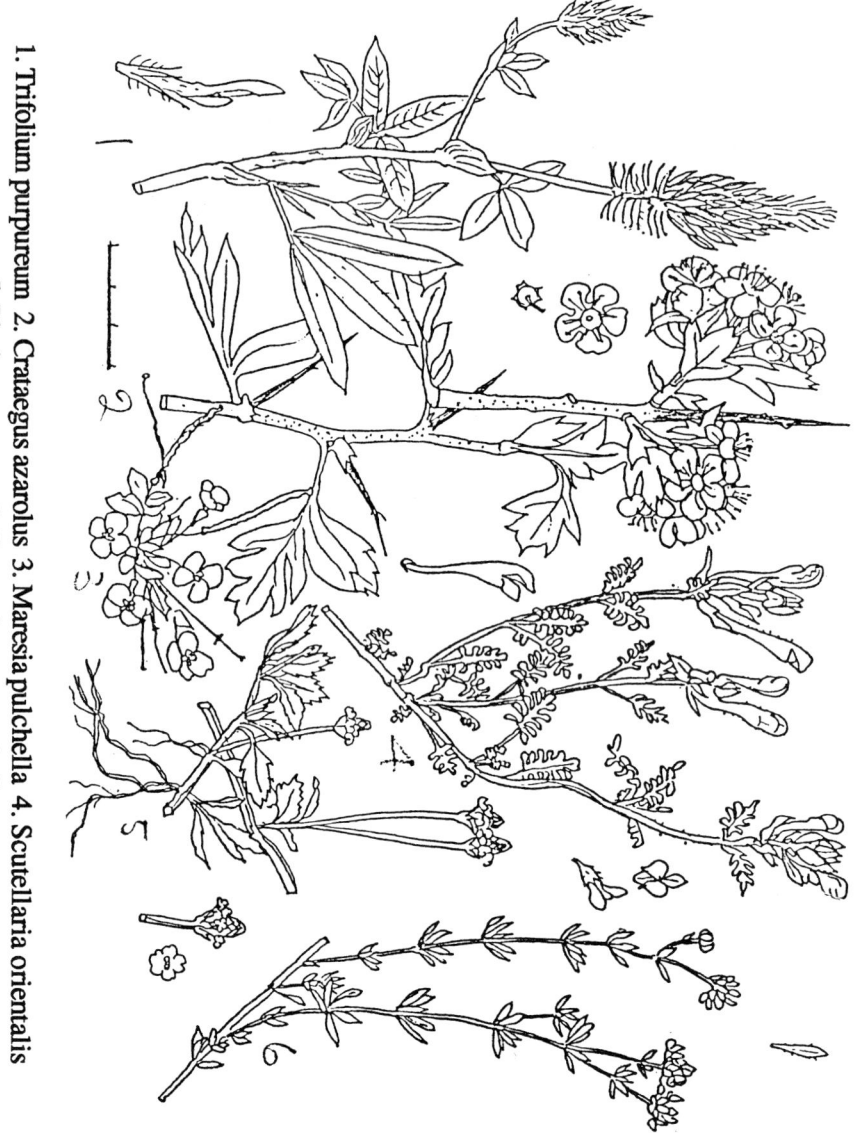

1. Trifolium purpureum 2. Crataegus azarolus 3. Maresia pulchella 4. Scutellaria orientalis 5. Phyla canescens 6. Dorycnium pentaphyllum

26. CONSERVATION

26.1 The human population of the world, including that of the Mediterranean region, is increasing at an alarming rate and exerting considerable pressure on non-domesticated plants and animals. Many species are now at risk of a great reduction in their numbers and even their eventual annihilation. Most people nowadays live in cities (Athens, for example, houses about fifty percent of the population of Greece) and whereas they appreciate the problem facing the larger warm-blooded animals such as the panda, elephant and whales, they have relatively little contact with plants and smaller animals and do not feel the same need to preserve them. To understand the importance of conservation one must first be able to recognise those species under threat and here is where readers of this book have a large part to play. Botanists, whether they are professionals or amateur enthusiasts are involved with the identification of plant species and their distribution. We are the vanguard for conservation - we have an important moral duty to help preserve what we enjoy today for future generations.

The whole notion of 'conservation' is complex. Man has been interfering with the environment for thousands of years by cutting down forests, cultivating and draining land yet most wild plants have adapted to the new situations with some species gaining ascendancy over others. Even without any interference by man the situation does not remain static; the climate changes slowly and diseases and wild animals play their part. The problem is often not so much about saving a single species but rather an entire environment. Then the question arises as to what should be the boundaries of the area to be conserved and how any decisions can be enforced. Environmental matters, in the concept we are dealing with here, are often given a low ranking in political decisions and any legal enforcement decided upon is frequently overlooked when financial profits are at stake.

26.2 **Seashores:** The Mediterranean littoral is, perhaps, the area most under threat from tourism. Hotels and holidaymakers monopolise the sea shore, sand dunes and other areas near the coast and destroy much of the wild life that lives there. Not only do plants get trampled on; in some areas beach-cleaning machinery tears up and kills many of those that escape other forms of attention. Hotel managements build gardens around their premises and introduce foreign plants which may escape and compete with the inherent vegetation. One particular culprit here is the beautiful *Carpobrotus acinaciformis* from South Africa which is now so widespread and almost classed as a native of the Mediterranean. It is favoured because it consolidates dunes but it also effectively destroys much of the wild vegetation in its path. Sewage effluent from holiday resorts often damages the marine flora including the interesting flowering plant *Posedonia oceanica* which gives rise to the brown balls of felt on the sea shore resulting from the crushing and rolling of its leaves by the surf.

Fortunately, many of the plants which live in the sand near the Mediterranean shore line have a very wide distribution. *Cakile maritima, Calystegia soldanella, Crithmum maritimum, Eryngium maritimum, Euphorbia parialis, Glaucium flavum* are also found widely spread on the shores of northern Europe. Others, such as *Anagallis monelii* (5.6), *Centaurea aegiolophila* (22.5) *Nananthera pusilla* (11.10) and *Silene succulenta* (11.10) are less widely distributed and may merit special protection. Yet others like *Pancratium maritimum* are widespread around the Mediterranean shores but not elsewhere and they produce their beautiful flowers during the summer holiday season and are at risk of being culled in quantity by visitors and hotel staff who wish to enliven their dining room tables. This greatly reduces the amount of seed production.

The situation is exacerbated by the fact that sandy beaches are not very widespread round Mediterranean shores, probably because of the low daily tides, and are thus much in demand for tourist sites. The Atlantic coastline of Spain west of Gibraltar and that of southern Portugal has fine sandy stretches and carries an exceptional coastal and dune flora but is also being destroyed by building schemes. However, rocky shorelines are a relatively safe haven for plants, some of which are of special interest.

26.3 **Wet places:** Ponds, marshes and river estuaries are important habitats for many special plants and birds. Unfortunately, they are usually disliked by local inhabitants for they are potential breeding places for mosquitoes and the water associated with them can usually be 'better used' to irrigate cultivated land. Some plants such as *Orchis laxiflora* depend on wet ground conditions and have been severely depleted in recent years through drainage.

Artificially created resevoirs do not usually have shore lines which provide a habitat for plants that live in marshy areas and the surrounding land is often planted with foreign eucalyptus trees that support a very limited natural flora - though they are said to deter mosquitoes.

26.4 **Forests:** Many of the natural forests of the Mediteranean were cleared several hundred or thousand years ago to provide timber, charcoal and grazing land. Fortunately, there are a few precious protected relics such as that of La Sainte Baume (10.11) in the south of France. Old, self-regenerated deciduous forests still remain in some areas such as the Foresta Umbra in Gargano (12.6), smaller woods north of the Col de Vence in south east France (10.14) and the woodland of the north slopes of the Nebrodi Mountains in Sicily (13.10). These are refuges for rare ground plants that need protecting. Important wild coniferous forests also are still to be found in parts of Greece, Turkey and Corsica; again they provide refuge for uncommon species in the undergrowth. Amongst them the natural cedar forests of Turkey are particularly interesting. However, some special conifers, such as *Abies pinsapo* in Spain, and *Abies nebrodensis* in Sicily are under great pressure and have, at a very late stage, been given a measure of protection.

Most present day Mediterranean forests have been planted - chiefly with conifers and sometimes with native species under stress as *Cedrus libani brevifolia* (22.11) in Cyprus. Their undergrowth here is generally less varied than that of deciduous trees but they do provide a home for several orchid species and plants of other families such as paeonies. Non-native trees, such as eucalyptus, are generally less favourable as a haven for other plants but they occasionally harbour interesting species like *Scilla monophyllos* in Portugal (4.9).

Fire is a very considerable threat to woodland, notably in the south of France and in Corsica. Conflagrations may occasionally be caused by lightening and as such they are part of the natural process of regeneration. Indeed, they may sometimes play an important part in the well-being of the natural vegetation pattern through the temporary destruction of geriatric trees and shrubs in places and the opening up of the canopy for new tree seedlings and for bulbous and tuberous rooted species to flourish. Unfortunately many fires are started deliberately as arson, either to open up land for farming or building or simply to give pleasure in causing destruction. Much effort is now being employed to prevent this from happening and plant enthusiasts, like all other visitors, have a grave responsibility to take all precautions to prevent fires.

26.5 Crop Cultivation: The old, traditional systems of cultivation gave scope for a host of colourful and interesting annual 'weeds' to thrive. Many such as the poppy, cornflower, corn cockle, corn marigold and larkspur have added so much to our gardens. Indeed, cereal crops used to be a kind of international forum for such plants; their seeds having been spread with the sowing of the crop. Some may well have originated from the earliest regions of cultivation like Mesopotamia and what is now eastern Turkey. Bulbous and tuberous species such as tulips, gladiolus, *Anemone coronaria, Bellevalia romana*, have also flourished in cereal crops by producing their storage organs below the depth of ploughing that could be achieved by oxen, mules and horses. Modern techniques for agricultural 'improvement' with more efficient seed-cleaning machinery, deeper mechanised ploughing and, above all, selective herbicide spraying, have done much to reduce or eliminate these species which are often now confined to the edges of fields and waste places. Examples of regions badly affected in this way are to be found in parts of mainland Greece and the Mesaoria of Cyprus. Though similar damage was done in other parts of the Mediterranean several years ago.

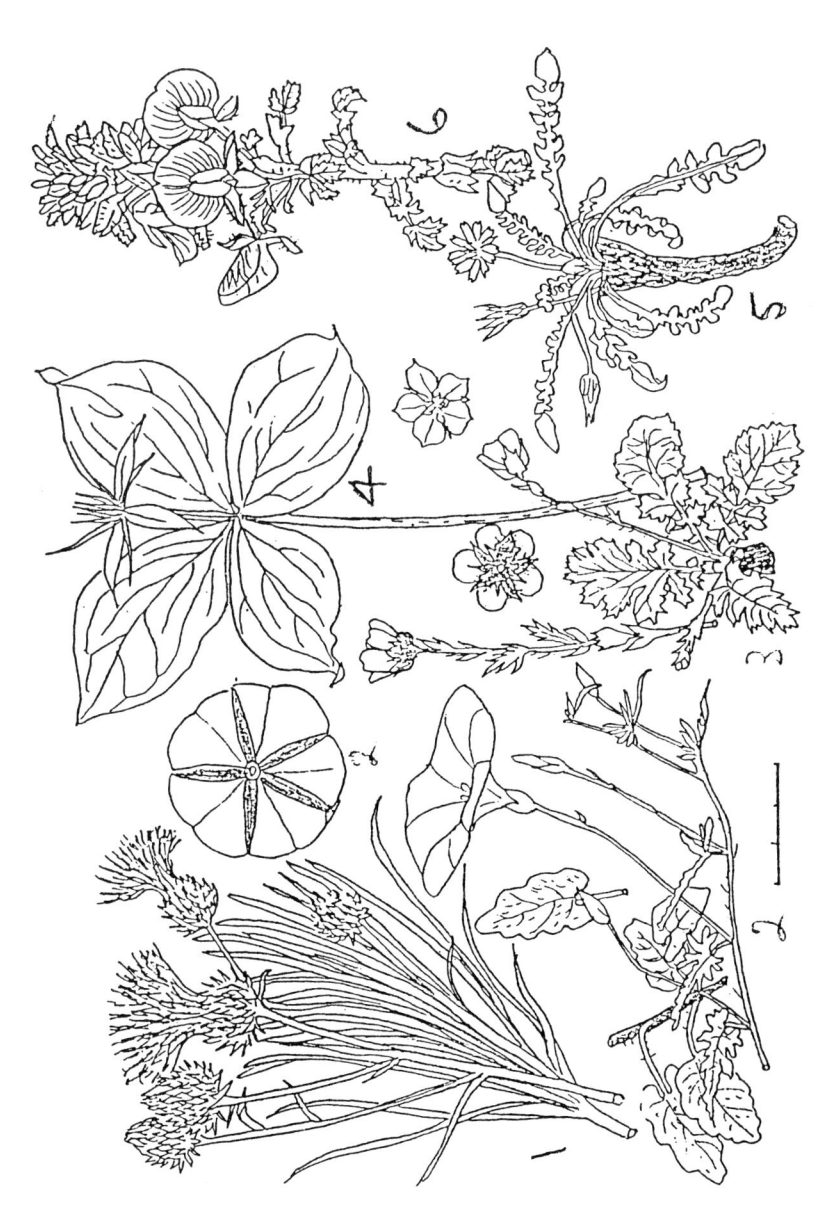

1. Ptilostemon chamaepeuce 2. Convolvulus althaeoides 3. Geum sylvaticum 4. Paris quadrifolia 5. Taraxacum aphrogenes 6. Ononis speciosa

1. Crepis incana 2. Urginea undulata 3. Valeriana tuberosa 4. Cistus clusii 5. Coronilla emerus 6. Centaurea solstitialis

Another practice which has reduced these 'weeds' is the introduction of new crops with different cultivation requirements. Irrigation, in particular, not only changes the crop environment, it often requires the sinking of wells that lowers the water table and adversely affects surrounding environments. Yet another practice is the widespread use of sheet plastic to produce early crops such as the area south of Almeria nick-named the 'Costa plastica' and that in the south of Crete in parts of the Messara plain.

It is to be hoped that, with the economic contraction of European agriculture, measures now being adopted to maintain traditional agricultural methods will help towards the survival of this interesting group of plant species.

26.6 Hills and Mountains: These extensive regions suffer most from excessive grazing by domesticated animals, especially sheep and goats. In some places such as the Cap Bon area of Tunisia (25.5), the highest parts of Mount Ida in Crete (20.10) and many parts of Turkey, grazing has practically destroyed all vegetation to produce large barren areas. Increasing human population is accompanied by an increasing demand for livestock and meat and it is important that some way of controlling this devastation should be found. This is not to say that domestic sheep and goats should be completely excluded for in limited numbers they help to maintain healthy garigue which is an important plant association for many rare and interesting species. Complete elimination of sheep would also remove a source of food for larger raptors such as eagles and vultures that are such an interesting and evocative element of the more remote mountains.

Visitors do not go to the mountains and hills in such numbers as those who congregate in the coastal areas but they can impose a threat to the natural vegetation. Excessive trampling in the mountains can certainly damage the environment; areas used for skiiing and the newfangled craze for mountain biking being under the greatest threat. Visitors in the summer may pick bunches of flowers which can greatly reduce the seed set and some dig up plants for their gardens. Dedicated plant enthusiasts will be well aware of their responsibilities in this respect.

Game shooting in the hills is, on the whole, favourable to plant conservation since the 'sporting fraternity' is keen to see that the plant communities that make up their hunting grounds are preserved and their demands usually carry more weight than those of plant conservationists. In a few areas where small birds are shot from static hides as in Malta, the activity is detrimental to plant life, because herbicides are often used to clear an area of vegetation around the selected area. The culling of plentiful larger game may be a definite advantage. Excessive numbers of deer and wild boar can lead to the destruction and even complete elimination of some plant species. Wild boar are especially destructive, digging up tubers and bulbous-rooted plants. Indeed, in some areas such as the Khroumerie district of Tunisia (25.8) they cause so much damage that it appears that the ground has been ploughed.

26.7 National Parks and Nature Reserves: The process of setting aside and conservation of areas of natural beauty and special interest for the flora and fauna is still at an early stage. Though some progress has been made since Polunin (1973) described the position in his 'Flowers of South-West Europe' it has hardly begun in the eastern Mediterranean.

The main intention of national parks is to conserve the environment and in some cases to control the hunting of game and this helps to maintain plant communities. Such parks usually provide ready access. Footpaths are often maintained and signposted and sometimes information on the flora is made available. Unfortunately measures to control building, farming and forestry in these parks are all too often overlooked and finances are usually too small to ensure adequate supervision.

Nature reserves play a more important part in the conservation of species for they are usually better documented and supervised and sometimes embrace a biological research unit. The main ones in the Mediterranean area are:

Reserve Marismas, Coto de Doñana - Spain
Parc naturel régional de Camargue - France

Parc naturel régional de la Corse - Corsica
Garat Ichkeul Nature Park - Tunisia

France leads the way in conservation of its natural flora and in addition to nature parks it has a series of 'Conservatoires Botaniques Nationaux' started in 1990 and now number five. The one relevant to Mediterranean France is the 'Conservatoire Botanique National de Porquerolles' at Le Hameau Agricole - Ile de Porquerolles. This is more than the usual reserve since it holds a bank of seeds and planting material of endangered species. Also, under the name of 'Conservatoires du Littoral', starting in 1976, some 260 sites have already been purchased and set aside for conservation, including La Camargue and three in Corsica. Somewhat similar conservation areas occur elsewhere, such as that for the preservation of *Abies nebrodensis* in the Madone mountains of Sicily, The Ghadira reserve of Malta and a number of sites in Israel. It is to be hoped that similar preservation areas may be introduced in many other sites such as the The Cabo de Gata of Spain, The Akamas peninsula of Cyprus and the mountains of Crete.

26.8 **Legislation:** Throughout this book the term 'plant enthusiast' has been used rather than 'plant hunter', for the latter connotes an intention to find plants and remove some of them for cultivation elsewhere as a kind of trophy. Whilst little harm may be done by collecting limited quantities of seed of the more common wild species or removing an occasional plant, it is nevertheless hoped that such activity will be kept to a minimum. Readers should also be aware that it is against the law in France to remove plants, pick flowers or collect seed of certain species except for approved scientific purposes. Some four hundred species are totally protected and thirty-six partially protected in France. Details may be obtained from Ministère de l'Environnement, Direction de la Protection de la Nature, 14 boulevard du Général Leclerc, 9254 Neuilly-sur-Seine, France or Associasion Française pour la Conservation des Espèces Végétales, Mairie de Mulhouse, 2 rue Pierre-Curie, 68200 Mulhouse, France.

BIBLIOGRAPHY

ALIBERTIS, C., ALIBERTIS, A. (1989) *The Wild Orchids of Crete*. Alibertis, Heraklion
BECKETT, E. (1988) *Wild Flowers of Majorca, Minorca and Ibiza*. A.A. Balkema, Amsterdam
BLAMEY, M. and GREY-WILSON, C. (1993) *Mediterranean Wild Flowers*. HarperCollins, London
BLANCHARD, J. W., (1990) *Narcissus, A Guide to Wild Daffodils*. Alpine Garden Society, Woking
BONAFE, F. (1977-80) *Flora de Mallorca*. Editorial Moll, Palma de Mallorca
BONNER, A. (1985) *Plants of the Balearic Islands*. Editorial Moll, Palma de Mallorca
BONNET, E., et BARATTE, G. (1896) *Catalogue Raisonné des Plantes Vasculaires de la Tunisie*. Imprimerie National, Paris
BOWLES, E. A. (1952) *Crocus and Colchicum* (Revised Edition). The Bodley Head, London
BRIQUET, J. (1910-55) *Prodrome de la Flore Corse*. Jardin Botanique, Geneva
BRUN, B., CONRAD, M., GAMISANS, J. (1975) *La Nature en France, Corse*. Horizons de France, Strasbourg
BUTTLER, K. P. (1991) *Field Guide to Orchids of Britain and Europe*. The Crowood Press, Swindon
CANDARGY, P.C. (1898) *Flore de l'Isle de Lesbos*. Bull. Soc. Bot. de France 45
CARLSTROM, A. (1987) *A Survey of the Flora and Phyteography of Rhodes, Simi, Tilos*. University of Lund
CONRAD, M. (1976) *Parc Naturel Regional de la Corse*. Production A.R.P.E.G.E., Clermont-Ferrand
COSTE, H. (1900-1906) *Flore Descriptive et Illustrée de la France, de la Corse et des Contrées limitrophes*. Paris
DAVIS, P., DAVIS, J. & HUXLEY, A. (1983) *Wild Orchids of Britain and Europe*. Chatto & Windus, London
DAVIS, P.H. (1965-88) *Flora of Turkey*. University Press, Edinburgh

DAVIS, P.H. (1949) *A Journey in South-west Anatolia I*. J. Roy. Hort. Soc. 74 part 3

DAVIS, P.H. (1949) *A Journey in South-west Anatolia II*. J. Roy. Hort. Soc. 74 part 4

DAVIS, P.H. (1951) *The Taurus Revisited*. J. Roy. Hort. Soc. 76 part 2

DAVIS, P.H. (1957) *The Spring Flora of the Turkish Riviera*. J. Roy. Hort. Soc. 82 part 4

DELFORGE, P. (1995) *Orchids of Britain and Europe*. Harper Collins, London

DEMIRI, M. (1981) *Flora Ekskursioniste e Shqiperise*. Shtepia Botuese e Librit Shkollor, Tiranë

FARRER, R. (1928) *The English Rock-Garden* (Fourth impression) T.C. & E.C. Jack, London

GAMISANS, J. (1985) *Catalogue des Plantes vasculaires de la Corse*. Parc Naturel Régional de la Corse, Ajaccio

GAMISONS J. (1991) *La Vegetation de la Corse*. Jardin Botanique, Geneva

GREY-WILSON, C. & BLAMEY, M. (1979) *The Alpine Flowers of Britain and Europe*. Collins, London

GEORGIADES, C. (1985-7) *Flowers of Cyprus - Plants of Medicine*. Georgiades, Nicosia

GEORGIADES, C. (1989) *Nature of Cyprus, Environment - Flora - Fauna*. Georgiades, Nicosia

GEORGIOU, O. (1988) *The Flora of Kerkira*. Willdenova 17

GIRERD, B. (1991) *La Flore du Département de Vaucluse*. Editions Alain Barthelemy, Avignon

GÖLZ, P. and REINHARD, H.R. (1984) *Die Orchideenflora Albaniens*. Mitt. BL. Arbeitskr. Heim. Orch., Baden-Wurtt.

HASLAM, S.M., SELL, P.D. & WOLSELEY, P.A. (1977) *A Flora of the Maltese Islands*. Malta University Press, Msida

HOLMBOE, J. (1914) *Studies of the Vegetation of Cyprus*. Bergen

HUXLEY, A., TAYLOR, W. (1977) *Flowers of Greece and the Aegean*. Chatto & Windus, London

KNOCHE, H. (1921-23) *Flora Balearica*. Montpelier (Reprinted in Holland)

LORENZ, R., GEMBARDT, C. (1987) *Die Orchideenflora des Gargano*. Mitt. Bl. Arbeitskr. Heim. Orch., Baden-Wurtt.
MAIRE, R. (1952-80) *Flore de l'Afrique du Nord*. Lechevalier, Paris
MARES, P., VIRGINEIX, G. (1880) *Catalogue Raisonné des Plantes vasculaires des Isles Baleares*. Paris
MEIKLE, R.D. (1977-85) *Flora of Cyprus*. Bentham-Moxon Trust, Kew
MURBECK, S.W. (18897-1900) *Flore de la Tunisie*. Collection of papers, Lund University
PHITOS, D., DAMBOLDT, J. (1985) *Die Flora der Insel Kephallinia*. Bot. Chron. 5(1-2)
PIGNATTI, S. (1982) *Flora d'Italia*. Edagricole, Bologna
PLITMANN, U., HEYN, C., DANIN, A. & SHMIDA A. (1982) *Pictorial Flora of Israel*. The Hebrew University, Jerusalem
POLUNIN, O. (1980) *Flowers of Greece and the Balkans*. Oxford University Press, Oxford
POLUNIN, O., HUXLEY, A. (1965) *Flowers of the Mediterranean*. Chatto & Windus, London
POLUNIN, O., SMYTHIES, B.E. (1973) *Flowers of South-West Europe*. Oxford University Press, Oxford
POTIER-ALAPETITE, G. (1979) *Flore de la Tunisie*. Paris (Incomplete)
PRIETO, P. (1975) *Flora de la Tundra de Sierra Nevada*. University of Grenada
RIX, M., PHILLIPS, R. (1981) *The Bulb Book*. Pan Books Ltd., London
RODRIGUEZ, L.G., NOGUEIRA, L.C., MIRALLES, J.M., NOGUEIRA, H.C. (1982) *Cabo de Gata, Guia de la Naturaleza*. Editorial Everest, Madrid
SAUNDERS, D.E. (1975) *Cyclamen, A Gardeners' Guide to the Genus*. Alpine Garden Society, Woking
SCHEMBRI, P.J., SULTANA, J. (1989) *Red Data Book for thre Maltese Islands*. Ministry of Education, Beltissebh
SCHÖNFELDER, I, SCHÖNFELDER, P. (1984) *Wild Flowers*

of the Mediterranean. Collins, London
SFIKAS, G. (1987) *Wild Flowers of Crete.* Efstathiadis, Athens
SFIKAS, G. (1990) *Wild Flowers of Cyprus.* Efstathiadis, Athens
STOCKEN, C.M. (1969) *Andalusian Flowers and Countryside.* Stocken, Devon
STRID, A. (1986) *Wild Flowers of Mount Olympus.* Goulandris Natural History Museum, Kifissia
STRID, A. (1986) *Mountain Flora of Greece, Vol 1.* Cambridge University Press
STRID, A., KIT TAN (1991) *Mountian Flora of Greece, Vol 2.* Edinburgh University Press
THOMPSON, H.S. (1914) *Flowering Plants of the Riviera.* Longmans Green & Co., London
TORNABENE, F. (1887) *Flora Sicula.* Catania
TORNABENE, F. (1889-92) *Flora Aetna.* Catania
TURLAND, N.G., CHILTERN, L. PRESS, G.R. (1993) *Flora of the Cretan Area.* HMSO, London
TUTIN, T.G. et. al. (1964-80) *Flora Europaea.* Cambridge University Press
VALDES, B., TALAVERA, S., FERNANDEZ-GALIANO, E. (1987) *Flora Vascular de Andalucía Occidental.* Ketres, Barcelona
ZOHARY, M. (1966-78) *Flora Palaestina.* Israel Academy of Sciences, Jerusalem

ALPHABETICAL LIST OF PLANT NAMES

Acer obtusifolium 360
Aceras anthropophorum 222
Achyranthes aspera 207
Adonis aleppica 430
Aethionema saxatile 314
Ajuga chamaepitys palaestina 348
Alkanna graeca 300
Alkanna tinctoria 169
Allium nigrum 190
Allium triquetrum 170
Alnus orientalis 338
Althea officinalis 138
Anacamptys pyramidalis 229
Anagallis monelli 43
Anchusa aegyptica 348
Anchusa variegata 314
Anemone apennina 202
Anemone coronaria 332
Anemone hortensis 207
Anemone palmata 32
Anemone pavonina 329
Anthemis cretica 374
Anthemis montana montana 211
Anthemis palaestina 362
Anthemis rosea var. carnea 402
Anthericum liliago 124
Anthyllis cytisoides 99
Anthyllis hermanniae 158
Anthyllis tetraphylla 94
Antirrhinum barrelieri 67
Antirrhinum granaticum 48
Antirrhinum hispanicum 85
Antirrhinum latifolium 134

Antirrhinum majus linkianum 43
Antirrhinum tortuosum 216
Arabis alpina 301
Arabis cypria 374
Arabis rosea 207
Arisarum vulgare 86
Aristolochia clematitis 133
Aristolochia maurorum 405
Aristolochia pallida 160
Aristolochia pistolochia 124
Aristolochia poluninii 405
Aristolochia rotunda 151, 244
Armeria pungens 17
Arum creticum 330
Asclepias curassavica 111
Aster sedifolius 66
Asteriscus maritimus 59
Astragalus caprinus 450
Astragalus lusitanicus 37
Astragalus massiliensis 17
Asyneuma limonifolium 263
Aubretia thracica 251
Ballis longifolia 337
Ballota pseudodictamnus 385
Barlia robertiana 222
Bellardia trixago 95
Bellevalia dubia 172
Bellevalia hackelii 14
Bellevalia nivalis 361
Bellevalia romana 172
Bellevalia trifoliata 360
Bellis annua 14, 151
Bellis margaritaefolia 182

Biarum carratracense 63
Biscutella vincentina 13
Blackstonia perfoliata 95
Brassica balearica 96
Brassica hilaronis 360
Buglossoides purpurocaerulea 144
Calendula aegyptica 454
Calendula arvensis 221
Calendula suffruticosa 17
Campanula drabifolia 313
Campanula hawkinsiana 251
Campanula lingulata 245
Campanula lusitanica 37
Campanula lyrata 385
Campanula rupicola 313
Campanula sparsa 230
Campanula spatulata 314
Campanula topaliana 287
Capsella grandiflora 237
Carduncellus pinnatus 197
Centaurea aegialophila 360
Centaurea aegyptica 454
Centaurea boiseri spaehi 80
Centaurea pindicola 246
Centaurea polyacantha 27
Centaurea pullata 463
Centaurea salonitana 270
Centaurea solstitialis 474
Centaurea sphaerocephala 102, 461
Centaurea triumfetti 121
Centaurium pulchellum 415
Centranthus junceus 246
Cephalanthera longifolia 158
Cerinthe major 172
Cerinthe minor 233

Cerinthe palaestina 438
Cerinthe retorta 244
Chamaecytisus supinus 251
Cionura erecta 425
Cistus albidus 96
Cistus clusii 474
Cistus monspeliensis 96
Cistus palhinhae 13
Cistus parviflorus 361
Cistus salvifolius 207
Clematis cirrhosa 86
Cneorum tricoccon 96
Colchicum lusitanicum 55
Colchicum pusillum 337
Coleostephus myconis 215
Comperia comperiana 229
Convolvulus althaeoides 473, Front
Convolvulus cantabrica 216
Convolvulus lineatus 454
Convolvulus supinus 448
Convolvulus tricolor 211
Conyza bonariensis 221
Coronilla emerus 474
Coronilla juncea 67
Corydalis solida 134
Cotula coronopifolia 14
Crataegus azarolus 468
Crepis incana 474
Crepis rubra 169
Crinthmum maritimum 332
Crocus clusii 43
Crocus corsicus 151
Crocus laevigatus 337
Crocus sieberi tricolor 302
Crocus veneris 338
Cryptostemma calendula 32

Cyclamen balearicum 86
Cyclamen graecum 330
Cyclamen hederifolium 158
Cyclamen repandum 157
Cyclamen repandum var.
  rhodense 194
Cymbalaria pubescens 182
Dactylorhiza romana 207
Daphne gnidium 38
Daphne laureola 134
Daphne oleoides 251
Daphne sericea 170
Dianthus crinitus 385
Dianthus haematocalyx 246
Dianthus lusitanus 63
Dianthus sylvestris 138
Digitalis viridiflora 277
Dipcadi serotinum 450
Diplotaxis erucoides 221
Dittrichia viscosa 48
Doronicum orientale 160
Dorycnium hirsutum 99, 263
Dorycnium pentaphyllum 468
Dorystoechas hastata 391
Draba azoides 127
Draba parnassica 304
Erica arborea 18
Erica lusitanica 18
Erica multiflora 221
Erodium gruinum 371
Erodium petraeum valentinum 80
Eryngium maritimum 55
Erysimum bonnanianum 182
Erysimum cephalonicum 322
Erysimum sylvestre 237
Euphorbia biumbellata 103

Euphorbia characias 99
Euphorbia pithyusa 152
Fagonia cretica 86
Fedia cornucopiae 14
Fritillaria carica 402
Fritillaria involucrata 124
Fritillaria lusitanica 23
Fumana ericoides 95
Fumana procumbens 68
Gagea fistulosa 304
Gagea graeca 371
Galanthus reginae-olgae 304
Galium canum ovatum 385
Genista corsica 152
Genista lucida 94
Genista umbellata 63
Gennaria diphylla 27
Gentiana cruciata 277
Geranium macrorrhizum 245
Geranium reflexum 277
Geranium sanguineum 170
Geranium versicolor 274
Geum coccineum 274
Geum sylvaticum 473
Glaucium grandiflorum 447
Globularia alypum 337
Globularia punctata 103
Gomphocarpus fruticosus 38
Gynandriris sisyrinchium 172
Halimium commutatum 13
Halimium halimifolium 152
Haplophyllum coronatum 244
Helianthemum almeriense 85
Helianthemum apenninum 127
Helianthemum cinereum 68
Heliathemum hymettum 301

Heliathemum obtusifolium 362
Helanthemum oelandicum 120
Helicodiceros muscivorus 94
Heliotropium europaeum 138
Heliotropium hirsutissimum 332
Hepatica nobilis 120
Hermodactylis tuberosus 180
Hippocrepis balearica 95
Hippocrepis unisiliquosa 362
Hippocrepis valentina 79
Hyacinthoides hispanica 23
Hyacinthoides italica 120
Hyocyamus aureus 374
Hypercoum procumbens 454
Hypericum aegypticum 221
Hypericum balearicum 95
Hypericum rumelicum 230
Iberis amara 133
Iberis candolleana 112, 121
Iberis crenata 37
Iberis sempervirens 338
Inopsidium acaule 32
Inula crithmoides 59
Iris haynei 392
Iris lutescens 103
Iris planifolia 180
Iris pseudopumila 112, 180
Iris pumila attica 313
Iris sintensii 230
Iris unguicularis 301
Jasminum fruticans 27
Juniperus oxycedrus macrocarpa 43
Kickxia commutata 48
Knautea drymeia 270
Lactuca perennis 124
Lamium bifidum 160
Lamium flexuosum 182
Lamium garganicum 144
Lamium moschatum 348
Lathyrus articulatus 197
Lathyrus digitatus 282
Lathyrus ochrus 18
Launea resedifolia 47
Lavandula dentata 95
Lavandula lanata 55
Lavandula multifida 68
Lavandula stoechas 18
Lavatera maritima 67, 86
Lavatera oblongifolia 85
Lavatera olbia 179
Lavatera trimestris 179
Legousia pentagonia 402
Legousia speculum-veneris 329
Leopoldia comosa 94
Leopoldia weissii 415
Leptoplax emarginata 251
Leucanthemum coronopifolium 79
Leucoium nicaense 120
Limodorum abortivum 329
Limoniastrum monopetalum 338
Limonium insigne 70
Limonium ovalifolium 47
Linaria algarviana 13
Linaria angustissima 233
Linaria chalenpensis 282
Linaria genistifolia 244
Linaria pelisseriana 322
Linaria peloponnesiaca 263
Linaria reflexa var. castelli 216
Linaria supina 120
Linaria triphylla 96

Linum bienne 170
Linum flavum 237
Linum maritimum 38
Linum narbonese 120
Linum pubescens 447
Linum punctatum 182
Linum thracicum 245
Lithodora diffusum lusitanicum 37
Lithodora fruticosa 85
Lithodora hispidula 374
Lobelia urens 55
Lobularia maritima 17
Lonicera implexa 79
Lotus ornithopodioides 152
Lycium intricatum 68
Lysimachia atropurpurea 47
Malcomia flexuosa 314
Malcomia littorea 27
Malva cretica 301
Malva moschata 144
Mantisalca salmantica 63
Maresia pulchella 468
Marrubium alysson 450
Matthiola fruticulosa 448
Matthiola lunata 70
Matthiola tricuspidata 329
Medicago arborea 313
Mercurialis annua 221
Minuartia baldacci 251
Misopates orontium 95
Moenchia mantica 237
Moltkia coerulea 415
Moricandia arvensis 85
Moricandia moricandioides 59
Muscari commutatum 313
Narcissus assoanus 122

Narcissus bulbocodium obesus 32
Narcissus cantabricus 70
Narcissus gaditanus 68
Narcissus poeticus 170
Narcissus serotinus 73
Narcissus tazetta 190
Neatostema apulum 94
Neotinea maculata 222
Nepeta reticulata 73
Nigella ciliaris 438
Nigella damascena 372
Nonea vesicaria 461
Oenothera drumondii 438
Omphalodes linifolia 37
Onobrychis venosa 362
Ononis speciosa 68, 473
Ononis spinosa 144
Onosma fruticosa 362
Ophrys araneola 124
Ophrys argolica 302
Ophrys argolica elegans 371
Ophrys bertoloniformis 169
Ophrys bertolonis 169
Ophrys bombyliflora 229
Ophrys bommuelleri grandiflora 371
Ophrys gerrum-equinum 244
Ophrys fusca 222
Ophrys garganica 169, 193
Ophrys kotschii 371
Ophrys lutea melana 169
Ophrys provincialis 121
Ophrys reinholdii 302
Ophrys sintenisii 371
Ophrys sphegodes atrata 229
Ophrys sphegodes mammosa 314

Ophrys sphegodes panormitana 190
Ophrys spruneri 302
Ophrys tenthredinifera 202
Ophrys vernixia 229
Orchis anatolica 371
Orchis branchifortii 179
Orchis caspia 430
Orchis champagneuxii 121
Orchis galilea 430
Orchis italica 222
Orchis lactea 215
Orchis longicornu 179
Orchis morio 329
Orchis pallens 304
Orchis pauciflora 169
Orchis quadripunctata 329
Orchis tridentata 222
Orchis ustulata 222
Orobanche ramosa 215
Orphrys carmeli 430
Osyris quadripartita 79
Otanthus maritimus 338
Oxalis pes-caprae 211
Pallenis spinosa 202
Pancratium illyricum 193
Papaver dubium 425
Parentucellia viscosa 322
Paris quadrifolia 473
Peganum harmala 59
Periploca laevigata 67
Petrorhagia glumacea 302
Petrorhagia velutina 302
Phagnalon rupestre 47
Phlomis italica 99
Phlomis lychnitis 18
Phlomis lycia 415
Phlomis purpurea 70
Phyla canescens 468
Plumbago europaea 138
Podonosma orientalis 447
Podospermum canum 300
Polygala nicaensis 103
Polygala preslii 197
Polygonum equisetiforme 48
Potentilla saxifraga 124
Prunus mahaleb 127
Psoralea bituminosa 202
Pteranthus dichotomus 448
Ptilostemon chamaepeuce 473
Pulsatilla rubra 127
Putoria calbrica 245
Pyrus amygdaliformis 330
Quercus aegilops 337
Quercus alnifolia 288
Ranunculus agyreus 405
Ranunculus asiaticus 348
Ranunculus bullatus 73
Ranunculus cadmicus var. cyprius 372
Ranunculus gramineus 127
Ranunculus millefolliatus 160
Ranunculus millefolliatus leptaleus 361
Ranunculus paludosus 362
Ranunculus rupestris 461
Ranunculus sartorianus 304
Ranunculus sericeus 385
Ranunculus subhomophyllus 301
Ranunculus tingitana 68
Retama monosperma 448
Rhamnus alaternus 80

Rhamnus lyciodes 70
Romulea clusiana 17
Ruta chalepensis 17
Salvia candidissima 274
Salvia dominica 447
Salvia fruticosa 47
Salvia lanigera 450
Salvia ringens 246
Salvia viridis 233
Saponaria calabrica 237
Satureja cuneifolia 66
Saxifraga chrysoplenifolia 282
Saxifraga flexuosa 282
Saxifraga pedemontana cervicornis 152
Scabiosa arenaria 461
Scabiosa argentea 300
Scabiosa cretica 190
Scabiosa maritima 211
Scabiosa prolifera 372
Scabiosa stellata 454
Scabiosa tenuis 270
Scilla autumnalis 38
Scilla cilicica 361
Scilla forbesii 402
Scilla liliohyacinthus 127
Scilla messenaica 304
Scilla monophyllos 14
Scilla verna 23
Scorpiurus muricatus 94, 144
Scorzonera undulata 448
Scrophularia lucida 332
Scutellaria altissima 233
Scutellaria orientalis 468
Scutellaria rubicunda 263
Sedum coeruleum 197

Senecio linifolius 66
Senecio ovirensis 122
Senecio rodriguezii 111
Serapias cordigera 157
Serapias lingua 157
Serapias neglecta 103
Serapias perviflora 157
Silene conica 230
Silene echinospermoides 415
Silene italica 121
Silene littorea 85
Silene quinquevulneraria 157
Silene sericea 151
Silene succulenta 158
Smilax aspera 133
Solanum bonariense 48
Spiranthes spiralis 338
Stachys glutinosa 151
Stachys oxymastrum 244
Stachys spruneri 313
Sterbergia lutea sicula 194, 330
Symphytum bulbosum 300
Symphytum gussonei 182
Symphytum ottomanum 230
Symphytum tuberosum 121
Tanacetum cinerarifolium 215
Taraxacum aphrogenes 473
Taraxacum gymnanthum 337
Teucrium fruticans 99
Teucrium pseudochamaepitys 14
Thlaspi praecox 122
Thymus capitatus 330
Trachelium caeruleum 66
Tragopogon porrifolius 216
Trifolium purpureum 468
Trifolium resupinatum 402

Tuberaria guttata 32
Tuberaria lignosa 122
Tulipa armena var. lycia 402
Tulipa cypria 374
Tulipa sylvestris 215
Tulipa sylvestris 215
Tulipa sylvestris australis 23
Ulex parviflorus funkii 43
Urginea maritima 38
Urginea undulata 474
Urtica atrovirens 158
Valeriana tuberosa 474
Verbascum macrurum 287
Verbena tenuisecta 438
Vicia sicula 197
Vinca difformis 94
Viola aetnensis messanensis 170
Viola aetolica 302
Viola arborescens 13
Viola cenisia 122
Viola graeca 302
Viola speciosa 244
Wiedemannia orientalis 425
Withania frutescens 73

Place Index
Only one salient page number is given for each entry.

Abunyola 107
Achaia 296
Acropolis 278
Acrotiri Peninsular, Cyp. 379
Adana 429
Adrano 199
Afantou 365
Afendis Hristos 357
Afendis Kavoussi 359
Afula 439
Ag. Eleftherios 318
Ag. Gerasimou Monastery 317
Agia Roumeli 342
Agia Varvara 353
Agii Deka 352
Agios Dimitrios 299
Agios Geogios, Cyp. 386
Agios Ilias 289
Agios Ioannis, Crete 350
Agios Nikolaos 356
Agios Vasillios 350
Agrigento 212
Aigosthena 281
Aïn Draham 463
Aïn Okter 456
Ajaccio 176
Ajiassos 327
Ak Daglari 411
Akamas Peninsular 383
Akko 441
Alanya 408
Albunol 89
Alcala de los Gazules 39

Alcolea 81
Alcudia 107
Aldarax 78
Alfambras 29
Algarve 15
Algatocin 57
Algeciras 34
Alhaurin 61
Alikianos 339
Almoriana 54
Altinkum 429
Altinyayla 426
Amanthus 376
Amfissa 258
Ammouda 345
Anamur 411
Anamurium 411
Anapoli 344
Anduze 137
Anoyia 352
Antalya 406
Antequera 62
Arahova 258
Arapköy 399
Archangelos 366
Argolis 280
Argos 281
Argostoli 317
Arhanes 355
Aronia 296
Arrifana 30
Asco Valley 164
Asomaton Monastery 353

Asos 321
Atajate 57
Athens 276
Atheras 321
Attiki 276
Avrakondes 357
Baba Dag 420
Badja Ridge 218
Bafa Gölü 429
Bahia de Alcudia 107
Balcón de Canales 83
Balerma 69
Banyalbufar 108
Barbate de Franco 41
Barrage de Ben Metir 465
Barragem da Bravura 29
Bastia 166
Baths of Afrodite 287
Bédarieux 132
Bedoin 140
Belice 25
Belvoir Castle 440
Beni Khalled 455
Benidorm 90
Bensafrim 29
Benyamina 61
Bey Mountains 406
Bizerta 451
Bobadilla 58
Bodrum 429
Bolonia 40
Bonifacio 167
Borgo 166
Bormes les Mimosas 148
Bou Fichu 458
Buffavento Castle 399

Bugibba 218
Bulla Regia 465
Butrini 239
Cabo de Gata 81
Cabo de Trafalgar 39
Caesarea 434
Cala de San Esteban 115
Cala de Santa Galdana 116
Cala Mesquida 119
Cala Ratjada 105
Cala'n Porter 116
Calas Coves 116
Caldas de Monchique 30
Caledonian Falls 394
Calenzana 161
Callosa 92
Calpé 91
Calvi 178
Calviá 107
Camargue 139
Campolato 185
Cannes 153
Cap Bon 455
Cap Corse 162
Cape Agios Ekaterinis 310
Cape Akrotiri Drapano 350
Cape Arnauti 387
Cape Drepanon 386
Cape Formentor 109
Cape Kormakiti 400
Cape Matapan 286
Cape Milianos 366
Cape Mirtias 366
Cape St. Vincent 22
Cape Sunion 279
Cape Tenaro 286

Carcassonne 128
Cargèse 177
Carpentras 142
Carpino 191
Carrapateira 26
Casta 165
Castaniccia 161
Castelejo Aguia 25
Castellar de la Frontera 54
Catania 201
Causse de Sauveterre 132
Causse de Sévérac 132
Causse du Larzac 132
Causse Méjean 132
Causse Noir 132
Cavaillon 141
Cedar Forest, Cyprus 295
Cefalù 204
Céreste 145
Cerro de Caballo 88
Cesarò 199
Chania 335
Chelmos 297
Chiancatapilone 189
Chionistra 393
Chios 331
Chora Sfakion 342
Chott el Djerid 466
Chrissoskalitissa Mon. 336
Chrysorroyiatissa Mon. 387
Clician Gates 429
Cirque de Navacelles 136
Ciudadela 113
Clermont l'Hérault 135
Çobanisa 412
Cogolin 150

Coin 61
Col de Gratteloup 148
Col de l'Ecre 155
Col de Larone 171
Col de Murs 142
Col de Rates 92
Col de San Stefano 165
Col de Vence 154
Col de Vergio 177
Col de Vizzavona 171
Col del Tap 132
Coll de Puig Major 108
Collesano 205
Collobrières 149
Combe de l'hermitage 142
Comino 217
Conca 161
Coral Bay 382
Corinth 280
Corte 177
Costa de la Luz 33
Costa del Sol 49
Cote d'Azur 123
Coto de Doñana 44
Courseqoules 155
Cuervo 88
Cueva de la Pileta 58
Cueva del Gato 58
Curium 381
Dalaman 427
Daliyat el Karmil 439
Dead Sea 446
Defile de l'Inzecca 171
Delphi 258
Dermil Geçidi 427
Dhavios 399

Dhiarizos 389
Didyma Hills 285
Digaleto 317
Dikti 357
Dikti Mountains 356
Dimisianata 319
Dingli Cliffs 220
Distrato 266
Djebel Abiod 457
Djebel Ichkeul 463
Djebel Zaghouan 458
Douze 466
Draguignan 153
Drapano 317
Driopi 285
Duilhac 130
Durres 242
Duzler Cami 407
Ebonas 369
Egosthena 280
El Aquian 81
El Haouaria 457
El Huerta del Cura 91
El Rocio 46
El Torcal 62
Elafonissi Island 336
Elat 446
Elbasan 242
Elche 91
Eleftero 266
Elmali 441
Elyaquim 435
Embalse de Guadarranque 56
Enfidaville 458
Enippeas River 250
Enix 78

Enna 212
Epidavros 284
Episkopi 350
Episkopi, Cyprus 380
Epta Pigai 366
Eressos 326
Erimanthos 296
Ermitage de la Trinité 175
Ersekë 234
Es Grau 119
Esporles 107
Estepona 49
Étang de Biguglia 165
Evisa 177
Evros Delta 249
Faliraki Beach 364
Fango Valley 178
Femmina Morta Pass 210
Ferragudo 284
Festos 353
Fethiye 413
Finike 411
Fiskardo 321
Florina 271
Fodele 355
Foia 30
Fondon 81
Fontana Amorosa 387
Foresta Umbra 188
Forêt de Aitone 177
Forêt de Sainte Baume 147
Formentor 105
Fornells 118
Fournes 339
Fraile del Valeta 88
Fréjus 148

Fuengirola 61
Gabes 466
Galatas 285
Galeria 178
Galilea 107
Galilee 441
Garat Ichkeul 462
Garrucha 72
Gazipasa 411
Getares 34
Ghadira 226
Ghar el Kebir 457
Giala 383
Giardini 198
Gibilmanna 204
Gibraltar 64
Girne 396
Githion 293
Gjirokaster 236
Golan Heights 444
Gölhisar 426
Golo Valley 177
Gonia 350
Gonies 356
Gorg Blau 108
Gorges de Galamus 128
Gorges de l'Arnette 131
Gorges de la Nesque 142
Gorges de la Restonica 177
Gorges de Loup 155
Gortina 353
Gourdon 155
Gouvia 308
Granada 83
Granbois 143
Grand Canyon de Verdon 153
Grand Plan des Canjuers 153
Grands Causses 132
Grasse 155
Gratteri 206
Gravia 259
Grazalema 60
Greolières 155
Grotta dell'Acqua 189
Guadalest 92
Guadalfeo River 89
Guadarranque River 54
Gyrka Pass 235
Haifa 435
Hala Sultan Tekke 378
Hammamet 453
Heraklion 353
Hisarönü 419
Horns of Hittim 441
Huelva 44
Hula Reserve 444
Hyères 149
Ideon Andron 352
Igualeja 53
Ioanina 262
Iouhtas 355
Iraklio 353
Isla Colom 119
Isles d'Hyères 149
Isles Sanguinaires 176
Isnello 206
Jaffa 434
Jendouba 465
Jerusalem 446
Jimena 56
Jimena de la Frontera 56
Jordanean Desert 446

495

Judea 445
Kalamata 289
Kalambaka 261
Kalavrita 297
Kalloni 326
Kandanos 335
Kantara Castle 399
Karakas 325
Karitsa 257
Kas 424
Kastellani 308
Kastoria 269
Kastritsi 299
Katara Pass 261
Kato Katelios 315
Kato Vermion 272
Katohori 321
Kaya 419
Kefor 440
Këlcyrë 235
Keliba 455
Kemer 424
Keramia 327
Kesme Bogaz 413
Khalki 367
Khirokita 377
Khroumirie 463
Kidirak 418
Kilani 389
Knossos 355
Kolibia 366
Kolpos Geras 324
Kolpos Kallonis 326
Konitsa 266
Korba 455
Korbous 456

Korçë 232
Korikos Peninsular 339
Korinthia 296
Korkuteli 407
Kouris Valley 388
Kritharakia 299
Kumluca 412
Kusadasi 429
Kykko Monastery 395
Kyrenia 396
Kyrenia Range 395
Kythrea 399
L'Isle Rousse 159
La Albufera 107
La Alcazaba 88
La Bambouserie 137
La Canonica 166
La Couvertoirade 136
La Garussière 154
La Granja 107
La Miquelette 150
La Revellata 178
La Sainte Baume 146
Lago de Varano 189
Lagos 26
Lake Antionoti 310
Lake Drakolimni 268
Lake Ichkeul 452
Lake Kinnaret 441
Lake Ohrid 242
Lambu Mili 327
Langada Gorge 289
Languedoc 125
Lanjaron 90
Lapithos 396
Lapta 377

Lara Bay 387
Larissa 261
Larnaca 376
Las Alpujarras 89
Lassiti Plateau 356
Latsi 387
Lazaros 357
Le Caylar 136
Le Châtelet 140
Le Lavandou 148
Lentini 201
Lepetimnos Hills 325
Les Agriates 164
Les Calanches 177
Les Corbières 128
Leskovik 235
Letoon 423
Levadia 258
Levka Mountains 344
Limassol 376
Limni Korission 312
Lindos 366
Linguaglossa 199
Lipades 307
Litochoro 250
Liveras 400
Lixouri 321
Llogora Pass 241
Llucmajor 104
Lodève 135
Logis-du-Pin 155
Los Caños 41
Loutra Eftalous 325
Lozari 165
Lykavitos 278
Lyso 383

Ma'agan Mikhael 435
Macinaggio 164
Madone Mountains 204
Mafa Ridge 218
Mahon 114
Makheras Monastery 393
Makounta 383
Malaucène 141
Malia 356
Mandraki Harbour 364
Manfredonia 181
Mani 293
Manoel Island 220
Manosque 141
Marais des Tre Padule 167
Marathos 355
Marbella 61
Marineo 212
Marmaris Peninsular 429
Maroulas 347
Mas Lomnbardo 189
Massada 446
Massif de l'Esterel 126
Massif des Maures 148
Matalscanas 44
Matmata 466
Mattinata 186
Mavri 352
Mazagon 46
Mazamet 131
Mazaugues 146
Mdina 220
Mellieha Bay 226
Menzel Bourgiba 462
Mercadel 118
Mersin 429

Mesaoria Plain, Cyp. 400
Messara Plain 353
Messina 198
Meteora 261
Methymna 324
Metsovo 262
Millau 135
Mirador de ses Animes 108
Mistra 286
Mohos 356
Mojácar 71
Mojon Alto 88
Molyvos 324
Monadendri 264
Monalithos 370
Monastery of La Rabida 44
Monchique 30
Mont Ventoux 140
Montagne de Lure 140
Montagne du Lubéron 141
Montagne Noire 141
Monte Carbonara 208
Monte Ciccia 195
Monte Cinto 168
Monte d'Oro 171
Monte Incudine 168
Monte Renoso 168
Monte Rotondo 168
Monte San Angelo 183
Monte San Salvatore 208
Monte Soro 210
Monte Stello 168
Monte Toro 118
Monte Tozzo 168
Montejaque 58
Monti dei Cervi 208

Monti Iblei 200
Montpellier 125
Morphou 399
Mosta 220
Motril Valley 71
Mount Akramitis 370
Mount Attaviros 370
Mount Carmel 435
Mount Enos 317
Mount Etna 200
Mont Filerimos 368
Mount Gilboa 439
Mount Herbetarim 444
Mount Hermon 444
Mount Hymettos 279
Mount Ida 351
Mount Kedros 353
Mount Meron 443
Mount Olymbos, Lesvos 327
Mount Olympus 249
Mount Osa 256
Mount Pantokrator 308
Mount Papingut 235
Mount Parnassos 258
Mount Parnes 280
Mount Pilion 256
Mount Profitis Ilias 327
Mount Smolikas 266
Mount Tabor 440
Mount Taiyetos 289
Mount Typmhi 265
Mount Vermion 272
Mulhascen 88
Munciarriti 208
Muzine Pass 238
Mycene 281

Mystra 286
Mytilini 323
Nabeul 453
Nafplio 284
Nahal Me'arot 439
Nahariyya 441
Naoussa 272
Narbonne 125
Nauplion 284
Nazareth 440
Nebrodi Mountains 210
Nefza 463
Negev Desert 446
Neokhorio 387
Nerja 44
Netanya 434
Nice 153
Nicosia 388
Nigde 429
Nijar 71
Odiáxere 29
Ölü Deniz 414
Omalos Plateau 339
Omodhos 389
Omolio 257
Orange 140
Orgiva 90
Ormos Ag. Georgiou 312
Ovacik 420
Pagi 308
Pahnes 340
Palazzolo Acreide 203
Paleohora 336
Paleokastritsa 306
Paleopanagia 291
Palma 107

Pamphylia 409
Panahaiko 299
Pano Panayia 387
Pap Rema 254
Paphos 382
Papingo 266
Pateras 280
Patras 299
Patrimonio 162
Peloponnese 276
Penal d'Ifach 91
Pentadactylos Mts. 395
Peristerona 388
Permet 235
Perpignon 128
Peschici 191
Petaloudes 368
Petra 325
Petra tou Romiou 381
Petreto-bicchisano 176
Peyia 382
Peyrepertuse 130
Piana de Albanese 212
Piano di Bataglia 208
Piano di Catania 201
Piano Zucchi 208
Piazza Armerina 203
Pic de Nore 131
Picota 30
Pindus Mountains 248
Pinos Genil 83
Pirio 178
Plage de Bussaglia 178
Plage de Pinetto 116
Plain of Jezreel 439
Plain of Sharon 434

Plakias 351
Plan d'Aups 147
Plateau de Vaucluse 141
Platres 390
Playa de las Negras 82
Playa de Matalascanas 44
Playa de Mazagon 46
Playa de Valdevaqueros 40
Plomari 328
Podradec 232
Poliana 291
Polis 383
Polizzi Generoso 209
Ponte Leccia 168
Poros 285
Poros, Cephalonia 319
Porticcio 176
Portimao 26
Porto de Lagos 30
Porto Germano 281
Porto-Vecchio 166
Pozuelo 89
Prado Llano 84
Prasies 347
Prespes National Park 269
Prines 350
Prinilas 308
Prionia 254
Profitus Ilias, Lesvos 327
Promentorio de Gargano 181
Propriano 175
Puerto Camacho 89
Puerto de la Virgen 76
Puerto de Soller 107
Puerto de Bujeo 36
Puerto del Cabrito 39

Puig Major 108
Punta Paloma 40
Punta Pozzo di Borgo 176
Pyrgi 308
Pyrgos 383
Qaf e Llogaresë 241
Qarne Hittim 441
Quéribus 130
Rabat 220
Ramla Bay 226
Randazzo 199
Ras el Drak 458
Restonica Valley 168
Rethymnon 346
Rhodes Town 364
Rhone Valley 123
Rio Aguas 71
Rio Barbate 41
River Guadalquivir 44
Roccapina 175
Rodi Garganico 191
Rodopou Peninsular 339
Ronda 57
Roquetas del Mar 77
Roussillon 143
Rozato 317
Sa Colobra 109
Sagres 16
Sahara Desert 466
Salakos 368
Salamis 401
Salina Bay 224
Samaria 445
Samaria Gorge 342
Sami 319
San Giovanni 184

500

San Pedro de Alcantara 50
San Roque 54
San Telm 105
Sant Elm 105
Santa Agata di Militello 210
Sault 141
Scaloria 185
Scanzano Lake 213
Segheria il Madrione 188
Seki 424
Seki River 424
Seli 272
Serandë 236
Serra de Arabida 16
Serra de Espinhaço do Cão 29
Serra de Monchique 15
Serrania de Ronda 52
Sfax 466
Sgombou 308
Sidari 311
Sierra Bermeja 50
Sierra Blanquilla 33
Sierra Cabrera 75
Sierra de Alcaparain 62
Sierra de Algibe 35
Sierra de Altana 92
Sierra de Bernia 92
Sierra de Filabres 76
Sierra de Gador 77
Sierra del Nino 35
Sierra del Pinar 60
Sierra Lujar 89
Sierra Nevada 83
Sierra Prieta 62
Sigrion 326
Silifke 429

Silves 30
Sisco Valley 164
Sitia 359
Skala 315
Skripero 308
Slima 220
Smyies Site 387
Solenzara 159
Soller 107
Sorbas 76
Soulatge 130
Sousse 466
Sparta 286
Spartilas 309
Spili 350
Spilios Agapitos Refuge 255
St. Barnabus Mon. 401
St. Florent 164
St. Guilhem-le-Desert 137
St. Hilarion Castle 398
St. Maurice-Navacelles 136
St. Maximin 146
St. Paul de Fenouillet 128
St. Paul's Bay 224
St. Pierre de la Fage 135
St. Pilon 146
Stomio 257
Suartone 167
Syracuse 203
Tabarca 464
Tahtali Dag 413
Taormina 198
Tarifa 39
Tébessa 458
Tel Aviv 431
Tepebasi 400

Tepelene 236
Termessos 407
Thessalian Plain 261
Tiberias 440
Tillissos 352
Tirana 242
Tombs of the Kings 382
Tomir 108
Toplou Monastery 359
Torremolinos 61
Torrent de Pareis 109
Touzeur 451
Trikala 261
Trimiklini 390
Tripolis 286
Troina 199
Troodos Mountains 388
Troumpeta Pass 307
Tsolou 308
Tunisian Desert 466
Uleila del Campo 76
Vacares 88
Val de Boi 29
Valetta 219
Valico del Lupo 187
Valledemosa 107
Vathia 293
Vejer 41
Vence 153
Venta de la Mena 78
Vicos Gorge 265
Victoria Lines 218
Vidauban 149
Vieste 186
Vikos Gorge 265
Vila do Bispo 25

Villa Monrepos 307
Villes sur Auzon 142
Vizzavona Valley 168
Vlorë 241
Volakias 340
Voukolies 297
Vrisses 345
Wadija Ridge 218
Xantos 423
Xilokastron 298
Xyloskalo, Crete 340
Yermasoyia Valley 378
Yithion 293
Ypsonas 388
Zaghouan 458
Zagoria 264
Zahara de los Atunes 41
Zefat 442
Zikhron Ya'aqov 435